Routledge Revi

Circulation in Population Movement

First published in 1985, this collection of essays deals with processes of population movement and how they have operated over time. It is also about people: Melanesians who number some five million and inhabit the region stretching from the Indonesian province of Irian Jaya to the Independent State of Fiji. Standard work on Movement in third world societies has emphasized migration, involving a shift in residence from one domicile to another, at the expense of the interchange of people between diverse places and different circumstances. Many moves, as from villages and towns, are circulatory: they begin at, go away from, but ultimately end in the same dwelling place and community. This book focuses on the full range of territorial mobility, especially circulation, and its meanings for the people involved.

This volume brings together indigenous scholars, foreign field researchers, and international authorities from many of the social sciences: anthropology, demography, economics, geography and sociology. It presents a set of multicultural statements about the mobility of particular peoples within a region of the third world. This collection about specifically Melanesian issues aims to stimulate broader visions among population scholars, and it underlines the pressing need for more theoretical and empirical work on a volatile, yet neglected, category of population movement.

Circulation in Population Movement

Substance and concepts from the Melanesian case

Edited by
Murray Chapman
and
R. Mansell Prothero

First published in 1985
by Routledge & Kegan Paul Plc

This edition first published in 2012 by Routledge
2 Park Square, Milton Park, Abingdon, Oxon, OX14 4RN

Simultaneously published in the USA and Canada
by Routledge
711 Third Avenue, New York, NY 10017

Routledge is an imprint of the Taylor & Francis Group, an informa business

Publisher's Note
The publisher has gone to great lengths to ensure the quality of this reprint but
points out that some imperfections in the original copies may be apparent.

Disclaimer
The publisher has made every effort to trace copyright holders and welcomes
correspondence from those they have been unable to contact.

A Library of Congress record exists under ISBN: 0710204515

ISBN 13: 978-0-415-52537-4 (hbk)
ISBN 13: 978-0-203-11848-1 (ebk)
ISBN 13: 978-0-415-52829-0 (pbk)

Circulation in
population movement

By the same editors

Circulation in Third World Countries

Circulation in population movement

Substance and concepts from the Melanesian case

edited by
Murray Chapman
and
R. Mansell Prothero

Cartography by
Alan G. Hodgkiss

Routledge & Kegan Paul
London, Boston, Melbourne and Henley

First published in 1985
by Routledge & Kegan Paul plc

14 Leicester Square, London WC2H 7PH, England

9 Park Street, Boston, Mass. 02108, USA

464 St Kilda Road, Melbourne,
Victoria 3004, Australia and

Broadway House, Newtown Road,
Henley on Thames, Oxon RG9 1EN, England

Set in Press Roman
by Hope Services, Abingdon
and printed in Great Britain
by T.J. Press (Padstow) Ltd
Padstow, Cornwall

Library of Congress Cataloging in Publication Data

Circulation in population movement.
Bibliography: p.
includes index.
1. Migration, Internal — Melanesia — Addresses, essays,
lectures. 2. Melanesia — Population — Addresses, essays,
lectures. I. Chapman, Murray. II. Prothero, R.
Mansell.
HB2152.55.A3C57   1985   304.8'0993   84-23801

ISBN 0-7102-0451-5

Smol samting nomoa fo se tagio long oloketa pipol bulong Melanesia hu i kaen tumas long mifala . . .

Olsem hu i talem stori bulong oloketa long mifala, tisim mifala kastom bulong oloketa, ansearem laif bulong oloketa witim mifala. Oloketa gud fereni hu i entaprit fo mifala, kuki fo mifala an meanim pikanini bulong mifala. . . . Everi taem nomoa oloketa tritim mifala olsem wantok.

Some small return to the people of Melanesia . . .

And especially to those who told us their stories, taught us their customs, shared their lives, questioned their fellows, acted as interpreters, carried our stores, cooked our meals, and minded our children. . . .

And at all times treated us with the greatest of civility.

Contents

Illustrations

Tables

xiii

Contributors

Ron Bastin is a freelance consultant based in London, England.

Murray A. Bathgate is Social Assessment Officer in the Town and Country Planning Division, Ministry of Works and Development, Wellington, New Zealand.

Richard Bedford is Senior Lecturer in Geography, University of Canterbury, Christchurch, New Zealand.

Joël Bonnemaison is Maître de Recherche, Office de la Recherche Scientifique et Technique Outre-Mer, Paris, France.

Murray Chapman is Professor of Geography, University of Hawaii, and Research Associate, East-West Population Institute, Honolulu, USA.

John Connell is Senior Lecturer in Geography, University of Sydney, New South Wales, Australia.

John D. Conroy is Head of the Asia/Pacific Economics Branch, Office of International Assessments, Canberra, Australia.

Richard Curtain is Postdoctoral Fellow in Sociology, Research School of Social Sciences, Australian National University, Canberra, Australia.

Ian Frazer is Senior Lecturer in Anthropology, University of Otago, Dunedin, New Zealand.

Sidney Goldstein is George Hazard Crooker University Professor and Director of the Population Studies and Training Center, Brown University, Providence, Rhode Island, USA.

Michael Hamnett is Deputy Director of the Pacific Islands Development Program, East-West Center, Honolulu, USA.

Shashikant Nair is Vice Principal, Ba Methodist High School, Viti Levu, Fiji.

R. Mansell Prothero is Professor of Geography and Head of Department, University of Liverpool, England.

Rejieli K. Racule heads the Secondary English Section of the Curriculum Development Unit, Ministry of Education and Youth, Suva, Fiji.

CONTRIBUTORS

Dawn Ryan is Senior Lecturer in Anthropology, Monash University, Clayton, Victoria, Australia.

Marilyn Strathern is a Fellow of Girton College, Cambridge University, England.

Sakiusa Tubuna is Headmaster, Naiyala Junior Secondary School, Wainbuka, Viti Levu, Fiji.

Morea Vele is Assistant Secretary for General Economic Policy, Department of Finance, Port Moresby, Papua New Guinea.

James B. Watson is Professor of Anthropology, University of Washington, Seattle, USA.

Elspeth Young is Research Fellow in the North Australia Research Unit, Australian National University, Canberra, Australia.

Preface

This collection of essays is about processes of population movement and how they have operated over time. It is also about people: Melanesians who number some five million and inhabit the region stretching from the Indonesian province of Irian Jaya to the independent state of Fiji. Standard work on movement in third world societies has emphasized migration, which involves a shift in permanent residence from one domicile to another, but fails to consider the constant interchange of people between diverse places and different circumstances. Many moves, as from villages and towns, are in fact circulatory: they begin at, go away from, but ultimately end in the same dwelling place and community. This book thus focuses on the full range of territorial mobility, whether circulation or migration, and its meaning for the people involved.

As a collaborative effort, this work arises from long association. In the 1950s Prothero (1957, 1959) examined seasonal movements of population from northwest Nigeria, particularly of those going to wage labour which at the time was termed 'labour migration'. During the second half of the 1960s, Chapman (1969, 1970) considered all forms of circulatory movement on south Guadalcanal in the Solomon Islands. Our professional contacts extend from that time and in 1975-76 Chapman was able to spend a sabbatical year with Prothero in Britain, supported by a small grant from the Social Science Research Council, United Kingdom (Grant HR3731/2). During that period both of us explored some of the crosscultural comparisons to be made between circulation in tropical Africa and the island Pacific. This effort reinforced our belief in the critical importance of its comparative study throughout the third world and we decided to assemble a multi-disciplinary collection of essays. That volume, *Circulation in Third World Countries*, is a companion to the present one.

The second outcome of our 1975-76 collaboration was an International Seminar on the Cross-Cultural Study of Circulation, sponsored jointly by the Population Institute of the East-West Center in Hawaii and the Institute for Applied Social and Economic Research in Papua New Guinea, with financial assistance from the National Science Foundation, USA (Grant SOC77-26793), which also provided a subvention towards publication. Held at the East-West Center, Honolulu, during 3-7 April 1978, this seminar was concerned with circulation in Melanesia and brought together indigenous scholars, foreign field researchers, and international authorities from most of the social sciences: anthropology, demography, economics, geography, and sociology. Such a specific regional focus, perhaps surprising at first glance, reflected an attempt to facilitate tight comparative analysis in studies of population movement and to encourage research workers to address some basic issues of concept and technique. The Honolulu discussions have been reported briefly (Chapman 1978, 1979; East-West Population Institute 1978) and their primary content forms the basis of this book.

Data from national censuses and regional surveys, the primary sources for movement analysis, fail to capture the circulation of people because they fix snapshots of behaviour frozen in time. Intercensal comparisons based upon questions about place of birth and of residence identify long-term relocations but are far less useful in considering the ongoing interaction between mobility dynamics, population redistribution and socioeconomic change. Heavy dependence upon census and formal survey data leads to a common assumption that recurrent and short-term movements represent transitional or compromise behaviour that will diminish steadily the more a society becomes modernized or urbanized.

In countries with systems of population registration able to record changes of address, as in parts of western Europe and east Asia, it is possible from the statistical record both to identify the various and complex kinds of mobility and to test the appropriateness of standard conceptual distinctions − as between movements that involve a definitive change in place of residence ('migration') and those that do not ('circulation'). For most parts of the world such objectives can only be attained through first-hand enquiries and the fieldwork reported in this book confirms this. What is required are interlocking sets of data derived from field censuses, ongoing mobility registers, family genealogies, retrospective movement histories, and oral historiographic reconstructions of residential transfers, supported wherever possible by archival, ecological, archaeological, and linguistic evidence.

To the incredulity of most metropolitan scholars, many such data do exist for a number of Melanesian societies. Since the early 1960s geographers, anthropologists, economists, and sociologists have undertaken detailed studies of mobility behaviour within Papua New Guinea, the Solomon Islands, Vanuatu (formerly the New Hebrides), New

Caledonia, and Fiji. These field data contain sufficient detail and depth in time to be able to question, rather than blindly accept, the conventions and assumptions of movement and analysis that derive from very different kinds of society and polity.

Participants at the Honolulu symposium were asked to address such broad concerns from the standpoint of the Melanesian group or people about whom they had collected information over lengthy periods, in one or two instances for as long as twenty years. Each received a set of propositions about circulation that reflected the African and Pacific experience, on which Chapman and Prothero (1977a) had collaborated in 1975-76 and the revision of which is the first chapter in this volume. The editors provided all participants with materials about the classification of African population mobility (Gould and Prothero 1975b), the nature of wage-labour circulation in south-central Africa (Mitchell 1961a), a life history of a Fijian migrant family in the capital of Suva (Racule 1974), systems of movement in two Solomon Island communities articulated as circulation (Chapman 1976), and the hypothetical links between modernization and changes in circulation over long swings of time (Zelinsky 1971). Although somewhat different from their seminar version, essays in this volume thus display a commonality of purpose refracted through the differentiating prisms of intellectual tradition, research philosophy, and individual predisposition. Collectively this book represents a set of multidisciplinary and multicultural statements about the mobility of particular peoples within a region of the third world.

Two papers have been added to those originally prepared: one by Shashikant Nair, the first Indo-Fijian to examine rural-urban connexions in his own country; and the other, a behavioural perspective on wage-labour circulation by Marilyn Strathern. This has become a minor classic in the Melanesian literature and appears here in revised form with permission of the Australian National University Press. Conference versions of some field papers were quickly published to make them more accessible to Melanesian audiences (Gipey 1978, Vele 1978, Bonnemaison 1979), while those commissioned from international authorities were placed for maximum impact upon the respective disciplines of demography (Goldstein 1978), geography (Prothero 1979), political economy (Breman 1978-9), sociology and social anthropology (Mitchell 1985). Sets of seminar papers were also despatched to key archives, national and university libraries in the region to ensure they were immediately available to Melanesian students, planners, and administrators.

The circumstances which produce circulation are many and complex, and circulatory movements have a varied impact upon cultural, social, economic and political organization. Somewhat arbitrarily the essays have been placed in five major categories to denote differences in topic and time-span: customary processes of circulation; transformations since European contact; the links between circulation, the money economy, and formal education on the one hand and its connexions with rural-urban flows plus urbanization on the other; and, finally, Melanesian circulation set within the context of other third world regions or broader theoretical concerns. Despite this grouping the papers inevitably overlap in approach, methodology, and substance. All refer, for example, to the indirect if not the direct effect of ecological or political factors on circulation and no essay is without substantial comment upon the role of social and/or economic influences.

Acceptable terminology is a particular difficulty in the study of population movement. As our introductory chapter in both volumes makes clear, this is not surprising, given the great range of disciplinary and theoretical goals that scholars have pursued (cf. Chapman and Prothero 1983). Thus all essays have been standardized. 'Population mobility' or 'population movement' is the general term used for the territorial flow of people. The critical distinction between 'migration' and 'circulation', the two major types of population mobility, denotes whether or not a return to place of origin is involved. Hybrid terms such as 'circular migration' are especially confusing. These have been retained and placed within single quotes only when they refer to a particular literature (as in economics) or are critical to the intellectual history of circulation. When first used, terms signifying particular kinds of circulation also are placed within single quotes: thus 'commuting', 'labour migration', and 'pendular migration'. The fact that such standardization has been possible, given the bewildering variety of terms in the literature, underlines the great forbearance of authors. Since essays range widely in time and place, dollars Australian and pounds sterling have been adopted as baseline currencies, but no conversions have been attempted before the early 1960s. Pidgin words, for example of Bislama in Vanuatu, are enclosed within single quotes and words in italics are from Melanesian and other languages not English.

We are most grateful to those who were originally part of the 1978 symposium, prepared papers for it, and then agreed to revise and correct them according to editorial suggestions. Bringing together work of authors who are geographically so widely spaced, by editors living mainly in another hemisphere and separated by 150 degrees of longitude, has inevitably involved longer delays than was anticipated. The first revised manuscript from an island author reached us, with great concern about tardiness, in June 1978, and the last, from an established academic, more than four years later. Along with its companion volume on the third world, we hope this collection about Melanesian issues will stimulate broader visions amongst population scholars, and will underscore the pressing need for more theoretical and empirical work on a volatile, yet neglected, category of population movement.

Like people written about in this volume, we have maintained contact through a constant stream of letters in both directions, cables, and the occasional telephone call. Yet our collaboration has been much assisted by being able to meet in Hawaii in 1978 and in England in 1978 and again in 1979, thanks to the National Science Foundation (Grant SOC77-26793), the East-West Center (Honolulu), and the Universities of Hawaii and Liverpool. These also facilitated discussion about cartography, which was designed and executed by Alan Hodgkiss, whose key role is recognised on the title page. The maps and diagrams are an important aid for readers unfamiliar with Melanesian society and territory, but spotty base-map coverage did subsequently lead to many demands being made over vast distance – demands that Alan Hodgkiss invariably addressed with good cheer and quiet efficiency.

For Chapman, editorial work was concentrated in particular periods, and he extends thanks to several institutions for their assistance: the Australian National University, Canberra, through the award of a visiting fellowship in demography and human geography (July–December 1979); the East-West Population Institute, Honolulu, for a summer grant that facilitated retreat to the island of Hawaii in the state of the same name (July–August 1980); and the University of Hawaii for a sabbatical year spent in New Caledonia, through the joint auspices of l'Office de la Recherche Scientifique et Technique Outre-Mer, Paris, and the award of a *bourse pour séjour scientifique de longue durée* from the French Ministry of External Relations (January–December 1983). Without such institutional encouragement at critical points this collection would have remained part of academic mythology.

Typing manuscripts through several drafts and helping with correspondence has involved a small band of heroines in diverse locales: Kathy Martinez, Lynette Tong, Mary Kehoe, and Beverly Honda in Honolulu, Geraldine Badham in Canberra, and Michelle Dubois in Noumea. References were meticulously checked by Geraldine Badham, thus eliminating errors in one of every five supplied by authors and demonstrating to some the lost art of how to cite correctly their own doctoral theses. The index has been prepared with great care by Lynne Garrett. Finally, particular thanks go to Norman Franklin, chairman of a notable publishing house, whose tolerance during this volume's long gestation has helped ensure that it did not become, in the words of one associate, 'a remaindered classic'.

Murray Chapman
R. Mansell Prothero

1 Circulation between 'home' and other places: some propositions

Murray Chapman and R. Mansell Prothero

Substantive work on migration . . . has yielded a far greater degree of diversity than prior theoretical formulations could have anticipated. Diversities have been uncovered over time, thus allowing us to formulate a number of models of historic types of migration and to 'unhitch', as it were, migration theory from modernization theory, an alliance which, while useful was also blinding. Diversities have also been discovered across space. Culture areas have been found to vary greatly with respect to the pervasiveness, acceptability, selectivity and mechanisms of adaptation to temporary and permanent population movements. . . . The monolithic category called 'migrant' increasingly has had to be broken down into a complex array of subtypes based on the variety of forces operating to stimulate the move, upon the periodicity and degree of permanent commitment related to the move, and upon the location of migrants within the social structures of both sending and recipient units (Abu-Lughod 1975:202).

Internal migration, conventionally defined, involves a shift in permanent residence from one place to another. Yet movements that involve moderately long durations of stay at a destination do not necessarily eliminate an eventual, and equally 'permanent' return to the places from which they originated. This fact is reflected in the qualifying adjectives used by scholars from many disciplines. The notion of 'return migration' is basic to formal demography (e.g. Feindt and Browning 1972); the economist Walter Elkan (1967) uses 'circular migration' to describe wage-labour movements in east Africa; in the Latin American literature 'pendular migration' is the standard term to denote comparable processes (e.g. Skeldon 1977); and Breese (1966:83), a sociologist focusing upon third world urbanization, speaks of 'floating migration'. Pierre George, the French geographer, considers the fluidity of wage-labour movements in west Africa to be so persistent and widespread that he dispenses with both the term 'migration' and its several adjectives in favor of one evocative word: 'turbulence' (George 1959: 200). Throughout the third world, this constant mobility has an impact on even the most dominant of primate cities. Thus the political scientist Joan Nelson (1976) talks of 'sojourners' and 'new urbanites' in a comprehensive study entitled 'temporary versus permanent cityward migration in developing countries'. This latter distinction is basic and echoes that made years ago by the geneticist Cavalli-Sforza (1963) between reciprocal flows as against the displacement (or relocation) of individuals – between those moves that begin at, go away from, but ultimately return to the same dwelling place, in contrast to those that are essentially away from one place of residence to another and are also relatively permanent (Fig. 1.1).

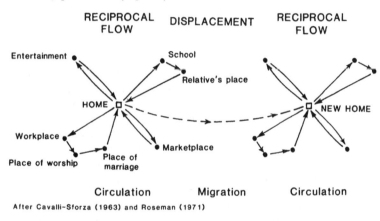

After Cavalli-Sforza (1963) and Roseman (1971)

FIGURE 1.1 *Reciprocal flow and displacement of people*

A clear implication in all these terms and distinctions is that the conventional concept of internal migration only faintly captures the full meaning of territorial mobility. Rather than constituting a one-way transfer, many movements involve the ongoing interchange of people between points of origin and destination, such as villages and towns, as individuals and in groups of varying size. This paper, first, briefly summarizes some attempts that have been made to articulate this constant ebb and flow; and secondly, states some propositions derived from the African and Pacific experience that may be tested against field data for insular and other third world societies. It is not a polished statement, interrupted by frequent qualification and citation; rather it is an attempt to present a perspective and to engage specialists in dialogue based on primary field experience.

The nature of repetitive movement

Serious attempts to capture the inherent meaning of repetitive movement in third world societies date from the 1920s and have involved scholars from most of the social and behavioural sciences (see Chapman and Prothero 1983). One of the best known occurred in the late 1950s when the British sociologist, J. Clyde Mitchell (1961a, 1961b), proposed the idea of labour mobility in south-central Africa as 'circulation'. Continuing the tradition of and drawing upon the extensive field research undertaken from the 1940s at the former Rhodes-Livingstone Institute in what is now Zambia, Mitchell recognized the continual oscillation of African villagers between their local communities and the agricultural estates, the mines, the commercial centres, and the sea ports to be an important characteristic of socioeconomic life. Although accurate estimates were not available, 'reports from individual areas in different parts of Africa have the same general story to tell: that something of the order of half of the able-bodied men are absent from tribal homes at any one moment and that almost all have been away at some time or another' (Mitchell 1961a:259). Absences were temporary, ranging from a few months to one or two years, and in west Africa often seasonal.

Such constant movement, or circulation as Mitchell aptly termed it, reflected peoples strongly tied to their tribal heritage but who, in the face of both economic need (head tax, land shortage) as well as their desire for some of the material items of a money economy, left their villages to engage in temporary employment. The wage labourer was thus responding to two contradictory sets of forces: centrifugal ones that induced him to depart his tribal domicile and centripetal ones that drew him back again. Assuming the need for money to be a necessary, but not by itself sufficient reason for movement to occur, Mitchell saw the interplay of centrifugal and centripetal influences as capable of producing three distinctive situations:

1   The negation of centrifugal (economic) influence by cash cropping or local employment so that the centripetal influences arising from the involvement in a social structure prevail. In this situation labour migration is absent or infrequent.
2   An oscillating balance between the two influences so that economic needs cannot be satisfied in the rural areas and people migrate to satisfy their needs. Social obligations in the rural areas, however, sooner or later force them back again when they once again feel the necessity to seek wage-earning occupations. Here labour migration is present.
3.   An involvement in a system of social relationships in town negating the social obligations in the rural areas so that the need to return is reduced. At the same time the ever expanding economic wants linked with the prestige systems in towns implies that the

economic pulls remain unaltered. Here labour migration and the circulation between town and country ceases (Mitchell 1961a:278).

Reflecting his own research and that of his colleagues, Mitchell's concept of tribal mobility as circulation referred only to the ebb and flow of wage labourers. In Melanesia, later in the 1960s, both Chapman (1970) and Bedford (1971) adopted this suggestive model but expanded its definition to embrace all movements that both began and terminated in the local community, on the grounds that villagers eventually returned to their homes no matter what the reasons for their departure nor for how long they had been absent. Within the Solomon Islands for example, the main commercial, administrative, educational, and medical services are concentrated in the government centres of Honiara, Auki, Gizo and Kira Kira (Fig. 5.1), and the dominant sources of cash during the decade 1960-70 were either in these or on coconut plantations sited on the coastal margins of various islands. Hamlet and village communities, in contrast, are far more widely dispersed. Unless a tribesman has completely rejected his local environment, as very few had done in the late 1960s, the continual shuttling back and forth from village to other rural villages, mission stations, plantations, district centres, and to Honiara is indubitably the lifestyle that Mitchell's concept of circulation aimed to illuminate. Throughout the Solomons and Vanuatu, as in south-central Africa, such constant movement reflects simultaneously the conflict and complementarity between the centrifugal attractions of and demands for commercial, social and administrative services and wage employment, and the centripetal power of village obligations, social relationships, and kinship ties.

About the same time the geographer Wilbur Zelinsky (1971), operating at a much higher level of generalization, formulated his 'Hypothesis of the mobility transition'. Persuaded by the impelling power of the modernization process, he argued that 'there are definite, patterned regularities in the growth of personal mobility through space-time during recent history, and these regularities comprise an essential component of the modernization process' (Zelinsky 1971:221-2). A society or people will pass through four unilineal 'phases' of mobility experience – premodern traditional, early transitional, late transitional, and advanced – during its transformation from a traditional-subsistence to an urban-industrial state, in the course of which there is 'vigorous acceleration of circulation'.

Basic to Zelinsky's hypothetical model is the concept of territorial mobility, in which there is both migration ('any permanent or semi-permanent change of residence') and circulation – defined as 'a great variety of movements, usually short-term, repetitive, or cyclic in nature, but all having in common the lack of any declared intention of a permanent or long-lasting change in residence' (Zelinsky 1971:225-6). The distinction between moves that are intentionally permanent and those that are not is crucial to third world studies and from the 1970s became incorporated into research on Black Africa (e.g. Gould and Prothero 1975b) and southeast Asia (Hugo 1975). Nonetheless the detail and time depth of data required to test the 'mobility transition' is awesome. Even so Bedford (1971), in a skillful reconstruction, grouped changes in niVanuatu movement to wage work over 150 years into three phrases: 1850–1900, 1870–1940, and post 1930. In sequence, these are seen to constitute a transition in 'circular migration', from labour recruiting for oversea cotton and sugar plantations, coconut estates, and nickel mines, giving way slowly to local and shorter-term contracts when villagers increasingly signed up to work on coconut holdings established by Australian, British, and French planters. Then, in 1942, these long-standing patterns were dramatically interrupted by labour needs at American military bases on Efate and Espiritu Santo (Fig. 4.1), out of which in the postwar era there emerged the port towns of Vila and Santo (Bedford 1971:Part II, see also 1973b).

Although Zelinsky was apparently unaware of Mitchell's pioneering statement, there are fascinating parallels in their perspectives on circulation, given different styles of formulation and levels of abstraction. Common to the thinking of both, but not to the elaborations by Chapman or Bedford, is the assumption that it is a transitory form of population movement linked to particular processes and phases of socioeconomic change – notably urbanization, modernization, and industrialization. For Mitchell, the circulation of labour between village and town would cease once a rising social commitment external to the rural areas converges with the town-based pull of ever-expanding economic needs (Mitchell 1961a:278). It is similarly implicit in Zelinsky's argument that all societies will follow the same sequence of mobility phases, from 'premodern traditional' to 'advanced' and probably beyond (cf. Zelinsky 1979; Goldstein this volume). Almost twenty years after his original statement, Mitchell (1976:personal communication) admitted that this inexorable process was occurring far more slowly in sub-Saharan Africa than he might previously have forecast.

The paradox of Zelinsky's 'mobility transition' is that, in being presented as a stage-type model, the discontinuities between each of its four cross-sectional phases are magnified. Similarly, in Bedford's reconstruction, the definitional disadvantage of separating interisland 'circular migration' from intraisland 'oscillation' obscures the ongoing linkages between pre- and postcontact modes of movement behaviour. Such schema also reflect an analytic bias. Namely that, under any conditions of socioeconomic change, the indigenous elements of a mobility system are seen to be of lesser importance and hence are examined more infrequently than whatever are the external forces operating upon that system. Such external forces, being historically more recent, are more apparent, more familiar, more easily recognized and, from the standpoint of contemporary modes of enquiry, relatively simpler to both comprehend and analyze. Consequently differences rather than similarities in people's mobility behaviour over time have been and continue to be emphasized. Since 1950, scholars have paid much greater attention to the rate, incidence, and differentials of internal migration, which were seen to be indicative of social and economic transformation, rather than to the persistence, reinforcement, and amplification of circulatory forms of movement. Circulation, far from being transitional or ephemeral, is a time-honoured and enduring mode of behaviour, deeply rooted in a great variety of cultures and found at all stages of socioeconomic change.

When first written in 1977 this last statement was based far more upon intuition and philosophical preference than upon widespread field evidence. Despite specialist studies since then in parts of Africa, southeast Asia and the island Pacific, along with methodological advances in oral history and ethnolinguistics (e.g. Miller 1980), there remains a broad lack of confirmatory data. Such derives partly from the fact that until recently most primary research has focused upon the structure rather than the process of movement; partly that most ethnographic enquiries have been couched in terms of a highly elastic 'ethnographic present'; and partly that towns and cities in the Pacific are sufficiently young to bolster the deterministic viewpoint that their belated but rapid emergence will inevitably follow western growth models. Much African research undertaken since the 1960s, for example, has pointed to a more even balance of females as a ratio of males in destination places, to the reduction in job turnover, the slow but durable rise in occupational skills, and the oversupply of unskilled workers (e.g. Masser and Gould 1975:3-7; Nelson 1976:724-32). Such evidence, from enquiries conducted at both the national and the regional scales as well as for particular urban centres, is cited as indicative that villagers are becoming more urban in their ways, more permanently tied to town and city places as their permanent domiciles and that, as a corollary, a diminution has occurred in the circuits of movement made between 'home' villages and other locations.

Some few, sporadic results were reported during the 1960s that questioned this prevailing doctrine. In a survey of 1,390 African students at senior high schools and teacher training colleges, Mitchell (1969a:176) found that very few conceived themselves to be 'definitely a townsmen' or 'definitely a countryman'. On the contrary: 'By far the greater majority (56.0% of both males and females) considered themselves to be "partly town and partly countryfolk". Even among those who have spent most of their lives in towns the majority look upon themselves as "part townsfolk and part countryfolk" ' (cf. Nair, Racule, this volume). Similarly Bellam (1963:99, 149) reported that, in 1962, only one out of a sample of 129 Melanesian males working in Honiara, the capital of the Solomon Islands (Fig. 5.1), considered that he would remain in town upon retirement. This finding was reinforced several years later by a study of two communities in south Guadalcanal, for whom it was suggested that 'a distinct increase in the proportion of people, particularly women, who oscillate between the village and Honiara has not been accompanied by a parallel rise in those who see it as their permanent domicile' (Chapman 1969:133; see also Chapman 1976).

It has been left to Walter Elkan, a most uncommon economist, to achieve a possible marriage between such apparently contradictory results. Focusing upon Nairobi, in Kenya, he has sifted both the statistical and the field evidence to address 'the question whether there are beginning to be substantial numbers in Nairobi who are wholly and permanently dependent upon wage employment and whose ties with their villages of origin have become severed' (Elkan 1976:695). It is not surprising that he ends this exercise 'with considerable uncertainty':

> If one were forced to reach a conclusion, it would probably be that nothing that could conceivably be described as a permanent proletariat is emerging in Nairobi – if by proletariat one means people who no longer have a farm income and who are totally dependent upon a wage income for their livelihood . . . On the other hand, the very temporary migration lasting typically no more than 2 or 3 years is also probably no longer the dominant pattern which it was in the 1950s. For the great majority the village is still home, but this does not preclude them from spending a considerable period of their lives in Nairobi, returning home during that time as often as they can (Elkan 1976:705).

In short, 'what is not in dispute is that people now stay longer in Nairobi while they continue to have close ties with their rural homes' (ibid:706) – and previously antithetical conclusions become plausibly joined! Persons continue to circulate during their working lives in Nairobi, but from a locus of residence that has shifted from village to town. With their departure from the capital following retirement from a profession or in wage employment, that locus is transferred back to the natal community, from which circulation continues to occur back and forth to those family, kin, and friends who live in other places (cf. Fig. 1.1).

Some basic propositions

We have reached the point where it is possible to frame some basic propositions about circulation as a long-established and enduring form of population movement, in the genuine hope that they will be rigorously tested on the ground. These propositions attempt to articulate the processes of movement and how they might have operated over time rather than, as with so many previous formulations, being overly concerned with the shape and structure of mobility behaviour.

1 In the third world, circulation as a form of mobility has not evolved from the impact of alien or western influences upon indigenous circumstances; as, for example, through the initiation of cash cropping into age-old systems of subsistence cultivation undertaken by tribal and peasant peoples. Rather, these and other externally-generated changes have reinforced customary circuits of mobility and added new ones. Western or other types of metropolitan contact introduced services for political administration, religion, exploitation, commerce, health, and formal schooling, the distribution of which were rarely congruent with the existing location of people. The discontinuities between the distribution of indigenous population and these introduced services resulted in the emergence of new forms of circulation between preexisting settlement (dispersed households, hamlets, villages, market centres, small towns) and the location of the new developments (towns, mines, plantations, ports, agricultural estates). Consequently it is necessary to think of the interrelations between circulation and these socioeconomic changes in terms of concepts like 'maintenance', 'modification', 'amplification', and 'accommodation' (Prothero 1973). Circulation has endured but has been modified; its incidence has been greatly magnified; but in turn this serves only to emphasize and reinforce customary patterns of mobility. Changes in the prevailing forms of population movement have been largely ones of degree rather than of kind. Although mobility basically remains circulatory, in some instances the social transformations have been sufficiently complex that 'modification' is no longer an adequate term to describe the process and 'amplification' or 'accommodation' are more appropriate.

2 Over centuries, circulation has occurred at the level of the village community and of its particular subgroups, the extended or nuclear family, and specified individuals. In some circumstances circulation was (and is) related to ecological regimes or natural hazards of the local environment; in others, it reflected basic features of customary lifestyles, like marriage, raiding weaker neighbours, barter and trade; in others it resulted from beliefs, values, and attitudes (as in the attribution of disease to sorcery); and in yet others from the decisions of the elderly, the prestigious, and the socially, economically and politically important (for example, the Melanesian 'big man'). Peoples' mobility behaviour can therefore be viewed as a system whose locus is the village, the local subgroup, the extended or nuclear family, compared with more open systems found in EuroAmerican societies where individuals or groups shift successively from one place of residence to another. In analysis, therefore, the village, the market centre, or the local town can be both the place of origin and of ultimate destination of completed

circulations. In form, circuits rather than vectors of movement prevail; in demographic terminology, return rather than linear migration is the dominant kind of mobility.

3 The basic principle of circulation, in both its customary and contemporary forms, is a territorial separation of obligations, activities, and goods. Among third world peoples, this separateness manifests two major influences. On the one hand is the security associated with the 'home' or natal place through access to land and other local resources for food, housing materials, and trading items; through kinship affiliation; through the presence of children and the elderly; and through common philosophies, values, and beliefs. There is, on the other hand, the wider locational spread of opportunities and associated risks involving local political and religious leaders; kinsfolk; marriageable partners; items for exchange or trade, ceremonials and feasts; and the introduced goods and services of wage employment, commerce, medicine, education, religion, politics, and entertainment. Following metropolitan contact, the establishment and continued growth of commercial centres, mines, plantations, ports and agricultural estates has intensified rather than diminished people's circulations. Certainly the circuits that are made between places of origin and destination have expanded spatially and temporally, and for sure they often involve greater numbers shifting for increased and a more complex variety of reasons. But, most crucially, no basic change in the form of movement is implied. As a result, it can be suggested that the rapid growth of urban places, post contact, has been a function of a dramatic increase in numbers of people who circulate into, through, and out of them, and is far less indicative of a rise in total urban commitment on the part of those who are in towns or of a markedly higher ratio who regard them as their permanent domicile. This will appear a provocative statement, but evidence for its support is to be found in an increasing number of specialized field or regional reports and similarly fugitive sources.

It is known that, irrespective of the length of time an individual spends away in town or other destinations, a continuing commitment to the natal community is manifest in the cross flow of letters, goods and gifts and the return flow of cash remittances; in local investments made by long-absent persons; and in the return upon retirement of those whose entire working life may have been spent in urban, agricultural, or mining settlements. Absent individuals make frequent visits to fulfil their social obligations in the family or kinship groups, and to participate in traditional or other ceremonies and festivals. Organizations known by such titles as 'Sons Abroad' and 'Home Improvement Societies' are found not only in the countries within which circulation occurs but also where members from a particular community have gone overseas (for instance, among Nigerians in the United Kingdom).

4 Both the incidence and durability of circulation is made possible by

the flexibility of indigenous structures in residence (change in settlement sites over generations, stable villages as an anchor to unstable hamlets); in social groupings (extended families, exogamy, polygamous marriage, customs relating to divorce and remarriage, practise of adoption and quasiadoption of children); and in leadership (achievement rather than ascription). Individuals belong to cultural systems that permit not only the exercise of choice but also the fulfilment of desired ends in locations beyond the natal place. The latter is possible because social structure is not localized in one community or household nor in several communities whose territories are continuous. Circuits of movement consequently were and still are one means to achieve a desired end and physical transfer need not be accompanied by social-structural displacement. Where sizeable numbers are away, regardless of whether the time involved is a day or much more and whether the destination is a neighbouring village, a distant town, a different country, or even another continent, the social structure is bi- or multilocal and these varied destinations become a sociospatial extension of the 'home' community.

5   The basic nature and deep historical roots of circulation are indicated by indigenous concepts. On Guadalcanal in the Solomon Islands, the word *lela* is common to eighteen dialects and translates in pidgin as 'go walkabout' − a practise found throughout much of Melanesia (cf. Frazer, this volume). *Lela* is acceptable and known as 'good custom'. Traditionally, it meant to go visiting, to travel on lineage business, to attend a feast, or to escape from a local quarrel or substantial humiliation − but always with the future intent of returning to the natal place. Since European contact through trade, missions, and colonial administration, the meaning of *lela* has gradually amplified to include many kinds of circular mobility. As a result, this word can now also refer to going away to wage labour, to the port town or administration headquarters, to mission stations, to schools, or to medical facilities (Chapman 1970:166-71). The existence of local concepts that underpin terms such as *lela* argues against the ethnographic image of the territorially immobile tribesman or peasant and the more recent stereotype of the disoriented, detribalized countryman in town. The persistence and amplification of age-old concepts of mobility through to the present, plus the flexibility of indigenous structures, indicates why villagers were ready and able to go away to the greater range of destinations that arose from the establishment of colonial economies. Such a reaction on the part of indigenous people, rather than being explained solely in terms of the political and economic pressures or legislative decrees to which they became subject, may be equally viewed as the exercise of choice according to traditional criteria and time-honoured practice.

What has been said about *lela* for Guadalcanal applies similarly to *masu cin rani* in northern Nigeria (Prothero 1957, 1959). Here, people have always moved during the dry season 'to eat elsewhere', not necessarily because this was forced upon them but also because they were free to do so. In travelling to other places, traditionally over relatively short distances, they were able to conserve food supplies within the local area for several months of the year at the same time as they acquired earnings from crafts or labouring, maintained social contacts, and met various kinds of obligations. By the 1930s, population growth and the resultant pressure on available land resources meant that, for some, *cin rani* became more of a necessity. In journeying to places of more conspicuous development that required wage-labourers, young adult males extended their movements beyond traditional pathways. Thus it was increasingly common for circulation to occur over long distances − as exhibited by the *yan tuma da gora* (men who travel with the gourd). Apparently this elaboration was not sustained during the 1960s, due partly to political disruption and perhaps also to greater opportunities for local employment. During the next decade, circuits of mobility were extended again under the pressure of drought conditions. Thus, in recent times, people's movement exhibits considerable flexibility in response to changed social, political, and economic conditions but with circulation persisting as a key characteristic (Prothero 1975).

Pointers to the future?

Recent forms of circulation may thus be seen as extensions of precontact modes of movement behaviour. Had mobility not been part of the lifestyles and thinking of indigenous peoples before western contact, then the territorial fluidity of wage labour demanded by externally-defined developments would not have occurred so easily nor in ways that had mutual advantages. Put another way, in areas of Black Africa and the island Pacific where population movement was spontaneous rather than forced or tightly controlled, the preexistence of local mobility and the flexibility of indigenous structures was as much conducive to subsequent economic change and social transformation as was the apparent power and efficacy of successive 'waves of modernization' (Gould 1970).

If this line of reasoning is considered, then there is also raised several thorny issues. The practical implications are obvious enough, for 'people who regard themselves as sojourners in the city will seek different kinds of housing, demand fewer amenities and services, behave differently with respect to making friends and joining organizations, use accumulated savings for different purposes, and respond to different political issues and candidates than will people committed to the city as their permanent home' (Nelson 1976:721-2). But in such situations, what is the meaning of 'home' and of 'urbanism', of 'commitment' to village and to urban domiciles? Why should the grandest settlements — the primate cities and intermediate towns — exclusively command so much theoretic and analytic attention when third world societies consist of bi- and multilocal populations, relatively stable in their demographic composition but comprised nowadays of individuals and small groups in constant motion between the natal or ancestral village and a variety of other places? Confronted with the rapid momentum of social change throughout the island Pacific, scholars should resist being seduced into thinking that its cities, towns, and market centres have begun a remorseless climb up an evolutionary ladder prescribed by prior world history. To rephrase slightly a classic remark made about Africans by the anthropologist Max Gluckman: If the Pacific island townsman is a townsman, then he is a townsman with a difference!

Acknowledgment

This paper is a revision of one presented by invitation in March 1977 to the Working Session on Urbanization in the Pacific, Association for Social Anthropology in Oceania, Monterey, California, USA (Chapman and Prothero 1977a).

Part I
Customary circulation

2 The precontact northern Tairora: high mobility in a crowded field

James B. Watson

To an observer on foot, traversing the country of the northern Tairora in 1925, there might seem little to shake the conventional view that subsistence-based villagers are stable folk, deeply attached to their land, even if not necessarily at peace with their neighbours. When asked the inevitable question where they belong, most people would say that they have always lived just where they are now. They are each other's kin and have common ancestors who lived there before them. They could truthfully add — but most likely would not — that certain of those ancestors still visit the ancestral ground. To emphasize their complete attachment, some might pat the ground with their palms.

Much else they might say or do could only suggest a fixed and isolated existence, the kind westerners like to think of as timeless. In separate territories as small as a few square kilometres, inhabitants proclaim the magic of their marvellous water, the strength of their soils, the depth and awe of a section of forest quite privately their own. Northern Tairora could easily seem part of *The Land that Time Forgot* (Leahy and Crain 1937), as so much of the outside world believes the Central Highlands of New Guinea (Fig. 2.1) to have been in 1925.

If, however, a traveller had been touring annually through the same country over five or ten years, and if every year a map was made of the inhabited sites and a census taken of the local populations, then something quite different would be learned about these seemingly timeless, territorially ingrown inhabitants. Traced through the maps of several years, their settlements would be seen to move, usually creeping but sometimes leaping across the landscape. Much of that movement would be distinct from the growth of individual settlements, a shifting of location for other reasons. Growth in the form of daughter settlements would also appear in the maps, suggesting that the country was becoming more crowded.

In comparison with the maps, periodic censuses would show more

15

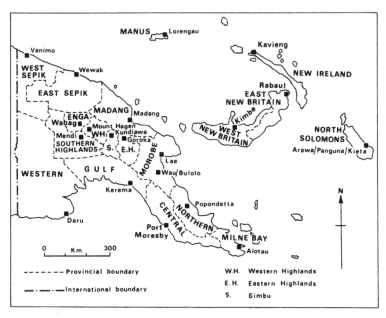

FIGURE 2.1 *Papua New Guinea: provinces and main towns*

striking changes, making the settled places seem stable. One could liken the settlements to sandbars in a slowly swirling pool of humanity. Each sandbar is frequently lapped by ripples and now and then washed by a wave that sweeps away some, occasionally all of the people who occupy it. Those dislodged then either construct a new sandbar or land on existing ones and join their occupants. Another wave meanwhile might wash up new people to add to those remaining on the sandbar that emigrants had lately quit. Though the swirl of human currents changes the locations of sandbars, inching them slowly along or whirling them to new locations, each new site can be usually traced to a former one nearby. In the flow of people some migrants move together and in relocating do not go far. But occasional waves carry people past the nearest sandbars onto others beyond. And some sweep away everyone, so that only the trace is left of a sandbar whose former inhabitants are now deposited elsewhere.

The fluidity of the regional population, especially in the face of its semblance of fixity and contradicting the claims of territorial attachment, could not fail to strike the observer. Yet the places to which people go are on the whole not basically distinct from those that they leave. The difference in resources and opportunities is either negligible or no more often better than worse. The pattern of people's lives scarcely changes after they settle down in new locations which are, after all, no great distance away nor in another sort of country. The observer

16

might therefore wonder whether mobility indicates something unusual about the northern Tairora.

As with all movement — migration or circulation — there is the basic question: why does it occur? What are its motives, especially if they are not readily apparent like better jobs or an easier life? What are its outcomes? What kind of society sustains such a mobility pattern? And why do people speak of their society as one in which such movement would not be common, if possible at all? The problem of explanation remains the same, regardless of whether movement is rare or commonplace, contradictory or consistent with local claims, or has obvious motives. It is doubtful that the northern Tairora are unique in their region; nor would they be unusual in a wider comparison with 'simple horticulturists'. Having recognized the problem, we now turn to the explanation of local mobility patterns in terms of general applicability, against enough of the essential background to make that feasible.

The northern Tairora

Considering precontact mobility patterns in Melanesia can only mean talking about small local groups, normally no more than two or three hundred people. Most often they are village horticulturists and always they are without knowledge of urban centres, formal government, markets, beasts of burden, roads, or wheeled vehicles. Peoples like the northern Tairora represent hinterland Melanesia, away from coasts and navigable rivers and without watercraft to connect them with distant points. Quite likely the northern Tairora represent more than Melanesia, since a larger portion of humanity's history is characterized by the above features. If there is any justification for a cross-cultural base in the study of mobility patterns, then peoples like the Tairora deserve a substantial place in it.

The local groups of the precontact Tairora, seldom more than two hundred residents, control small territories and are for most purposes the ultimate political units, the makers of war and of peace. Numerous such units crowd the country, some exceedingly hostile and some closely allied to each other, while the disposition of others lies at various points along a gradient between. Every group's history is foremost an ongoing statement of their connections and oppositions to other groups and this summarizes their present political stance. Although the claimed territories of residential groups are often adjacent, so bitter and so dangerous is the hostility between enemies of long-standing that their territories are likely to be separated by buffer zones. These empty spaces serve as no-man's-land, though adjacent groups may in fact claim them as their territory (Read 1954).

The contemporary Tairora are a series of such local groups, mainly the immediate descendants of the foregoing people (Watson 1967, 1979).

17

All speak one of four to six dialects of a single language and they number some 11,000. Their territories lie to the south, southeast, and southwest of Kainantu, in the Eastern Highlands province of Papua New Guinea, at altitudes between 1,220 and 3,350 metres (Fig. 2.2). For practical purposes the Tairora may be considered as belonging historically and ecologically to the central highlands of New Guinea. since their staple crop is the sweet potato (*Ipomoea batatas*), much of it used to feed pigs, an intensive husbandry compared with non-Highlands people nearby (Brookfield '961, 1962). Use of the term northern Tairora recognizes distinct dialect and social differences between the Tairora-speakers of the north and the south. By confining discussion to the more populous northern section, these differences are of no concern.

The ethnographic present which this paper treats is the immediate precontact period, when open warfare still prevailed and the movements of men and women beyond their local territory were often hazardous. That period began to end in the early 'thirties though still today there is little difficulty in plotting the 1930 map of inter-local enmity and amity, for those sentiments are not forgotten. My first contact with the Tairora was in 1954, followed by a brief revisit in 1959 and a further year of field work in 1964.

Migratory movement is not only common but also of pivotal concern for Tairora chroniclers. Once the true character of local life is recognized, it is easy to obtain information covering major movements of the remembered past, to collect accounts of the same movement from different sources, and thus to compare different sources for agreement. To understand the context and meaning of movement for the Tairora is a key concern of the ethnographer, unless there is justification for treating an essentially fluid and mobile regional society as if it were fixed and its local units static. That would be tantamount to saying that the mobility of the regional population is not worth investigating.

The field, or practical range of mobility of the Tairora has a distinctive scale, in both distance and cultural variety. For a given Tairora group it is very small when compared with those more familiar, like the hunter-gatherers of Australia (Wilmsen 1973). The field is minutely differentiated in the number and small size of communities into which it is divided, but differentiation is at a very low cultural level. The distinctiveness of units is personal or political: degrees of friendliness or hostility to others. Inter-group differentiation does not afford a radical option in lifestyle or mode of existence, like the contemporary plantation, town, mine, store, or mill. Using today's terms, there is nothing at either end of a move but the village. A move can affect the security of a person for better or worse but it cannot usually change the quality of opportunities, the activities that fill and define one's existence. Between origin and destination point the rules of life are much the same, but

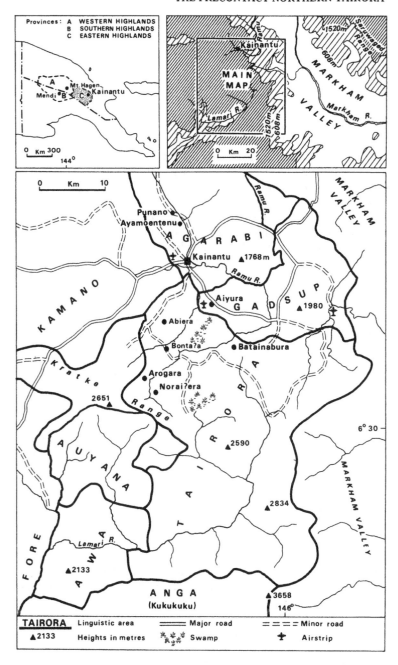

FIGURE 2.2 *Northern Tairora: linguistic and social areas*

there may be more or less room, more or less security – magically, socially, spatially – for the activity of a given individual at one place than another. Movement is everywhere; the question is only the purpose and strategy.

Movement in Tairora history

It is no exaggeration that the recent history of northern Tairora peoples is principally one of migratory movements. Even the origin song-cycle, *hakori*, recounts the long traverse of the world of Elder Sister and Younger Brother. In this society, history itself is movement. A certain personnel, X, have the identity their name betokens because they are believed to be descended from the immigrant founders of what is today accepted as the first settlement in a country thought previously to have been empty. The remembered name of one such settlement, and hence of the original settlers, also denotes a widening personnel of descendants, of subsequent refugee immigrants, and of other recruits, now all together purported descendants-in-common of the founders. In the case discussed, such a name is Tairora, whose immigrant accretions have been submerged in this common designation, deeded the present membership of Phratry Tairora by those who founded the initial, or stem settlement.

Movement is meaningful to the northern Tairora in connecting a personnel with a place, and in both distinguishing one place and personnel and connecting them with other places and personnels. The arrival or departure of people is their ultimate political statement. In the conventions of this history there is scarcely any interplay of ideas, formation of institutions, rise or fall of custom or practice. The probable introduction of the sweet potato in the last several centuries, for example, does not seem to figure in local histories. A general movement from valley-bottom lands to drier slopes may, however, have been ecologically inspired by the adoption of the sweet potato. Whatever its true scale and character, this movement is possibly remembered in the story of a people fleeing from sickness, a magical danger somehow associated with the cry of a bird that inhabited the swamps and canebrakes.

As as result of the emphasis upon movement, the ethnographer both needs and is easily able to obtain through interview much information regarding its extent and significance. This information can supplement direct observation and support the inferences drawn from genealogies, residential lists, marriage histories, records of births and deaths, and biographical material. An insight into the ethnic ebb and flow is crucial. Otherwise, contrary to the emphasis upon statics in the recent past of social anthropology, the existing social system will seem more rent with contradiction and inconsistency than need be.

Given the obstacles and hazards in this region, it is all the more notable that movements of various kinds are a major fact of life. Being village horticulturists and not hunters with large areas regularly exploited in quest of animal or plant food, the Tairora have compact territories closely bounded by others in which the interloper is presumed to be a threat and at risk unless recognized as friend. Among the neighbouring Agarabi (Fig. 2.2), deliberately to enter adjacent enemy territory is a recognized form of self-destruction. Were the Tairora equally given to suicide, the practice would work just as well. All the more emphatically, then, do the many sorts of movements in this region stand out. Since no-one lightly travels abroad, the mere fact of coming and going is the more worthy of recollection by the local chronicler, as indicative that key contacts and initiatives have been made. For the same reason, certain movements are concealed as far as possible. When no movement has been observed, it is nevertheless assumed to have taken place if suggested by some important change in intergroup relationships.

The precontact Tairora scale

The world of the precontact Tairora people is exceedingly small. A radius of ten or twelve kilometres would probably exceed the territory traversed and settlements visited, or visited often, by most inhabitants of a given place. The most obvious exceptions are first, the flight of refugees to more distant asylums; second, the distant visiting patterns of leading men; and third, the trading partnerships of these same men or others with hosts in particular villages at the edge of the mobility range. It is not safe pathways that make such journeys possible, but rather that a secure destination exists at the end of the path and that leading men are expected to be sufficiently intrepid to undertake them.

The size of the Tairora world is further suggested by the people's unawareness that it is contained within a sea-girt island. The abode of the dead is visible within that world in the ranges that rise to the north of the Markham Rift (Fig. 2.2). The moons that successively light the Tairora region expire in a valley perhaps no more than a few days' walk to the west, a place few are believed to have seen.

Unless the perils of travel and the small range of mobility are recognized, the emphasis given here to precontact circular and migratory movements will seem misplaced. How should one regard the relocation of a people from ten or fifteen kilometres away? What meaning should be attached to the defiant departure of an Abiera native to take up residence among his wife's people — traditional Abiera enemies — five or six kilometres distant? Here the local scale is obviously important. It surely matters, in the case of both the refugee migrants and the departing native, that the alien destination is closer to the edge of the known than to its centre. The ethnographic estimation of that ratio is critical to appreciating the motives and consequences of movement.

21

What matters is not simply the distance or time or effort but the amount and kind of risk, for we must assess the nature of opportunity perceived by sojourners or migrants against their notion of risk, and we must judge the territory covered against the farthest distance they know (cf. Chapman 1978:564).

Types of precontact movement

It is possible to turn to the kinds of movements that people, in units of one or more, make within this field, to some of their reasons for doing so, and to the consequences of movement. Fourteen headings describe the circular and migratory movements of the Tairora (Table 2.1). Some of these same headings would be needed, in fact, to describe life in the 'fifties and 'sixties — with several new ones to include local and long-distance labour migration, travel, trade, and visiting. The kinds of pre-contact movement are each marked with a C (circulation), or an M (migration), or both where either may occur.

This summary calls attention to the number and variety of move-ments as defined by destination, distance, duration, frequency, partici-pation, cost, and purpose. Purely daily shifts such as between house and garden or firewood and water supply, or from one house to another within a settlement or local territory, are not listed. Visits to pig stations are included, although they occur almost daily within the local territory, because the location of pigfeeding stations sometimes presages other movements, like the siting of a new settlement.

Migratory movements

There are four kinds of migration, of which one involves the short-range transfer of dwellings and people from one residential site to another (Table 2.1: type 1). Such site changes occur frequently, normally within the same local territory, which at least suggests the fluidity of Tairora society. A residential group is known by the name of its site, so that even a relocation nearby means a change of name.

Since territories are small and suitable sites limited, residential shifts are restricted and some involve no more than a few hundred metres. These reflect in general dilapidation or infestation of houses. More specific is the feeling that a site, once satisfactory, may accumulate a magical or ghostly contamination. The sickness or misfortune of the current occupants is one manifestation, especially a series of illnesses or mishaps within a short period. The conviction grows that the place is unfavourable, and one by one households begin to accumulate new building materials on some nearby spot. A varied deployment of residents sometimes accompanies a move to a new site, since some inhabitants may select alternatives that suit them better. So the community moves through space and time. Three kinds of movements involving residential shifts to outside territories and groups are those of refugees, of brides, and of estranged or disaffected men, the first often with dependants (Table 2.1: types 2, 3, 4). Each of these may be migratory or circular. Only the movement of brides has migration as its normative aim and typical outcome. But brides sometimes run away, back to their recent homes, so the possibility of a circular movement remains.

Blocs of refugees are known to be unstable, especially during the early years of asylum, in that they probably wish to return to their own lands, even after some years, and sometimes manage to do so. It is a question of their unity and the outside support they can muster against the opposition of their likely enemies. The first asylum, in any case, may not be the last. Restlessness or anxiety, the qualms or failures of their hosts, can cause them to move elsewhere, even if not back to the point of origin. Some may return while others remain in exile. The refugee cycle is prominent in the life of northern Tairora communities. Few if any local groups in either 1954 or 1964 were without a trace of immigrant blocs in their membership. Whether a community was explicitly considered 'mixed' reflected the recency or size of the latest bloc of incomers, whether their speech was alien to that of their hosts, or how politically extraneous or incompatible they appeared.

The endless flow of refugee migrants produces some of the most distant transfers, some of them astonishing. They bring to lodge in a territory a bloc whose origin, in kilometres, in degrees of ethnic unfamiliarity, or in the number of local groups separating origin from destination, unquestionably precludes prior kinship ties. Preparing the way for such movements are the circuits of far-flung trading partners and distant fighting allies, as well as the pathways pioneered or maintained by the wide-ranging strong man (Table 2.1: types 5, 6, 8). That distant asylum is sometimes favoured over others at closer hand says much about the character of regional society and the meaning and making of kinship in it (Watson 1970).

A people do not capriciously abandon ancestral territory, houses, and gardens, and suffer the loss of pigs and crops that flight entails. Their motivation — doubts of their own capacity to defend and sustain themselves or to secure the help of sufficient allies — must normally be great to justify leaving so much behind. The contemporary Abiera recall that their forebears once stood on the ridge above the settlements they were obliged to abandon, shaking their shields and promising each other to return.

The Tairora-speaking Batainabura (Fig. 2.2) can show the specific site where a sizeable contingent of Kamano-speaking refugees settled among them for about 'two or three' years, before returning to their original territory. This departure was allegedly precipitated by

TABLE 2.1 Precontact Tairora movements characterized by purpose and participation

Type of movement	Circulation (C) Migration (M)	Purpose	Participation	Estimated Frequency
1 Intra-territorial shift of residential site (a) personnel integral (b) personnel redeployed: fissioning or budding off.	M	Need for new houses. Unfavourable magical influences. Avoidance of hostile or untrustworthy neighbours. Proximity to supportive neighbours. Separation of a segment or larger group, a possible response to a leader's ambition.	One to ten households.	5 to 15 years.
2 Refugees from outside the local territory.	C, M	Flight to asylum in foreign territory to evade internal or external enemies.	Blocs from one to perhaps a dozen families.	25 to 50 years.
3 Connubial flows.	(C) M	In-marrying women given by outside groups.	Females. Usually one at a time: occasional two-way sister exchange.	Up to 1 year or less.

4 Disaffected, disgruntled, or ostracized males.	C (M)	To avoid angry kinsmen or local retribution for injury, e.g., by taking up residence with affines. To take advantage of opportunities elsewhere.	Males, usually one at a time, or brothers if both oppressed by same kinsmen on same account.	? Highly variable. 8–10 years.
5 Strong man's travels for intelligence, politics, trade, reciprocity, network maintenance.	C	For intelligence gathering, developing partnerships, giving/receiving hospitality and prestige, military deals. Keep up contacts and liaisons with various nodes of a network.	Outstanding male leaders of renown or aspirants to same role. With entourage or solitary.	Distant: 3 to 4 times a year. Close at hand: much more often.
6 Visits to trading partners.	C	Maintaining contacts, exchange of hospitality, information, plus trading goods. Differs mainly in degree from strong man's travels, except maximum distances may be less and destinations fewer.	Similar to strong man but men of less renown, usually in small parties.	? Once or twice a year.

Type of movement	Circulation (C) Migration (M)	Purpose	Participation	Estimated Frequency
7 Intra-phratric visits, in multi-local phratries.	C	Visits to kin in closely connected local groups, often without special occasion such as ceremonies.	Variety of individuals, obviously the better connected and more active. Brother visiting sister a common pattern.	Once a month or more often.
8 War parties (a) for own account (b) on account of allies	C C	(a) To inflict damage on an enemy, kill persons or destroy property, make show of force, keep enemy off balance. (b) To repay a debt or make a debtor, build a reputation, strenthen an alliance.	Various-sized parties of men, from 6 to 8 or more.	Several times a year (informants say 'constantly') but rarely in rainy season.
9 Ambushes: lurking by frequented paths to water sources or gardens.	C	Similar to above but victims sought include women and children.	Very small parties; perhaps most commonly individual stalkers.	As above.

10 Sorcerer's travels (a) black (b) white.	C C	(a) To collect exuvia (especially faeces, semen), to administer poisoned food to unsuspecting children, or set traps with contact poisons for passers-by. (b) To respond to summons when certain types of illness are indicated.	(a) Magical specialists, usually singly, sometimes with help of invisibility magic. Always covertly, because of great risk. (b) Magical specialists known as *bure*. Overtly.	(a) Uncertain (fear of such visits is perennial). (b) Variable, according to illness.
11 Hunting in forest: both own or, for grassland dwellers, that of other groups, with their permission	C	For birds, marsupials, eggs, other forest products. For regular fare or ritual use (especially marsupials).	Men in ones and twos.	Intermittent: especially when game needed for ritual purposes. Varies widely among individuals.
12 Affinal visits, at times of initiation, *ihálabu, órana,* funerals, infant rites.	C	To receive their due from local and kin groups into which their women are married; to help mourn dead, name the new born, initiate the youth, and celebrate other *rites de passage.*	Family and lineage groups.	One to several times a year, depending on age of the married couple, distance involved.

Type of movement	Circulation (C) Migration (M)	Purpose	Participation	Estimated Frequency
13 Visiting pig stations (intra-territorial).	C	To carry food to pigs, keep under surveillance, maintain pig shelter; possibly as a domicile with adjacent gardens. Conceivably scout future location for establishment of a new colony.	Family units (man, wife, or wives), occasionally two or three inter-connected families. The potential nucleus of fissionary shift of residence.	Daily or if domicile established and safe, on longer cycle, with overnight stays if inclement weather.
14 Miscellaneous special purpose.	C	Eeling; lime burning in forest; gathering forest lianas and other construction material for housing, fencing, magical materials. Some of these purposes are combined with hunting, or vice versa.	Individual men or family units, sometimes parties of 10 to 12 with children. Forest tends to be considered strengthening and healthy. Lime burning requires a number of men.	Irregular. Frequent at times of new construction of fences or houses.

disagreement over the marital assignment of one of the young refugee women. Whether the 'two or three years' is accurate, the Kamano immigrants obviously did not stay sufficiently long to put down roots. We can conclude that if strong motives exist to put a bloc of refugees to flight, then equally strong motives can draw them back to their point of origin, provided that a return seems feasible and too long a time has not been spent in exile.

What is the proportion of refugee episodes whose outcome is circular rather than migratory? Migrations, more easily recognized, may appear more numerous. Of the two local groups whose history is best known to me, one has gone through at least one remembered cycle of circular movement and the other has passed successively through several phases of asylum, each followed by the return of some refugees. If these two groups are representative, then circulation may be as common for refugees as migration.

The individual male emigré (Table 2.1: type 4) differs from the inmarrying bride in that his motives are more purely idiosyncratic and variable. Some cases indicate disaffection, others ostracism. Moves also vary greatly in distance and duration. In 1953, a Tairora originally from Bonta?a had been living for several years in Agarabi-speaking Ayamoentenu (Fig. 2.2). His children were all born in the adopted territory, there seemed little prospect of return to his original group, and he said he planned to stay. For this migrant it is a distance of about sixteen kilometres and perhaps six or seven ethnic steps between points of origin and of destination. Ayamoentenu was his wife's natal group, which most Tairora-speaking peoples feared and with whom they had little connection. His movement appeared destined to end as a migration, although contact was kept with close kin and I encountered him when visiting Bonta?a ten years after our first meeting at Ayamoentenu.

At Arogara in 1955, I was startled one day to see a house ablaze a short distance from mine. People were not much disturbed by this, even though it threatened a nearby dwelling. Their calm reflected the fact that the owner had put the torch to his house, doing so as a last act of defiance to the community before decamping for Norai?era, a group about five kilometres distant from Arogara (Fig. 2.2). It had all happened before, for this man was impetuous. He would surely return in a year or two, several ventured, quite likely after firing the new house he would now build at Norai?era!

During my stay at Abiera in 1964, a native son sheepishly returned from a sojourn of perhaps two years at Arogara, his wife's group. The returnee was subjected to open teasing and ridicule on account of having chosen to live abroad with his wife's people, with whom the Abiera are bitter enemies. He was said to be welcome but, at least in the early days of his reappearance, was reminded repeatedly albeit humorously of his disgrace.

At Batainabura there lived in 1964 a Gadsup-speaking man married to a woman whose father had no sons. It was generally accepted that he would make Batainabura his permanent home and become the heir of his father-in-law. But his anomalous position in the group, not helped by his diffidence, produced a distinctly uneasy individual who seldom spoke or smiled in public. He appeared a permanent migrant, like the solitary examples found in other local groups. Speaking generally, a good deal of suspicion characterizes the relations of Batainabura and the Gadsup-speaking people nearby (Fig. 2.2). In ethnic terms, the latter were not a likely source of migrants, suggesting again that the emigrés are best described by their individual motives.

Brief periods spent with five groups, one Agarabi and four Tairora, indicate that individual male emigrés are prevalent if variable phenomena. Their significance may outweigh their total number, for they often connect divergent or otherwise unlikely places of origin and destination. This is additional to the fact that neither leaving one's natal group nor entering an adoptive one is wholly free of adverse implications. The frequency of the individual male emigré indicates a web of contacts cutting across the embattled field and demonstrates the *de facto* openness and mobility of local society, which can only be one of the fruits of circulation.

It is true that the data and observations described are nearly contemporary, from the 'fifties and 'sixties, while the implications being drawn allude to precontact conditions some twenty or thirty years earlier. Although open warfare had ceased by the time of the particular movements reported, inter-group hostilities remained. There was fear of physical aggression or sorcery and generally speaking an active distrust of outside persons whose loyalties could be challenged. In other words, the circulation and migration of individual males is not simply a post-contact phenomenon.

Circular movements

Of the fourteen kinds of movement identified in precontact Tairora, ten are circular and six exclusively concern destinations beyond the local or phratric territory. In the case of strong men, trading partners, and war parties (Table 2.1: types 5, 6, 8), some destinations are at or near the periphery of movement for the group of reference. Fighting at the behest of allies against their adversaries can involve warriors with a foe they might not otherwise face, far-flung groups that are simply beyond range of the primary network. Quite possibly, such involvements characterize some groups more than others, particularly those whose reputation and vigour are ascendant. Distant movements of war parties should be considered much the same as those for trading. Both activities are transactions that express and sustain an inter-district network and fighting supports these linkages quite as much as does the transmission of imported goods or the exchange of women. The involvement of individuals in circular mobility varies substantially. Some persons are relatively immobile. Their journeys beyond the local territory are to points closer at hand, are less frequent, less varied, or less protracted. Men more than women participate in the wider travels, except for inmarrying brides at time of marriage or for women accompanying their husbands. In 1954, an elderly resident of Agarabi-speaking Ayamoentenu described a journey of Matoto, in recent times probably the Tairora man of greatest renown. He had met Matoto about twenty-five years before, en route to Punano (Fig. 2.2). He remembered Matoto's boasting about the number of women in his entourage, stating they were all his wives and inviting the local man to count them. From Matoto's point of origin to Punano, about thirteen kilometres, was in those days a distance few undertook to travel, yet indications are that Matoto's network stretched for comparable length in other directions. The Ayamoentenu informant, once a man of influence, had never visited Matoto's base at Kambuta and Bahi?ora, which today are part of the Abiera community (Fig. 2.2).

What may be called Matoto's circulation pattern is one of the defining traits of men of renown, as Salisbury (1964) and Chapman (1976) have suggested and as was recognized in my detailed account of his career (Watson 1967). By itself, the wider and more frequent travel is impressive for the risks involved and the prowess and success implied each time the traveller returns safely. Matters of style apart, the conduct of business, gathering of intelligence, and making of contacts abroad are the very substance of prominence and local leadership. It is assumed, no doubt correctly, that such a man as Matoto depends upon sources of knowledge and allies which his fellows may only learn if he chooses. It is said that he makes private agreements and receives secret payments for bringing warriors to help a distant partner, for the sudden betrayal of one ally on behalf of another who are their enemies, for habouring refugees, or for permitting the killing of a murderer or sorcerer within his district. Such suppositions indicate that the movements of a strong man are strategic to his role and serve broader ends than demonstrating his prowess and fearlessness.

The travels of the strong man may be used to measure the circulation pattern of the region. If his movements involve the maximum range and risk, as both logic and evidence suggest, then they define the limits of local circulation. No one travels farther.

Movements that generate movement

Some of the kinds of circular mobility generate other movements, including migration. The points and pathways established and maintained by the journeys of strong men, traders, brides, affines, and war parties prefigure potential migrations for individuals or blocs fleeing untenable local conditions. Within this regional system, circulation goes far beyond the short-term purposes of trade or hospitality and of individuals who move temporarily from their home base to stop briefly at other points in space. Women from one place who marry into another group are said to make the road their kin follow. So likewise does the male sojourner make roads or keep them open. In either case, inmarrying women or mobile men must reduce the facts of physical distance and separation so as to ameliorate their personal isolation.

The effect of such movement is to superimpose a map of social triangulation upon one of simple points, spaces, distances, and physical difficulties. Transformed into an order of social distances, the farthest points may become socially nearer than others close at hand or merely equidistant. Allies may be secured; enemies outflanked or held at bay.

The circulatory lifelines not only extend nearly to the ends of the known but also are the reasons that most of it is known. The country beyond can be viewed from within: like the Saruwaged range (Fig. 2.2: inset), considered the abode of the dead; or it is reported by partners who live at the periphery: peoples whose own periphery naturally extends another reach beyond. Fantasy holds sway in beliefs about this beyond: naked peoples utterly promiscuous and incestuous, nesting in trees and eating only bananas; folk undaunted by menstrual blood, who deliberately collect and consume it; cannibals; or a poisonous land whose very vegetation is deadly to those who stray there.

The circulatory lifelines discriminate, for otherwise the known territory would be subdivided only by distance and quarter. Their discrimination is largely political; there is much less reason for economic differentiation. True: salt, shell, pots, plumes, bowstaves, and stone axes must reach the centre by some path from the periphery, for no one possesses their source. But the same items are available to all centres both one's enemy's and one's own, and the particular routes and circulatory pathways do not appear critical. Partners may be necessary to ensure imports but there are no monopolists and little variety in the items traded. The uniqueness of a group lies less in some particular production than in its local reputation and strength, its particular alliances and enmities. These are the details with which the traveller in precontact Tairora must conjure.

Conclusion: hypothesis

This paper opened with a synopsis of the mobility patterns of northern Tairora. The regional field was represented macroscopically as a great flow of people recurrently lodged in or being dislodged from the various residential bases that comprise the social landscape. For ordinary comings and goings as well as for farther journeying, the individual types of movement within northern Tairora space correspond to many of the social roles of local society: males, brides, strong men, trading partners, fighters, refugees. People are defined by where they go. The features that stand out in this mobility system are the risky, long-distance contacts of strong-man and trading partners, the sometimes distant migration of inmarrying women, and the intermittent disruptions of local communities, resulting in the temporary or permanent dispersal of their personnel as either founders of new settlements or refugees among existing communities. As refugees, the dispersed sometimes merge with their respective hosts, sometimes return to their point of origin. This rooting and uprooting of people is presupposed by the regional flow described.

Compared with the stock image of individuals indelibly fixed by kinship in a certain personnel and anchored by birthright in a certain place, the northern Tairora reality is distinctly transitory, even precarious for some individuals. Situations arise in which the individual has no choice but to move, because of either the hostile action of outsiders or internal friction. When dislodged, he moves most safely in concert with others who both share his fate and with whom he expects the best chance of success. Whether he is proven right or wrong partly depends on those with whom the migrant bloc choose to cast their lot and on how closely compatible are the two sets of people: hosts and refugees. In a field of such numerous partnerships, varying in degree from amity to extreme enmity, a perfect match between resident and immigrant is perhaps impossible but prospects may be improved if the refugees flee to the edge of their social range, rather than seeking asylum close at hand. At the periphery may be found peoples known through prior contacts but perhaps relatively neutral toward their particular friends and enemies. Greater harmony is thus possible and the danger of betrayal reduced.

We can of course distort the picture of northern Tairora by dwelling overmuch on the mobility of its population. Not everybody is at a given moment on the move. If they were, it would scarcely be possible to speak of residential bases or of personnels as standing with others in a field of partnership. Indeed, some residential communities succeed notably in remaining identifiable and continuous, indefinitely riding successive waves of immigration and emigration by being strong, holding their territory, and avoiding dissolution through expulsion and the refugee process.

How do some individuals manage to spend all or most of a lifetime in one territory when a significant number are obliged to move, sometimes more than once? The most satisfactory answer to this question will be framed in ways that advance the cross-cultural study of circulation. The following argument, drawn from a larger analysis of partnership

33

(Watson 1980), invites comparison between northern Tairora mobility and that of other fields. Here, for brevity, definitions and the supporting argument are kept to a minimum.

In the search for a general explanation of Tairora mobility, one problem is that the economic rubrics most familiar in studies of population movement do not apply. The flow of individuals and blocs of people in this field obviously cannot be accounted for by an outside market for labour or goods, opportunities for schooling, or the lure of towns. It is not some differential of earning power, a higher price for products, access to better resources, a less arduous life, or even a more exciting one that produces the mobility of the precontact Tairora. Because this case does not fit rubrics now acceptable, it is worthwhile to discover how it might be interpreted.

The northern Tairora region is a field of open partnerships. Intergroup relationships fluctuate and are negotiable; yesterday's friend or kin can be tomorrow's enemy, and vice versa. Such a condition arises in a field in which the effective, top-of-the-ladder, social units are numerous. In northern Tairora, as noted, these social units are personnels defined by co-residents. Besides being (1) numerous, they are also (2) small, (3) near to each other, and (4) similar in production and reproduction — indeed, similar in culture. These four conditions, which may be compared with mobility fields other than northern Tairora, appear to be systematically related to the local fluidity of its population.

The similarity of origins and destinations in the northern Tairora field contrasts diametrically with the heterogeneity of culture and opportunity that typifies mobility in modernizing populations. With respect to partnership, similarity means that complementarity is lacking as a possible element of regional stability. If partners were differentiated as to production and hence less readily substituted one for another, changing them would be more difficult and less frequent. Circular and migratory flows might also be slowed if the same knowledge were not adequate nearly everywhere.

Local personnels in Tairora engage in substantial transactions with each other: transactions described synoptically as women, wealth, and warmaking. The intensity of this traffic is predictable from the crowding of personnels or, if one prefers, from the lack of a wider polity to unify different personnels. Wider polities would reduce the number of top-of-the-ladder personnels, which in turn would substantially change the dynamics of Tairora partnership from numerous and open to few and closed.

Small personnels are more precariously situated than large ones for sustaining any sizeable or steady flow dependent upon their own production and reproduction. To maintain sufficient capacity to support partnerships, a personnel generally requires three strategies. The first is to intensify production and at least keep pace with rivals. The second recognizes that it is efficient to develop friendly partners or allies, upon whom one can depend. That there are numerous other small personnels nearby makes the engagement of friendly partners both feasible and mutually advantageous.

Every additional partner means an extra obligation to sustain the flows of women, wealth, or warmaking upon which northern Tairora society depends. This pressure underscores the need for the third strategy within the local group: to maintain a sufficient population. Insofar as possible, a personnel must keep approximate parity of production and reproduction with its rivals or potential rivals, with friendly as well as hostile partners. Since the partnership of friends is a key to survival in this field, a personnel must develop the capacity to be beneficial in order to keep the sort of counterparts who can reciprocate in like manner. The need to maintain full ranks means that, from time to time, local groups will attempt to recruit additional members, whenever their numbers have been thinned by death or the departure of disgruntled insiders.

Enough has been said to place in brief perspective a number of implications of mobility systems like northern Tairora. Intelligence is critical concerning the plans and manoeuvring of all partners, friendly as well as hostile. Conflict of intention or a surprise betrayal by friends is no more acceptable than a surprise attack by enemies. Prudence requires that a personnel's intelligence matches that available to rivals and greater intelligence is obviously advantageous. Thus the key position of strong men, who are willing for personal advancement to make dangerous journeys and develop various distant contacts. From these contacts may emerge further partnership, entailing the exchange of women, help in fighting, and even now and then a haven for refugees.

Rivalry is costly and intense, a condition which escalates in partnership systems like northern Tairora. Inevitably there will be losers among the groups of such a field. This chance is multiplied by openness and the fact that local groups, apart from production, strength, or luck, are fairly interchangeable. Thus rapid shifts of loyalty may occur as one's partners seek to optimize their alliances. Such a process cannot fail to have its occasional victims: personnels who have failed to keep pace in production, reproduction, recruitment, or beneficial partnership. Such personnels may be catastrophically weakened through the sudden desertion of partners for more promising ones. Losers are the recurrent source of refugees, the potential recruits for undermanned local groups whose amities and enmities are reasonably congruent.

In the recruitment of outsiders may be served both the personal ambitions of strong men and the needs of the community to match its rival. While leadership is not formalized in northern Tairora but based rather on charisma or credibility, it is only the wide-ranging strong man whose contact with the refugee community may antedate their expulsion. If so, he will be in an advantageous position to pose as local spokesman and protector of the new immigrants. As outsiders, they are beholden to him but equally as clients help strengthen his hand. If the recruitment of refugees is sometimes the salvation for a depleted local group, it is also the seed or sometimes the culmination of internal tension and fission. Factionalism, usually present between strong man rivals and their respective contingents, becomes exacerbated by differences between residents-by-birthright and immigrants. The latter have no choice but to assert equivalent rights to resources and security. The system presents, in the case of refugees, a recurrent loosening of ties to the residential base. In recruiting the uprooted to help maintain its foothold the local group may find, paradoxically, that some of its original members later become disaffected. Thus, in practice, refugees are sometimes the replacement for departing natives.

This, then, is the outline of a mobility system associated with particular cultural and demographic circumstances. The argument is that top-of-the-ladder personnels, which are numerous, small, and proximate, will be intensively engaged as partners in an open field. When they are essentially undifferentiated in type of production, that openness will be greater and we should expect to find fragile partnerships and a regional mobility cycle similar to the precontact northern Tairora. This proposition is offered as being testable and hence potentially useful in advancing the comparative study of mobility.

Northern Tairora does not fit the stereotype of a static, preurban world nor the complementary notion that mobility is a modern or urban phenomenon. Were there a weighted scale to permit comparison, the circular and migratory flows of northern Tairora might in fact exceed those of some highly mobile contemporary fields. The comparison, however, could do little but dramatize the misconception of static, preurban society. Unless one is content with gross distinctions, different systems of population movement must be compared in more respects than magnitude. We can at least be clear that the transition from pre- to postcontact Tairora is not the mobilization of a once 'sessile' population or one with 'little genuine residential migration' (Zelinsky 1971: 222, 230).

Although the single exception challenges the general claim, evidence may count for little if considered unique. Over two years of ethnographic exploration in the immediate area and extensive consideration of the wider, regional literature (Hays 1976) make it most unlikely that the higher mobility of northern Tairora is the outcome of singular local conditions. That we are considering a phenomenon of at least regional proportions is attested in the independent findings of a genetic survey:

Under conditions of extreme isolation and low population size, heterozygosity decreases in succeeding generations and loci tend toward fixation in both theoretical and experimental populations . . . . However, in the Eastern Highlands heterozygosity does not decrease and genes do not tend toward fixation. Probably conditions of more extreme inbreeding than found in the Eastern Highlands are necessary before a tendency toward gene fixation would be seen . . . . Careful collection of ethnohistory and anthropological data over long periods of time may prove as valuable as genetic surveys in unravelling the nature of human heterozygosity. The mixing and exchange of genes is probably far more extensive and isolated populations far less static than previously believed in the Eastern Highlands (Wiesenfeld and Gajdusek 1976:188).

Once these data yield to general explanation, the case for a substantial, preurban mobility is no longer limited to genetic or ethnohistorical data, and a systematic account of one such system strengthens the prospect for both regional and broader generalization. The theory sketched here is not a particularized version of northern Tairora or Eastern Highlands history; it is a testable prediction of mobility in specified preurban circumstances. Not only can these circumstances be recognized and defined but their equivalents are surely also to be found among populations besides those of the Eastern Highlands province. Crowded fields, in short, make for a markedly mobile population and suggest the wider occurrence of preurban systems of high mobility.

To the extent that present mobility theory presupposes static preurban societies, a stronger and more comprehensive formulation is required. Nor is this need only prompted by the highly mobile, crowded field. Although not dealt with in this paper, one readily thinks of the theoretical extension required to describe the inherent mobility of the *un*crowded field, sparsely populated with hunter gatherers, such as those of Australia or Athabaska (Wilmsen 1973). These preurban populations can no more be called 'sessile' than the village horticulturists of the Eastern Highlands. To the extent that a 'mobility transition' (Zelinsky 1971) presupposes a shift from sessile to mobile societies, it will be limited to cases in which truly static, pretransition populations can actually be found. But to be adequate, such a theory of transformation must account for more than gross contrasts of magnitude, however these might be shown.

Magnitude is probably not the greatest difference between pre- and postcontact mobility ·of the Tairora. The age and sex distribution of migrants and circulants has surely changed, as well as their postcontact destinations, distances, and directions. The precontact movements of refugees, for example, comprise representatives of both sexes and all ages whereas in modern labour circulation young males predominate. More basic are the contrasts of purpose and motive. A life-and-death urgency and high risks often attend pre-contact movement. Particular mechanisms — notably ideology — are required to provide a modicum of local stability and *esprit de corps* for residential groups in the midst of such a fluidity of personnel. The ideological aspects of the situation have only been hinted at here and more by way of implied paradox than understandable connection. In the mobility field of the precontact northern Tairora, gross magnitudes can be identified to compare with the modern movement of their descendants who respond to markets for labour and goods, the tug of town life, or the monotony of the village. But we have no such differences between origins and destinations as occur with modern flows.

3 Precontact movement among Eivo and Simeku speakers in central Bougainville

Michael Hamnett

Anthropologists, many of whom purport to describe traditional societies, have given little attention to movement behaviour among groups other than hunters and gatherers, nomads and pastoralists. The great emphasis on patterns of post-marital residence in studies of the precolonial condition of tribal and peasant societies implies that all agriculturalists are sedentary (see Fox 1967:77-96; Keesing 1976:242-265; Murdock 1949:79-90). Where population movement is addressed (eg. Murdock and Wilson 1972:255-265), the dominant assumption is that residence groups transfer as a unit.

The geographer Ward (1971:82) claims that in precontact Papua New Guinea there was some residential mobility, primarily in response to population pressure, warfare, and natural disaster. Like the ethnographers, Ward presents no substantive data nor suggests the incidence of movement or distances travelled. Similarly Zelinsky (1971), in his hypothesis of the mobility transition, characterizes 'premodern traditional societies' as having 'little genuine residential migration'. He notes that 'some circulation (i.e. short term absences) might occur, . . . normally within well-trodden social space', but admits the evidence for premodern mobility is patchy and imperfect (Zelinsky 1971:234-5).

This paper considers precontact movement for a group of Eivo and Simeku speakers on the island of Bougainville, Papua New Guinea (Fig. 3.1). The reconstructed distribution of settlement and individual households reveals that both migration and circulation were a feature of precontact life and also demonstrates how past movement patterns may be quantitatively described even where no written record exists. Among the Eivo and Simeku, households changed their usual place of residence about every six years when 'usual' was defined as the settlement where persons generally ate for at least a year. Most such moves involved not only the relocation of households but also changes in the composition of settlement groups. Most residential transfers occurred over short distances and the decision to remain was made gradually, especially if households maintained multiple domiciles. Circulation, or temporary changes in residence of less than six months, was frequent and also mainly over short distances, although shifts of up to nineteen kilometres were recorded.

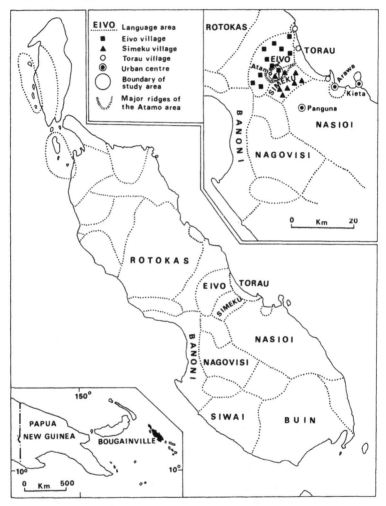

FIGURE 3.1 *Eivo and Simeku speakers and the languages of Bougainville island*

Collection of field data from August 1974 to May 1976 focused on the village and hinterland of Atamo, inhabited by speakers of both Eivo and Simeku (Fig. 3.1). From before European contact (c1925) until the mid-seventies, settlement in the Atamo area has changed from several small hamlets, comprising one to ten households, to a village of fifty households with a series of satellite hamlets. Reconstructing the precontact pattern was thus necessary to establish a baseline from which to describe the contemporary position as well as the changes which occurred both during and after the advent of Europeans. Following Bedford (1973a), reconstruction was achieved by means of life-time residential histories. Throughout the Atamo area, only thirty-six out of a *de jure* population of 403 (December 1974) had been born prior to European contact, so that their various places of residence had to be identified as well as other contemporaneous settlements. Use of retrospective data to establish past events is admittedly fraught with problems, especially in tribal societies, but Chagnon (1964) has shown in research with the Yanomamo that a high level of consistency can be achieved in reconstructing patterns of settlement and the composition of residential groups. In addition, the advantages of using the resultant diachronic data to understand process and change far outweigh the disadvantages of relying upon synchronic, but more recent information.

Before collecting residential histories in the Atamo area, genealogies were obtained for every individual. These included not only names of and relationships among all persons, but also places of birth, death, pre- and post-marital residence. For those born outside, it was estimated approximately when they entered the population, with whom, and why. Baptismal, marriage, and death registers established by the nearby Catholic Mission in 1929, four years after European contact, provided sufficient documentation to interpolate the likely dates of each vital event recorded in the genealogies.

Life-time residential histories were obtained from the oldest male and female of each sibling set represented in Atamo, including at least one adult from every household. Questions were asked about the establishment and dissolution of settlements, the places where people were born and died, and the names of inhabitants at frequent intervals in a settlement's history. Moves were dated by reference to the genealogical information previously collected, the accuracy checked of details about who was born and died at particular settlements, and information added about residential changes not contained in the life histories. The resultant data were sufficiently consistent to permit the nature of and changes in residential groups to be reconstituted, the completeness of which were further checked against site and feast histories.

Individual mobility was more difficult to trace, since most changes of residence were made by households as long as the founding couple remained intact. Independent adults, whether previously married or not and especially males, were quite mobile; some effectively had no place of residence for several years. Moves for marriage were moderately easy to discover but information on other kinds were not and the data are incomplete. The primary focus of this paper is upon the period 1920-24, the five years prior to sustained European contact, although the three post-contact periods have been discussed elsewhere (Hamnett 1977). Placed within the context of movement patterns from 1920 until 1976, those described for 1920-24 seem representative of the years 1900 to 1925. Data on movement prior to 1900 are inadequate to determine the degree of fit with the 1900-25 patterns.

When reciting residential histories, individuals might comment: 'We sometimes spent time at this other settlement', or 'We used to gather almonds or hunt or go there for feasts'. Such statements reveal that the Atamo were making temporary, circular moves in pre- and early contact times similar to those of today. Individuals had multiple residences, which they occupied at quite regular intervals of up to several months. It is possible to plot the distances travelled in the course of such moves, catalogue the reasons given, and identify the tasks performed during such periodic absences. While the distances travelled during circulation from a person's usual residence may have increased somewhat since European contact, there is no evidence of a decline in either their frequency or duration of absence. More permanent changes of residence were similarly consistent from 1920 through to 1974 (Hamnett 1977).

The Atamo area

In the mid 1970s, Atamo village and its twelve affiliate hamlets were located in an inland valley at the Eivo Census Division, central Bougainville (Fig. 6.1, cf. Fig. 3.1). The 403 inhabitants of this village and hamlets are descendants of Eivo and Simeku speakers who occupied the valley long before European contact and have shared it with Kanavitu villagers (Figs. 3.1, 3.4). At contact, Atamos claimed approximately seventy-two square kilometres of land. Most lies in the valley bounded to the south by the Crown Prince range where altitudes reach 1280 metres, and flanked by two major ridges trending north towards the coast. The Atamo area is separated from that of the Simeku-speaking Kanavitu by a smaller ridge that divides the valley. Two major rivers, which join to form the Arakabau just north of where Atamo village now stands (Fig. 3.2), dominate the valley, whose steep ridges are cut by small streams.

Since before contact, Atamos have been shifting cultivators. Until the Second World War the principal root crop was taro, with fallow periods which ranged from twenty to twenty-five years. Plantains, yams, and tobacco were also grown. Despite the amount of land required for gardening, given this long fallow, agriculture was not a significant factor in limiting settlement size or influencing the decision to move (Hamnett

1977:46-53). After contact, the Atamo cultivated Chinese taro, bananas, pineapple, sweet potato, and European garden vegetables. Protein was provided by pigs, both wild and domestic, opossum, birds, river fish, and prawns. Hunting, done primarily in preparation for feasts, was secondary to pig husbandry in terms of both food obtained and effort expended. Since pigs require new forage, owners would shift their place of residence to provide 'fresh ground' for them (Hamnett 1977:53-55).

Of the two languages spoken by the Atamo, Eivo belongs to the Rotokas family of the northern stock of Bougainville's non-Austronesian languages (Fig. 3.1) whereas Simeku is a sub-language of Nasioi, which is of southern stock (Allen and Hurd 1963). They share about seventeen percent cognates. Before Europeans came, although less so today, members of both groups inter-married, exchanged pig feasts, and occasionally lived in the same settlements (Hamnett 1977:31-4). Most adults precontact spoke only their own language but some were bilingual.

Throughout history, the cultures of Eivo and Simeku speakers may be considered congruent: standards of expectation, of perceiving, believing, evaluating, communicating, and of acting were sufficiently similar for both groups to live within the same community and interact more frequently than with language mates in other communities. Equivalents exist for almost all Eivo and Simeku concepts, with language, kin-term categories, mourning song styles, and norms for choosing feast recipients the only cultural differences cited.

Both language groups shared an ideology of matrilineal descent. Each individual belonged to one of nine matrisibs, or clans. Atamos describe most social interaction, including the formation of residential groups in terms of clans, which are said to belong to moieties. Through membership of moiety and clan, kin relationships include all Eivo and Simeku speakers as well as those of other languages. Rights to land transferred through women. The most secure were held by the matrilineal descendants of each tract's original inhabitant. Access to and use of land for subsistence was freely granted through gift, sale, exchange, or as compensation for aid in feasting and warfare.

Members of both language groups recognized three major types of spirits: 'land spirits', 'ghosts', and 'malign spirits', the first two of which were thought to have places of residence just like humans. These spirit shrines were the sources of supernatural power and consequently precontact settlements tended to be sited on or near them (Fig. 3.2). In addition, 'land spirits' protected the land rights for their true owners and breaches of norms of land tenure, as of any other norms, were punished by supernatural sanction (Hamnett and Connell 1981).

Major pig prestations, the traditional means for males to achieve renown and become 'big men', occurred at 'memorial feasts' which

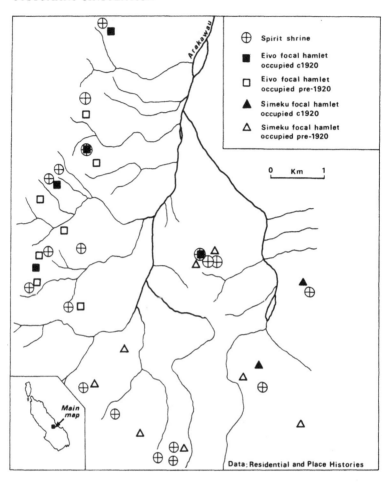

FIGURE 3.2 *Spirit shrines and focal settlements in the Atamo area, c1920*

were obligatory for a deceased father, mother, sibling, spouse, or child (Hamnett 1977:104-9). Prestations ranged in size from one to at least a dozen pigs, each of which was accompanied by cooked vegetables. 'Big men' were the group leaders, managers of land, and actively involved in most aspects of Atamo life. They attained their position through competitive feasting and actively recruited co-residents to aid in preparations which, for large prestations, took about two years and much labour. To attract followers, 'big men' granted rights to land, gave away trees, and used the rhetorical resources of kinship and descent. Although they could and did use sorcery, strong sanctions existed against its practise on co-residents, followers and close kinsmen. Warfare was infrequent, so that possessing its skills did not warrant renown (cf. Hamnett and Connell 1981).

Patterns of settlement and movement before c.1920

Prior to European contact, Atamos occupied the western, southern and southeastern portions of the valley they shared with Kanavitus. Settlements were located on inland ridges at altitudes between two and six hundred metres, but ridges nearer the coast, the coastal plain, and alluvial land on the valley floor were unoccupied, although used for hunting and gardens.

Compared with north and south Bougainville, this area was sparsely populated (cf. Brookfield with Hart 1971:69; Hamnett 1977:36). Genealogies and settlement histories yield an estimated *de jure* population, in 1920, of 165. This represents about 2.3 persons per square kilometre for land claimed by Atamos and about 4.2 persons per square kilometre for that used for habitation and subsistence gardens. However, this estimated population is undoubtedly too low, with the elderly and very young the most likely individuals omitted from household reconstructions.

Movement into the Atamo area from the south and east reflects the migration of groups and of individuals to marry. Some migration histories trace the clans of both Eivo and Simeku speakers to Siwai and Buin (Fig. 3.1) while others originate in Nagovisi, some of whom moved through Nasioi-speaking areas and others directly into Eivo territory. The more recent source of most Eivo and some Simeku speakers is the western edge of the Crown Prince range. Only one of the nine clans represented in Atamo at contact did not have a spirit shrine, which indicates its recency. Informants knew of migration histories for seven clans, three of which were considered to have recently arrived about five generations ago.

Although the details of only one migrating clan are remembered, they are probably indicative. As a result of an arranged marriage, a woman migrated 7.5 kilometres from a Simeku to an Eivo settlement. Before this occurred, she was given land to legitimize her transfer to Atamo and some members of her clan subsequently followed. It is also likely, as in this case, that clan histories summarize individual acts even though reported as group movement. Similar circumstances surround the recent immigration of sub-clans of those clans represented at contact. In some cases, 'big men' emphasized common descent categories and extended land rights to immigrants; in others, arranged marriage and a gift of land explained their presence.

Even before contact in 1925, the location and composition of settlements changed from time to time. Genealogies and settlement histories indicate that previously inhabited ridges near the coast were abandoned

45

about 1900, perhaps as a result of fighting. Torau speakers (Fig. 3.1), who colonized the coast north of Atamo in the late 1800s, may have attacked Eivo and Simeku settlements, whose survivors moved inland to join the core area of Atamos.

Within the Atamo valley itself, changes in the distribution of people can be seen when the 1920 location of focal settlements, larger hamlets with resident feast-givers, is compared with those occupied earlier (Fig. 3.2). Explanations vary of why this occurred. In the southeast quadrant of the valley, the border between Eivo and Simeku habitation, focal hamlets were occupied by Simeku speakers of a clan that died out in the first decade of this century. The decline of mixed language (Eivo/Simeku) settlements located on ridges between what now constitutes Atamo and Kanavitu territory (Fig. 3.4) followed the death of a 'big man' and surviving inhabitants either joined other groups or established new settlements in different areas. Despite the difficulties of dating settlement change, the distribution of Atamo people was in flux before 1920 and the underlying processes paralleled those occurring immediately prior to contact.

Atamo settlement in 1920

For the Atamo area, 1920 is the earliest date for which the reconstruction of settlement and population is possible (Fig. 3.3). In that year, settlements ranged in size from one to ten households with a mean of 2.5 households, which in turn contained an average of 3.8 persons. Located on or near spirit shrines (Fig. 3.2), these hamlets were composed of individuals related through primary kin ties and one or more members of most households held secure rights to land. Its feast organizers, 'big men', usually lived in the focal settlements, whose larger size may have been enhanced from the late nineteenth century by the impetus given to prestation by the introduction of steel.

Although such a summary indicates the pattern of settlement and residential groups, it provides no indication of the degree of population mobility. Precontact, households and individuals moved in two ways: through short-term cyclical changes in location, or circulation, which did not involve abandoning the original settlement; and by more permanent residential shifts (migration) which led to the redistribution of people. In general, cyclic shifts of less than six months' duration were made to harvest almonds, to hunt game, or to clear land for food gardens.

Most subsistence gardens of the early 'twenties were within two kilometres of the hamlet of residence and located on or adjacent to the same ridge. Temporary houses and shelters of palm leaf were constructed at garden sites for the week or two spent there. Occasionally, garden houses were more substantial if household members decided to leave the hamlet and reside in them.

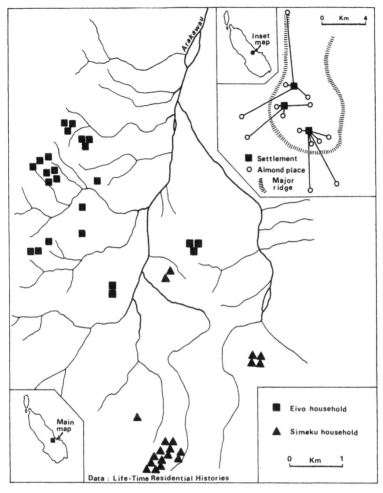

FIGURE 3.3 *Distribution of Atamo households, 1920*

Almond harvests drew individuals and entire households to bush areas for up to two months at a time. Information from three focal settlements for the period 1920-24 (Fig. 3.3; inset) indicates that trees were picked both within the Atamo valley and beyond the major ridges, thus involving return journeys of up to twelve kilometres. Almond areas were harvested regularly, perhaps annually and biennially, and if a feast were in preparation then some residents would remain at the settlement while others left to gather nuts. Hunting also occurred when feasts were being prepared and occasionally in conjunction with almond picking. Trips to hunt opossum and pigs might involve whole households

47

relocating to the almond and other bush areas for as long as a month or two.

For the Atamo area as a whole, the most complete reconstruction of patterns of precontact movement refers to the years 1920-24 (Fig. 3.4). Despite a great deal of mobility, there is no indication that the underlying processes or resultant distribution were exceptional. The only singular event was the arrival about 1921 of a punitive expedition to the Atamo area, which affected but two settlements. Apart from this, fluctuations in the size and distribution of settlements reflect the changed status of households: their establishment, generally through marriage; dissolution, usually through the death of either spouse; and movement.

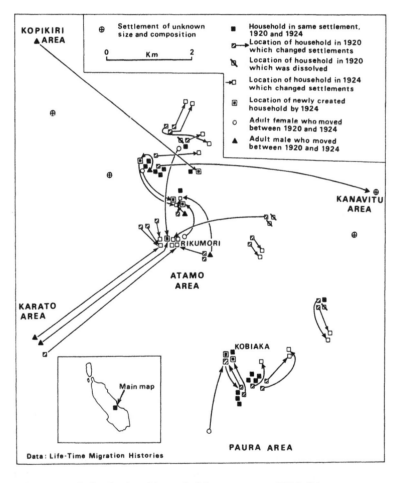

FIGURE 3.4 *Individual and household movements, 1920-24*

Although the numerical increase of settlements and households between 1920 and 1925 was small, two in both instances, fifteen out of seventeen hamlets underwent changes in their size and composition. Establishment or dissolution of households had a minor impact compared with relocation: out of forty-three households present in 1920, four were dissolved as a result of death whereas twenty-three shifted their usual place of residence. Eight of those moves were within and fifteen across a land-tract boundary. In addition, one of five households newly established through marriage relocated across a land-tract boundary, raising to twenty-four the household moves made during five years.

Not only was household movement directed into some settlements and out of others, but also most occurred between adjacent areas and over short distances (Fig. 3.4). In ten out of sixteen cases, such transfers contributed to either growth or decline of hamlet size. Why should movement occur at all, and why did it result in the consolidation, dispersal, or relocation of residence groups? Such questions are most effectively answered by examining case materials.

Case I: Rikumori hamlet

Rikumori hamlet (Fig. 3.5) illustrates most dramatically the processes of settlement change between 1920 and 1925. About 1920, there was no settlement at Rikumori (Fig. 3.5a:e11). By 1923 Katepato, a man who held fairly secure title to Rikumori land, had established a hamlet. There had been sickness at his former settlement (e6) which caused the death of an old woman and Katepato wanted a place, free from misfortune, in which to make a pig feast. He was married to two women, one from the Atamo area and the other an inmigrant from Karato (Fig. 3.4).

Besiatuan, son of Katepato's mother's mother's brother, also lived at e6 and helped establish Rikumori. Besiatuan had migrated to the Atamo area as a child after his parents were killed two valleys to the north. Having married two women, he moved to e6 and was given rights to use land which his father's clan obtained two generations before.

Each year the two men, Katepato and Besiatuan, travelled to the Karato area to gather almonds from trees which Besiatuan had acquired through his parents. While at Karato, Katepato convinced a man (Penuma) whose wife belonged to his clan to move to Rikumori and share the clan land. This Penuma and his wife did, by the way of e7 (Fig. 3.5a). Katepato next arranged a marriage between his mother's sister's daughter and another man from Karato, who inmigrated as a result. During the next almond harvest, Katepato visited clan relatives at a settlement in the Karato area which was in decline following a 'big man's' death. While there, he convinced another man (Akamoko) and his wife, who was reputed to hold title to the Rikomuri land, also to relocate.

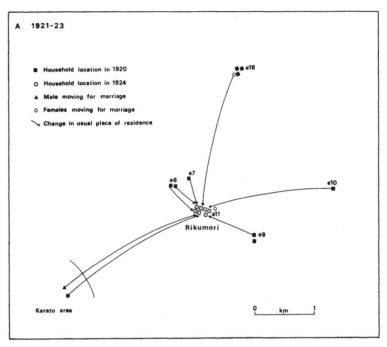

A 1921-23

■ Household location in 1920

□ Household location in 1924

▲ Male moving for marriage

○ Females moving for marriage

↘ Change in usual place of residence

Rikumori

Karato area

0   km   1

FIGURE 3.5a *Household movement into Rikumori, 1921–23*

Apukapunua, brother of both Penuma and Akamoko, resided on his wives' land at hamlet e9, less than a kilometre southeast of Rikumori (Fig. 3.5a). He was living with a relative by marriage with whom he was at odds, and when his brothers moved to Rikumori, Apukapunua followed to help Katepato with feast preparations. Akupekato, yet another brother of Penuma and Akamoko, resided at e10 on his wife's land. This hamlet had been in decline for some time and when its 'big man' died, Akupekato began spending more time with his brothers to help Katepato organize his feast, for which ability Akupekato was well known in Atamo. With the death of the only other male at e10, Akupekato also migrated to Rikumori.

Within three years, Katepato had established Rikumori and expanded it to seven households. During consolidation, individuals circulated into and out of Rikumori which, at its height, contained only one person born and raised on the sole tract of land in Atamo to which Katepato's clan had claim. Two of the three females drawn by ties of kinship belonged to Katepato's subclan and their parents had lived at Rikumori before they migrated upon marriage to Karato. Their husbands were all brothers, as were the heads of the additional households that transferred to Rikumori, even though their wives held no title to its land. All inmigrants who arrived after Katepato and Besiatuan originated from hamlets which were either in decline or small and without resident feast givers. Marriage accounted for only one out of five households which shifted to Rikumori, and the dominant reasons were to help with Katepato's feast or claim land to which their adult female members had some rights.

The decline of Rikumori illustrates how rapidly a hamlet can disperse when the occasion for consolidation has passed. After the Rikumori feast had occurred, Apukapunua abandoned his house and established a new hamlet at e19 (Fig. 3.5b), to raise pigs for his own ceremony on land to which his two wives held quite secure title.

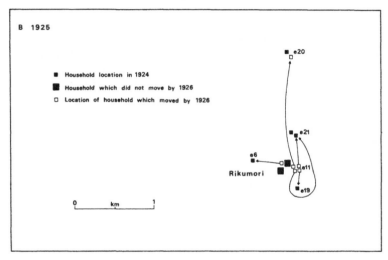

B 1925

■ Household location in 1924

■ Household which did not move by 1926

□ Location of household which moved by 1926

Rikumori

0   km   1

FIGURE 3.5b *Household movement out of Rikumori, 1925*

Katepato, the hamlet founder, now preparing for a second feast and feeling that his pigs were not doing sufficiently well, established a new settlement (e21) on land to which his first wife held full title. While there, Katepato married a third wife, sister of his second, who thus joined him. Penuma, of the same clan as these two sisters, accompanied Katepato in the move to e21. After Katepato had left Rikumori, Besiatuan returned to e6, where they both resided previously and where Besiatuan had continued to maintain a house and his rights to land which he had been given.

Akupekato departed Rikumori at the same time as Katepato and joined a man with whose clan relatives he had previously lived. Some time was still spent at Rikumori but primary residence shifted to e20 (Fig. 3.5b). There now remained only the households of Akamoko and Nesinua, whose wives were told by Katepato that the hamlet was sited on their land and they should therefore stay while he went to use his wife's land at e21.

Within two years Rikumori, having expanded from one to seven households over three previous years, was reduced to two. Both households which remained after about 1925 contained inmigrant women who had been settled on 'their land'; conversely, those without such rights left when Katepato moved to his wife's land. During dissolution, individuals moved into and out of Rikumori just as they did during its consolidation. Even after the residential group dissolved, those households which remained at the hamlet continued to aid Katepato and were considered his followers.

Case II: Kobiaka hamlet

Narea was born and raised at the Simeku hamlet of Toriana (Fig. 3.6:s1), the largest and most stable settlement in precontact Atamo (Hamnett 1977:123). Since his parents had died when he was a young adolescent he, his brother and sister became members of their classificatory father's household. This was unsatisfactory and, after an argument, Narea and his brother established their own household. About 1919, Narea married a Toriana woman and the couple lived in this house, of which his brother continued to be part. After being married about a year, Narea founded a new settlement at Kobiaka (Fig. 3.6) to prepare a memorial feast for his father, even though he did not have full title to that land.

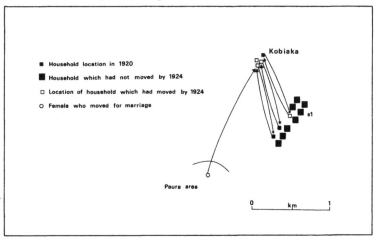

FIGURE 3.6 *Household movement into and out of Kobiaka, 1920-24*

In 1920 Kengkara, father of Narea's wife and with whom Narea associated most frequently, built a house at Kobiaka but continued to reside at times in Toriana. Also his brother Ben married a woman of the Paura area (Fig. 3.6), from which their mother had originated, and the newly-wed couple moved to Kobiaka. All three men (Narea, Kengkara,

Ben) convinced another clan relative living at Toriana to join them and help prepare the feast. This man built a house at Kobiaka but continued to reside mainly at Toriana.

During two years, Narea was thus able to attract two households from Toriana, in addition to his brother and newly-married wife. However, this residential group did not endure. Once the feast was over, the clan relative left his house at Kobiaka; a year later Narea's wife died in childbirth and the hamlet was abandoned as everyone returned to Toriana. Soon after, Narea married another of Kengara's daughters and re-established Kobiaka with his brother's help. Kengara also returned about 1925, when the memorial feast for his daughter and Narea's first wife was in final stages of preparation. Although he participated in the feast and stayed for short periods, his primary residence was at Toriana. Ben and Narea remained a further two years, after which Kobiaka was abandoned because the pigs were doing poorly, and founded another hamlet less than a kilometre away on the same tract of land.

Factors in household and individual movement

The case materials for Rikumori and Kobiaka, complemented by field data for other Atamo hamlets during the period 1920-24, indicate that different conditions favoured the stability of settlements and of migration into and out of them. The use of self-reported reasons to analyze behaviour performed by individuals has been much debated by social psychologists and is more heavily criticized than the use of retrospective records of events (e.g. Festinger 1964; Aronson 1968; Walster and Bercheid 1968). Reconstruction of processes that underlie precontact movement force some dependence upon reasons given by the Atamo for past actions (Table 3.1), checked and supplemented however by the attributes of those involved and the residential groups of which they were members.

In general, settlements increased their size from inmigration when a 'big man' was rising to power and gathering a following or, more specifically, preparing a feast. Certain households and individuals were capable of being dislodged, either locally or from areas with which the 'big man' and his followers had close kin or frequent interaction. These conditions were offset by others favouring dissolution. When misfortune, usually in the form of illness, death of a 'big man', or pigs doing poorly, befell a settlement, then outmigration rose. Departure of a 'big man' or a lull in feasting activity produced the same result. In addition, a settlement's stability was affected when few residents held secure title to garden land or shared close kin ties and when there was competition for renown or political following.

Movement into and out of settlements can be viewed as the observable

TABLE 3.1 *Reasons for household movement in the Atamo area,*
1920-25

	Eivo speaker	Simeku speaker	Total reasons
+ Establish new hamlet for pig husbandry	1	1	2
− Pigs doing poorly	0	4	4
+ Aid in feast preparations	4	1	5
+ Presence of relatives (consanguine or affine)	4	0	4
+ Seek use of other land	4	0	4
+ Move nearer food gardens	0	0	0
+ Harvest almonds	0	0	0
− Illness in settlement	2	2	4
− Death of a settlement resident	2	2	4
− Anger or social discord in settlement	0	0	0
− Punitive expedition	4	0	4
Unknown	2	0	2

− Reason for leaving settlement.
+ Reason for joining settlement.

result of particular decisions on where and with whom to live. As long as the core nuclear family of a household remained intact, its members moved as a unit. Household heads made this decision, as reflected in the fact that female informants could provide few reasons for residential shifts and most denied having ever known them. Women may have been consulted or offered opinions, but for a household to migrate was and is considered a male decision.

The destination of a migrant household was most determined by land tenure, kinship, and leadership. As a rule, Atamo households moved to sites where they held secure land rights, especially if the head aspired to gain a following and became a 'big man'. Since most households had rights to at least two parcels of land, where to relocate was greatly influenced by the quality of relations with land owners or managers at potential destinations. Number of close kin was similarly influential, with the most important being mother's brother/sister's child and parent/child links between the head or his spouse and an alter. Leaders also actively recruited followers, most successfully among households with a high propensity to move. As demonstrated by the Rikumori and Kobiaka cases, kin relationships were used as rhetorical resources and land rights granted to secure followers; consequently it is difficult to discern whether kinship, land tenure, or the persuasive powers of a political leader had the greatest influence on a household's choice of destination.

Whenever the core of a household was dissolved through the death of a marriage partner, there was a high propensity for the remaining members to move. A male surviving his wife generally maintained an independent household and even with remarriage was less likely to relocate, unless his relations with co-residents were strained or more status could be achieved by transferring to his wife's household. Females surviving their husbands were absorbed into an existing household along with the younger children, who remained with their mother even if she subsequently migrated to marry. If the father were the surviving parent, children often moved into the household of a mother's brother, unless their father had more than one wife, in which case the original household remained intact; soon married the sister of his deceased wife; or did not subsequently migrate to marry. The adolescent children, by contrast, were among the most mobile segment of the entire population and males, in particular, attached themselves to various households in different settlements until they eventually married.

Postmarital changes of residence, like other individual movement, had little impact upon the composition of settlements or residential groups. All six marriages between 1920 and 1924 resulted in new households. Three males moved to their wife's hamlet, one female transferred to her husband's settlement, and two couples established residence at a completely different site. Newly-married couples tended to live in settlements in which 'big men' or aspiring feast givers were resident, many of whom had arranged the marriages. The one polygamous union in 1920–24 saw the third wife move to her husband's hamlet.

Not all moves, of either households or individuals, entailed changes in usual place of residence. Households quite often maintained dwellings at two hamlets, even while spending most time at one. Multiple residences and frequent movement between them created a locational ambiguity and allowed Atamos to exercise options about the residential group with which they chose to affiliate. In addition, short-term mobility for hunting game and gathering almonds provided political leaders the opportunity to maintain contact with a territorially-dispersed network of individuals. From this network, 'big men' could recruit followers and co-residents from outside the Atamo area. Consequently these flows are not independent of but closely linked with permanent changes in the distribution of people.

Conclusion

Among the Atamo, residential change was an integral and unexceptional part of precontact life. If, in the course of fieldwork, people had only been asked where forebears had been born, had lived just before and after marriage, and had died, then most movement reported for the

1920-24 period would have gone undiscovered. By asking about successive places of residence, it can be shown that in most cases migration occurred between birth and marriage, marriage and death. Similarly, if it had been assumed that the Atamo maintained only one place of residence and could not make temporary shifts, then the existence of circulation and of multiple residence as a characteristic of precontact mobility would have remained unknown.

This paper demonstrates that, for at least one 'premodern traditional society', population movement was more frequent than has been implied (Zelinsky 1971) and that precontact behaviour can be described and quantified by using the methods of oral history. Shifts in place of residence, some quite permanent and others distinctly temporary, occurred to areas outside the Atamo and both within and across land tracts. Circulation, or short-term changes, generally involved transfers between hamlets, reflected the ambiguity of having more than one 'usual' domicile, and in some cases led to more permanent relocation (migration). .

Both temporary and long-term shifts in residence occurred for various reasons but, most frequently, were associated with the activities of 'big men': feasting, pig raising, the recruitment of followers. Land tenure and kinship were also influential, as were the misfortunes of death and illness. For precontact Atamo, there is no indication that population growth, warfare, or agricultural condition were directly related to the movement of either individuals or households and thus no support for Ward's (1971) statement about the primary reasons for precontact migration among the mainland societies of Papua New Guinea.

Acknowledgments

Field research on which this paper is based was funded by the Research Corporation, University of Hawaii and the Culture Learning Institute, East-West Centre, Honolulu. Of numerous individuals in Hawaii and on Bougainville who lent personal and professional support, special thanks are due Judith Hamnett, Douglas Oliver, and the people of Atamo.

4 Territorial control and mobility within niVanuatu societies

Joël Bonnemaison

Mobility among the traditional societies of Vanuatu, formerly the New Hebrides, can be gleaned through old people's recollection, through traditions that still survive, and especially through the functioning of local groups that have maintained their customary environment. Based upon fieldwork carried out in east Aoba, central Pentecost, Tongoa, and Tanna (Fig. 4.1), Melanesian societies appear to comprise social groups or micro-groups that are clearly demarcated and divided, independent of one another, and mutually suspicious. The first discoverers or missionaries to visit the archipelago noted this fragmentation of social space. Based upon a tour in 1850 the Presbyterian missionary, John Inglis (1851:29), wrote:

> The principal permanent difficulties to be encountered in prosecuting Missions in the New Hebrides are the number and smallness of the tribes . . . Wars appear to be universal . . . No intercourses could have taken place among the inhabitants of the different islands, or very rarely.

An observer's first impressions are that man in traditional society is fixed to his own ground by bonds that permit only a meagre degree of geographical mobility.

The first impression needs quickly to be modified. For among these territorially-fixed groups, traditions relate everywhere to a multiplicity of contacts, networks of alliances, and cycles of cultural exchange that unite widely separate groups speaking different languages. Within each group, largely cemented by common territory and basically stable, there has always been an attitude of curiosity and openness towards the outside world, a need for exchange and mobility. Traditional societies seem built around two apparently contradictory concepts: the ideal of fixity and the desire for mobility. Factors such as time, location, and especially social and political organization have determined the

57

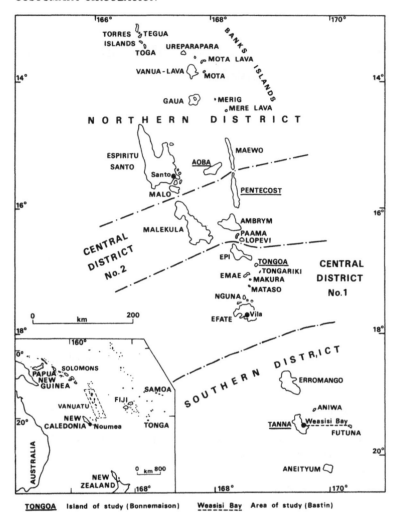

FIGURE 4.1 *Vanuatu*

prevalence of either fixity or mobility, but never to the extent of jeopardizing the overall coherence of a territorially-fixed society where inhabited space appears fragmented and where initially the types of mobility were strictly formalized.

Traditional society: territorially-fixed groups with formalized mobility

The concept of territorial fixation in traditional society did not necessarily imply a low degree of mobility but rather that any movement outside the territory was considered dangerous. For people who had come by sea and arrived in scattered parties after an unsettled journey, attachment to a particular piece of ground formed — and still forms — the essential bond. This territorial anchor assumed a mystical dimension which is often expressed in a myth of origin. Territory is lived as a space of safety but more deeply as a space of identity, and of communion with magical powers and local divinities.

The unity of local groups is defined, especially in northern and southern islands, by places and territory. Each group is related to one or several *nakamal*, places where people socialize and generally drink *kava* (*Piper methysticum*), the traditional beverage. The political arena is formed by the totality of territorial rights owned by its local groups. Often, common reference to some 'tabu ples' (sacred or magical places) gives to each group its unity and identity. It appears difficult to apply terms such as 'tribe' or 'tribal group' to such social units. However, they appear similar to what Raison (1976:192) calls a 'truly geographical society': 'Dans de véritables sociétés géographiques, l'appartenance à un lieu donné exprime l'appartenance à une unité sociale définie par l'espace autant que par la parenté.'

Place and space are thus personified through a mythical cycle, which explains the arrival of the human group (Shepherd Islands: Fig. 4.1) or else its magical appearance. On Tanna, for example, and indeed on most other islands of the group, men arose from certain unusual stones, rocks, mountains, or caves. Each clan and lineage has its own sacred places and legends, which are often but part of a more important mythical cycle. Myth defines a magical space from which the local group draws its cultural identity and justifies the laws and land rights of its component lineages. Each group does much more than simply coincide with its territory; the group is its territory, and customary ideology requires that each person lives and dies wherever he has his 'stamba' (roots). A man conceived as a plant can only exist wherever his roots; consequently trips to the outside world must be short and relatively infrequent. The strength of this bond is expressed in traditional society by the concept of territorial fixation.

For each local group, this territory was closed and specifically corresponded to a safety zone bounded most often by natural limits: watersheds, rivers, breaks of slope. Men were free to move wherever they had territorial rights but their movements were restricted as soon as they crossed the limits of their own group. No doubt this partially explains the cultural and linguistic fragmentation of Vanuatu. Each island, each area, and even each local territory was a closed environment: independent, with its own traditions, privileges and magical powers, and very often with its own dialect if not a completely different language. Within the archipelago, already geographically fragmented, it was as though each local group strove to reproduce an insular structure for itself and its allies.

Specific cultural norms governed the relations between these mutually insulated groups. There were taboos on movement along pathways and tracks linking one territory to another, which in time of war were absolute. In time of peace, such constraints were lifted at precise moments and for particular occasions, such as exchange cycles and marriage. This in turn explains the importance of the messenger in the central and southern islands (Fig. 4.1) – one who takes the word of the chief beyond local frontiers, which then allows neighbouring groups to travel beyond their limits and to meet in safety.

In the same way the littoral groups, 'man sol wora', and interior groups, 'man bush', were radically divided in every island. The 'man bush' were only given access to the beach after delicate negotiations, since it amounted to conferring the right to freely cross coastal territories. For this the 'man bush' had to pay in the traditional currency of pigs and women. At best, successful negotiations granted the right to descend along a track from which no deviations were permitted to a specific beach location several metres wide, where they could collect salt water, bathe, and engage in spear fishing. Every case of trespassing over defined limits was extremely serious and considered an act of war. In times of tension, coastal pathways were usually made taboo, thus restricting the hill clans to the interiors of islands and accounting for many wars that raged between 'man sol wora' and 'man bush'. Interior groups always insisted on having free access to the coast and to maritime routes, but this conflicted with the determination of littoral groups to deny crossings of their territory and to retain their monopoly on the maritime environment.

Within traditional society, the concept of locational fixity was so strong that *l'espace vécu* was forcibly confined to the limits of the original territory. This concept of *espace vécu* refers to the known space of an individual or group, including spiritual, social and physical space (Frémont 1976; Gallais 1976). It is analogous in English to the term 'lifeworld' used by Anne Buttimer (1976:277): 'the culturally defined spatiotemporal setting or horizon of everyday life'. In traditional niVanuatu society, the lifeworld was a zone of both safety and identity; it was impregnated with magical bonds, both benevolent and malevolent. Each parcel of land had its own name and an identity which linked it to a particular clan title, either patriclans or matriclans depending on the island. Each rock, mountain and spring had its own history and its own spirit. Similarly the name of every plant growing within the territory was common knowledge, together with its medical or magical power and preferred ecological site.

This anchorage to the local territory underpinned the concept of stability or fixation. Awareness of the outside world was slight. Mobility beyond local frontiers was dangerous, just as the external world was at the same time a source of anxiety and an object of curiosity. The external world could represent a means of enrichment, thanks to the exchange networks that could be established outside one's group; it could guarantee security if these customary networks developed into means of political alliance. The history of Melanesian societies seems to have continually oscillated between the ideology of fixation and of exposure to the outside world, between the desire for cultural and economic autonomy set against the need to trade or forge links with far-off places.

Traditionally, then, mobility could be very variable and, in the case of certain coastal groups, markedly intensive. Yet it was always regulated. Only a few individuals held the privilege of being able to travel to the outside world or could take the necessary initiatives. Nor was movement unrestricted. It always had to proceed along well-established pathways that led to precise places known as meeting points by the local groups involved. The flexibility and pragmatism inherent in Melanesian concepts thus permitted a reconciliation of two apparently contradictory ideologies: on the one hand the principle of territorial stability and, on the other, the attraction of mobility or the 'custom journey' beyond the local environment.

Socially-controlled mobility in the northern islands: travellers and 'big men' of a graded society

It is well known that the hierarchy of social grades in such northern islands as Aoba, Pentecost, Malekula, Ambrym and Malo (Fig. 4.1) concentrates political and economic power within the mutually co-opted members of a more or less closed society (Deacon 1934; Guiart 1956; Allen 1972; Bonnemaison 1972; W. Rodman 1973; M. Rodman 1979). Although the ceremonial rituals and cycles of this graded society are rather complex, its principles of social organization are fairly simple.

The society is divided into a number of hierarchical grades; a person who takes a particular grade must follow a ritual cycle whose costs and requirements increase with each rank. A variable number of pigs must be offered, given and sacrificed, some of which were valuable tusked pigs. Thus a person who is candidate for enrolment in a high grade must not only gain the sponsorship of its members, but also have access to pigs and mats that constitute traditional wealth. In parts of Vanuatu, this system of grades was abandoned as a consequence of conversion to christianity (Allen 1968). In other islands the system continued to function, especially in east Aoba and north and central Pentecost (Bonnemaison 1972; W. Rodman 1973). During the late 'seventies, in fact, there has been a strong revival of the grade system in these regions.

High rank in a grade system confers political status. To achieve a given rank, the candidate will have already skilfully manipulated the various rituals for exchange, loan and counter-loan for pigs that mark the great events of customary life. Before becoming a man of high grade he will have to become a business man; that is, in Bislama, the lingua franca of Vanuatu, a 'man blong bisnis'. The wealth of the high-graded man does not stem from accumulated goods but rather from systematically lending them to his social partners, who need pigs to achieve rank or to participate in 'bisnis'. The game consists in controlling the flow of goods (pigs, mats, currency) and in manipulating the different exchange networks.

In spite of first appearances, the capitalistic spirit is absent in the graded society. The system is not materialistic: all material values are created by the act of giving or exchanging and business means good social relationships more than profit. To have such relationships in a large geographical circle is the first goal; wealth and goods are simply a means to a socially defined end.

In this way, one can understand why the prospective 'big man' must give evidence of constant mobility. A man of high grade has often been defined by the political authority acquired through the use of economic power; he can just as well be distinguished by his geographical mobility and the control exerted upon the movements of others and the traffic of goods. The higher the grade, the greater his degree of territorial mobility. Every important chief in traditional society needed a large canoe, his own team of oarsmen, a firekeeper for long voyages, and a captain who was also the coxswain and often his spokesman. The chief's territorial sphere of influence depends in fact upon exchange relationships, and often kinship ties, which link him to other high-grade men, his equals in the graded society. By definition, the high chief ('big man') is one whose fame reaches far beyond the limits of his territory. Through him, relationships of exchange and trade can be built between groups that are geographically differentiated and often very distant.

A monopolistic tendency also reinforces the control over mobility acquired by men of high grade. Ordinary warriors, servants, and relatives do not initiate movement; it is the high-grade man who knows the routeways and who ventures along them. If several men accompany him, it is by virtue of being bodyguards or servants and not in their own right. Therein lies a basic feature of traditional society: both external relationships and the right to travel beyond the local territorial limits are the privileges of the social elite. For common people, fixity and stability are the norm: mobility is both a privilege of power and a means of acquiring it. Mobility allows for trade and access to a higher rank, but it is recognized power that permits mobility. Thus the principal lineages retain for themselves and their descendants a monopoly on power and on the advantages gained from travel.

At the level of *l'espace vécu*, traditional social organization based upon the grade system is reflected by concentric circles of unequal size:

the higher a man's rank, the greater will be his circle of mobility and the wider his perception of space. In general, the hierarchy of grades in the northern islands comprised ten or fifteen levels, depending upon cultural traditions and local groupings, which could be grouped into three major social categories. The ordinary men, 'smol man nomo', usually had access to the first three or four grades, which took place within a familiar social circle. Everyone had to try and achieve these grades if one wished to be considered a man: in other words, to have a normal social life and have access to death to the sacred places of one's ancestors. For these common men, the degree of mobility remained uniformly low and rarely extended beyond the limits of the home territory. Their sphere of movement never exceeded one to three hours on foot, according to the size of the territory, and they remained rigidly fixed to their home ground unless invited by a man of higher rank to be oarsman or bodyguard on an external voyage. Similarly, they married within their own local environment.

Above the lowest category could be distinguished an intermediate group, comprising mainly the elders of family lines and the 'big men' of medium rank who dominated the system of *nakamal*. In Vanuatu *nakamal* has wide, sociological significance. It represents the basic social unit formed by the clan or hamlet and also reflects political hierarchy. In general, men of medium rank have their own *nakamal* and dance places (*nasara*).

These men of middle rank generally attained the sixth or seventh grades of the hierarchy. Holding elevated rank, they were able to enlarge the circle of their mobility and thus had connections with those of equal stature in neighbouring groups, were very often linked with them in marriage, and participated in their main ceremonial events. However their sphere of mobility rarely exceeded a full day's journey on foot away from their *nakamal* and included only those neighbouring groups already linked through bonds of friendship and exchange.

Medium-rank chiefs controlled the circuits and exchange of people and goods both within their own group and with neighbours. In the case of coastal groups, the range of territorial mobility was much greater than for 'man bush'. Thus the 'big man' of north Pentecost had constant ties with counterparts of the Longana area of northeast Aoba or with south Maewo (Fig. 4.1). Each portion of sea coast, for example, had regular connections with the seafront of the nearest islands that could be seen on the horizon. Indeed, the relative security of maritime routes contrasted with the insecurity of land routes, which often had to cross territories where hostile clans would lie in ambush.

At the highest level of the hierarchy, the social and geographical landscape changed in dimension. From the tenth grade to the ultimate point of the hierarchy, a 'big man' was very important. His prestige and status exceeded the framework of his group of origin and encompassed the whole region where he held right of free travel. Such men carried the title *tanmonok* ('the end of the earth') or *mariak* ('that which is beyond'). They possessed wives and raised pigs in most groups within the region under their control and spent their lives in continual movement, from village to village and from one *nakamal* to another. Traditionally, there were only four such highly-graded men in central Pentecost among a population of four to six thousand.

The constant wandering of the *mariak* or *tanmonok* was the privilege of rank and also a means of enlarging their power. Pigs and mats involved in most customary transactions were procured through their good offices, since the highest 'big man' were required to be present at each traditional feast, if only to control the transfers of wealth that took place, to gain information on the size and value of pig raised, or to weigh the possibilities of future transactions. This mobility was also inter-regional and they were expected to respond to invitations from those holding equal power. On such occasions, the highest 'big man' often acquired wives from distant groups of different islands. The *tanmonok* or *mariak* was condemned by rank to continual movement. Sometimes this gave rise to gibes from men of lower rank. By having to be everywhere, those of highest grade ended up being nowhere: 'Hemi wokabaot oltaem mo hemi no kat wan stret haos blong em' (He is always on the move and does not even have a house of his own).

The lengthy journeys made by men of high grade expanded the horizons of groups that were territorially fixed. By establishing a common atmosphere of cultural contact and exchange, they enabled a meeting of people with different traditions, customs and languages. Consequently, in the case of the archipelago, it is difficult to speak of traditional societies with either a high or low degree of mobility. There were considerable variations according to geographical position ('man bush' and 'man sol wora') and to the sphere of influence of the highest 'big man'. Similarly, the range of territorial mobility and the perception of space increased or declined according to one's position in the social hierarchy. Even in groups where the territorial bond had an almost biological connotation, intergroup mobility was never absent, though always subject to control exerted by the underlying political organization.

Territorially-fragmented mobility: the routes of exchange

Just as mobility was a privilege, so likewise were the routes and pathways of exchange rigidly controlled. Once fixed, they very quickly tended to become exclusive. Beyond the home environment, men of high grade acted as intermediaries in the connections developed between different groups. Such relationships were expressed by the opening of one or more precise trade routes along which all goods flowed. From then on, such routes would pass through a number of gateways carefully controlled by groups located along the itinerary.

For example, all tribal groups under the jurisdiction of the chiefdoms of Ambanga and Lolossori in the Lolopuépué area of northeast Aoba had two principal routes of exchange (Fig. 4.2). The first, a maritime route, followed the northern coast of Aoba and afforded a link with groups in the extreme south west (Vilakalaka). The east Aobans came ashore at a precise anchorage at Nabutikiri, where they could push their canoes on to the beach, be assured of friendly contact, and barter fringed red mats for expensive tusked pigs.

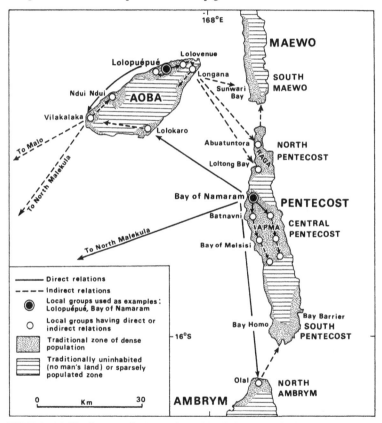

FIGURE 4.2 *Exchange relations of two local groups in the northern islands, Vanuatu*

Their partners from Nabutikiri would next search for pigs in the Nduindui area, renowned for its wealth in the ceremonial, curved-tusk variety, and subsequently exchange them with the easterners. Thus Nabutikiri functioned as a gateway through which the Lolopuépué groups could trade with those of Nduindui, when direct contact was impossible. The liaison with Vilakalaka also permitted indirect relations with the people of Malo and north Malekula (Fig. 4.1), with whom there were constant links from southwest Aoba (Bonnemaison 1972). A second major axis of exchange linked the people of Lolopuépué with those of Lolovenue in northeast Aoba, through whom indirect relationships were possible with Longana, extending from there to south Maewo or north Pentecost.

Very often, the end of an exchange route was marked by an unusual tree or rock. The people from the Bay of Namaram (Fig. 4.2) had thus planted a banyan tree on the beach so that their Aoban visitors could orient their canoes when approaching what to them was a strange island. This precise arrival point belonged to the Aobans and they felt secure as long as they remained there. Similar arrival points existed for overland transactions. Each group controlled the particular exchange route that passed through its territory and intending travellers would first check their friendliness.

Hence mobility in the traditional order was formalized both socially and spatially. The geographical space was fragmented, even divided, into a certain number of friendly territories linked by precise routes, along which all exchange had to proceed. Outside these territories and secure routeways existed an unfamiliar and dangerous world. The man who risked a journey beyond them was like an explorer; he faced many dangers, but if successful in creating a new connection, he won great prestige. Yet because of the perceived dangers of leaving safe routeways, the cycle of mobility appeared permanently fixed, covering a certain number of territories and being reserved for the prestigious family lineages who held political control.

The control of territory: the Tanna example

Traditional society on Tanna appears to have carried the principles of territorial fixation and formalized mobility to their logical conclusion. Social and cultural organization in the southern islands (Fig. 4.1) is not based on the concept of social competition found in the system of grades, but rather on a hierarchical system of titles handed down by generation along family lines according to principles that may be either hereditary or elective. Whereas in the northern islands trade arising from grade-taking leads to a certain freedom of movement in both men and goods and to a quest for the benefits of exchange, in the south by contrast the traditional bonds of alliance between different groups constitute a balanced reciprocal relationship in which each person must return what previously has been received (Sahlins 1972).

These formalized exchanges only occur during the ceremonial period, at the time of the cool season when yams are plentiful. They are initiated by a local aristocracy whose highest ranking members are known as *yremera*. The *yremera* are linked to one another by an alliance network which criss-crosses the entire island. Their power is expressed by the possession of the *kweriya*, or sparrow hawk feather. During traditional rituals, only certain members of given patrilineages have the right to wear a ceremonial head-dress bearing this feather. *Kweriya* is the emblem of the customary aristocracy. It is attached to a customary title which is handed down within certain family lines at a precise locality: the *nakamal*. Most of the large *nakamal* on Tanna have two *kweriya* (long and short) as well as several *yremera* titles. For example Yanemwakel, one of the most prestigious *nakamal* on eastern Tanna, has six *yremera*. The *yremera* are masters of the exchanges and networks of alliance between *nakamal* possessing the *kweriya* and they share authority over the journeys and routeways of the traditional world.

The island of Tanna is covered with several customary paths (*swatu*) which follow the major features of relief: crests, natural landforms, coastal fringes. These *swatu* link together the principal *nakamal*; they have personal names and some carry a mythological origin related to the exploits of some civilizing hero. For example, the *naokonap* pathway created by the hero Kalpapen crosses the island from north to south and follows the main crestline. Similarly, some maritime routes formerly linked certain littoral groups with neighbouring islands (Aneityum, Erromango, Aniwa, Futuna). The longest of these led directly to Mataso, in the Shepherd Islands (Fig. 4.1), which was the end of the known world for Tanna society.

Such routes are said to have preceded the appearance of man on Tanna. According to myths, the island was first inhabited by stones (*kapiel*), birds, and plants. These stones were alive and, just like birds, had the gift of speech and were constantly on the move. Most of them arrived by canoe (*niko*), bringing with them a range of various magical and cosmic powers that today reside with those men of Tanna who hold knowledge of secret plants.

The movements of these *kapiel* around the island and from coast to coast created both the *swatu* and the itineraries of alliances between political groups. One day the stones became silent, affixed themselves along the tracks, and then gave birth to people and clans whose descendants now inhabit Tanna. These *kapiel* are still powerful and are scattered throughout the island; some are considered sacred. Thus people arose in a world created by *kapiel*; the spirits which before resided in stone slipped into men. Space is divided into territorial groups and is centralised around particular places from which men have emerged. The Tannese, linked by their cosmology to living stones, are born from *kapiel* that have ceased to move. Like the stones, men are implanted into the ground and bound to particular places. Thus men are unable to leave their places without alienating both their identity and the powers inherited from the *kapiel*. To leave one's territory is therefore abnormal. Because travelling is important and serious, it must be strictly controlled. It can only be expressed within the network of *swatu* that traverses the island space and must remain under the political authorities of customary power.

In traditional society, this customary power was theoretically in the hands of *yremera*. These men could control the flow of goods and circulation of people along every section of the pathway passing through territories owing them allegiance. In the past, they held the right to impose taboos and to kill any unknown traveller who might venture along the route without authorization. Similarly, they could either allow or arrest the progress of a message from a neighbouring group. In Bislama, they were 'ket', a gate: something that closes or opens according to the wishes of its owner. The power of the *yremera* was also stationary. Unlike the high-grade men of the northern islands forever engaged in travelling, the aristocracy in Tanna could only in exceptional cases move away from the sacred places from which it drew its power.

Another dignitary travelled in the name of the *yremera*. Since each territory of Tanna is thought of as a canoe, this person was called *yani niko* (master of the canoe). The *yani niko* is the chief's spokesman and must broadcast the word of the *yremera*. He proposes and organizes the exchanges and meetings between different groups and, where necessary, leads the operations of war. The *yremera* remains in his *nakamal* at the fixed seat of a power that is held not by men but by places and *kapiel*. He says little and must not intellectualize ('hemi no mas tink tink tumas'); this role is delegated to the *yani niko*.

The rigidity of this structure, canalized through a pre-established network of customary itineraries and controlled by as many gates and *nakamal* as there are places and *kweriya* dignitaries on the island, left little room for innovation. It restricted the spatial configuration of potential relationships between local groups. But although geographically confined, such relationships were nonetheless intense for exchanges and gifts have always been, and still are, one of the principal preoccupations of Tannese.

In these territorially-fixed societies, each local group had its own itineraries, at the end of which lived its allies and preferred ceremonial partners (*niel* or, in the language of central Tanna, *napang-niel*). The *niel* exchanged women and sealed their political alliance during extremely important and formalized cycles at which large quantities of food and customary wealth (pigs, *kava*, mats, grass skirts, yams, taro) passed from one to the other, according to different position on the network of customary pathways. These ceremonial cycles between allied groups, linked together through distinct and named itineraries, resemble somewhat the grade-taking feasts in the northern islands: they create an external alliance and, through the division of both labour and food, they reflect the internal social hierarchy of the communities. For littoral groups, the cycle of sharing out and giving the sea turtle was one of the most prestigious and formalized types of exchange. Today it rarely takes place, doubtless because of the scarcity of turtles. On the other hand, other exchange cycles continue: those that follow the lifting of seclusion imposed upon young boys after circumcision (*kawür*); those concerned with exchange of special food between different groups (*niel*); and the complex *nekowiar* cycle, in which one or two thousand people participate in a grand ceremony lasting without interruption for a minimum of twenty-four hours and involving exchange of dances (*toka, nao, napen napen*) and of large quantities of customary wealth. All these rituals are currently experiencing renewed vitality throughout Tanna.

On Tanna, the sea turtle is considered a magical food as it is embodied with spirit and endowed with 'mana'. Only the *yremera* have the statutory right to consume the turtle and share out its cooked flesh among their community. An example drawn from the Whitegrass area, northwest Tanna (Fig. 4.3) provides a clear illustration of the network of relationships that is involved. If the *yremera* from the grand *nakamal* at Lenara'uya (Rakatné group) wishes to share a turtle with his group, he will ask his *yani niko* to take a message to the *nakamal* at Loomia (Ya' uné group), where his closest *niel* is located, to forward in turn the message to Katinuwiken. This last *nakamal*, situated close to the shore, will then select from amongst its own *niel* a group or family line holding the right to capture the sea turtle.

Certain coastal lineages hold the *kapiel* (magical stones) through which the capture of the turtle may be undertaken but, being considered affiliated with this animal, are not allowed to consume it. Conversely, lineages associated with other large *nakamal* may cook and share out, but not capture the turtle. It is upon this complementarity of rights and functions between different *nakamal* that the exchange system is based.

When the turtle has been captured by the littoral group who owns the necessary *kapiel* and holds territorial rights over the fishing ground, it will be sent to Katinuwiken, which will return a gift of pigs and *kava* and transfer the animal to Loomia, and thence to Lenara'uya. At each great *nakamal*, the live turtle is resold to the next one, with payment being accompanied by dances and the exchange of minor gifts (*kava*, pieces of yam, cooked pig) between the neighbouring partners. This process continues until the sea turtle reaches its destination.

Thus Lenara'uya, which has no direct connection with Katinuwiken, must pass through its *niel* at Loomia, who in turn must contact Katinuwiken in order to reach Loaneaï or another *nakamal*. This axis of connections materializes through a path, the *Numwanamkenem*, and then along the *Kwoteren*, a littoral route which encircles the island (Fig. 4.3). A customary itinerary has a certain number of *niel* spread along it, with

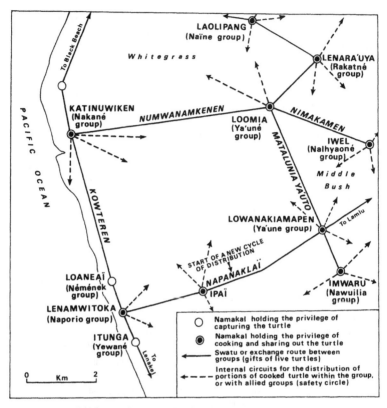

FIGURE 4.3 *Traditional cycle of sharing out the turtle: Katinuwiken, Nakané group, northwest Tanna*

the latter linked together in succession. For a relationship to exist between two distant groups, it is essential that there be harmony between all the *niel* located along the chosen route, for each one must agree to leave its territory open to messages and to the subsequent passage of men and goods.

Exchanges between distant *niel*, between partners belonging to different groups, are concerned with raw foodstuffs: live turtles, live or dead pigs, yam tubers, *kava* roots. On the other hand, cooked food is exchanged between close partners belonging to adjacent groups. Once the turtle has been cooked in the stone oven, it is divided and redistributed in the name of *yremera* within the safety zone that surrounds each great *nakamal* and comprises its membership. The difference between the gift of raw food and cooked food indicates the distance or closeness of social ties and reflects the structure of political allegiance between *nakamal*. Ultimately, each person's place and rank within the customary hierarchy is expressed through the sharing of the cooked turtle.

This example demonstrates the complexity of these systemic relationships, the multiple types of spatial control, and the hypersensitivity of the whole structure of exchange. Any violation of customary law or disrespect for a territory or its chiefs almost invariably led to war. At the end of last century the Great War of Whitegrass, still known in oral tradition as *Kamaguin-Nikou*, was waged for this very reason. The Nissinamin group captured a turtle and sent it directly to the Middle Bush without passing through its *niel* at Loomia (Fig. 4.3). The *yremera* and their *yani niko*, the 'big men' of this *nakamal*, decided to fight. They sent a request for warriors that was transmitted from one *niel* to another as far as the east cost of Tanna, with each *nakamal* agreeing to provide military support. The war took the form of an extremely violent raid by warriors during the course of which two local groups (the Nissinamin and their allies the Nakané, at Katinuwiken) were exterminated. This war largely explains the current land dispute that reigns in Whitegrass.

Informal mobility in traditional society

All systems of social control and cultural formalization have their limits and shortcomings. Each crisis within traditional society appears expressed by a weakening of social control and by greater freedom of mobility.

Outcasts were a common problem, although not a threat to the established social order. Intra-group conflicts, particularly such domestic ones as adultery and theft, could not always be resolved by fines paid in pigs or *kava* roots. When in serious trouble, a man had no option but to flee and try to join affines outside his own group. Often no such relations existed, for most marriage occurred within the immediate environment. Thus he would venture far from his home, by canoe or on foot. Upon passing the crests of inland mountains, often the no-man's land or buffer zones between hostile groups, the fugitive would cross to the other slope of the island. When he met with strangers, there was only one alternative: he was either killed on the spot or warmly welcomed, often being given an adopted rank and a new wife. Stories abound throughout Vanuatu, particularly in the northern islands, of an exile fleeing his local group, being welcomed by another, and later acceding to supreme power amongst them, usually by replacing his adoptive 'big man' upon death. These outcasts were the marginal men of territorial societies; they might succeed, but the perils were great.

More significant crises arose when those defeated in tribal wars had to flee their territory. They would try to join allied groups, particularly those in which they had kin, and wait until they could re-occupy their territory. The history of Tanna contains many examples of wars causing the displacement of whole populations from one part of the island to another. Such population movements were especially important at the end of the nineteenth century, when the arrival of Europeans, the trade in firearms, and the brutality of blackbirders resulted in a grave disruption of customary institutions. These events weakened the ability of traditional society to face the new conditions represented by the possession of guns and money and the presence of christian missions. Suddenly the social order became unstable and violent and merciless wars broke out almost everywhere. Entire groups took refuge in the Presbyterian missions that already occupied certain coastal points.

These periods of grave crisis recurred periodically in traditional society, particularly in response to influxes of foreign populations. They also arose during times of famine, sudden demographic expansion, natural catastrophe, and cultural mutation. At certain points in the history of Tanna, traditional society found itself incapable of functioning and the island was shaken by random movements which defied control. These were days of warriors, fugitives, of hiding places in the mountains, often of famine, and they continued until a fresh equilibrium was restored by the victors and a return made to the normal condition of territorial fixation.

Mobility in traditional society clearly varied according to group location, type of political organization, and cultural milieu. In some northern islands, several autonomous marine zones were created by the multiple axes of exchanges and connecting trade routes. These marine zones were small *Mediterranée*, characterized by the intensity of movements and circuits of exchange. The grade system, similarly carried along by its own dynamism, tended to produce more and more people who reached higher ranks, even if this simply resulted from the natural extension of exchange networks. Thus was created a momentum conducive to mobility.

By the time Europeans arrived, certain coastal areas were noted for their mobility, dominating the routes of maritime trade, their ports, and most circuits for the exchange of pigs. The best known of such highly mobile groups were those of Malo, north Pentecost (Raga), western Aoba (Ndui Ndui, Vilakalaka), the off-shore islands of north and south Malekula (Vao, Atchin, Wala, Rano, Toman) and probably the Banks Islands (Fig. 4.1).

In the southern islands by contrast, the absence of grade systems together with the presence of less flexible conceptions of space and territorial rights produced societies that were geographically more fixed, or at least where mobility was far more formalized. The transformations that followed the arrival of the White world ran deeper here than anywhere else in Vanuatu.

Circular mobility and 'custom journeys'

The first European contacts and the labour migration to commercial plantations outside Vanuatu – first to Queensland, then to Fiji and Samoa, and in lesser degree to the nickel mines of New Caledonia – is well known (Shineberg 1967; McArthur and Yaxley 1968; Bedford 1971). Traditional society was profoundly changed by successive waves of sandalwooders, blackbirders, and self-confident, enterprising missionaries, together with the settlement from 1870 of planters and traders selling firearms and steel tools. At first, mobility was imposed and organized along authoritarian lines, as blackbirders removed variable quotas of coerced or voluntary workers from different islands. The total numbers were considerable. Between 1863 and 1906 some 40,000 niVanuatu were recruited to Queensland and another 10,000 left for Fiji, Samoa and New Caledonia (McArthur and Yaxley 1968:17; Bedford 1973b:196).

By 1880, with the growth of plantations and the increasing number of white colonists, an internal recruiting system had developed within Vanuatu. Plantations being established throughout the archipelago, notably on the islands of Efate, Epi, Espiritu Santo and Malekula (Fig. 4.1), required cheap manpower to clear bushes, construct fences, plant trees, and husk and prepare copra. The nascent Condominium of the New Hebrides imposed tentative legislation to limit local abuses of recruitment which, although of questionable effectiveness, nevertheless stopped the excesses of the preceding period (Scarr 1967).

The termination of internal warfare (M. Rodman 1979) and the weakening of traditional power structure brought about a dramatic increase in personal mobility. Most taboos on unauthorized journeys disappeared; each person became free to go wherever and with whom he wished. Formalized movement thus became unshackled, opening out towards new poles of attraction formed by European plantations, recruitment of boat crews and, later, the development of urban centres.

Parallel with the opening of these new routeways was the progressive disappearance of the grade system in many regions of the northern islands, particularly among the first coastal groups to be christianized, and a general disintegration of the old structures of exchange. The former trade routes and canoe journeys from one island to another fell gradually into disuse. So did a number of customary rituals, which were sometimes prohibited by the new religious authorities.

From the meeting of these two elements – the end of formalized mobility caused by a transformation within traditional society and the attraction of salaried labour on European plantations – there appeared the phenomenon variously known, with some justification, as 'circulation', 'circular mobility', and 'circular migration' (Chapman 1970, 1976; Bedford 1971, 1973b; Bonnemaison 1974, 1976). If this circular mobility is studied in terms of flow, periodicity and statistical indices, it does not appear fundamentally different from that described elsewhere in Melanesia. But to be fully understood, it must first be placed within its specific cultural context.

Between the end of last century and the immediate post-war years, labour movements to European plantations constitute the dominant flow. These initial movements were regarded as part of normal mobility, as conceived by traditional society; in fact, the first men leaving for plantation work were evocative of former customary journeys. In this way niVanuatu society showed that, despite many vicissitudes, it had the capacity to adapt to new situations within its own terms and cultural models. Five major components of this circular mobility can be recognized.

Brief and repetitive movement

Before the second world war, most movements to plantations were successive journeys of short duration whose regularity was controlled by the pressures of village life. Wage-labourers usually left at times of rest in the traditional agrarian cycle, just after the April/May harvest of yam or following heavy clearing and planting work during August/October. The average time spent away varied from three to six months.

This brevity of stay and the instability of a labour force under the constant pull of its village domicile figured among the planters' greatest difficulties. As a result, the planters obtained the right to import an overseas work force of Tonkinese, Gilbertese, and Wallisians. The Tonkinese reached a total of 6,000 in 1929 and the development of large plantations in Vanuatu is mainly due to their labour. The Melanesian worker remained fundamentally attached to his territory and his time way from home was always of short duration. He continued to maintain his food garden and to plant coconut palms near his village, to which he would return whenever called for a meeting, a customary feast, or by the illness of a relative. Even though under stress, traditional society retained an emphasis on its territorial bonds and kept some sort of control over its absent members.

Spatially-controlled movement

Just as movements are limited in time, so are they confined in space. The random nature of first contacts with Europeans gradually evolved into a network of specific relationships between certain local groups and plantations. The people of a particular village on Pentecost, for example, always worked on the same plantation in Santo, and still do, while those from Mataso (Shepherd Islands) travelled to a neighbouring plantation near Vila (Fig. 4.1). This was not simply a matter of habit or convenience but more like the opening, in the traditional sense, of a new routeway to the outside world. By this means, the local group enlarged the dimensions of its lifeworld, ensured the recognition of a particular place (plantation or enterprise), and directed there most of its reserve of young men.

Wage-labourers consequently went to known territory which, by definition, was a place where they had relatives and friends. The planters, by maintaining continuous connections with a particular area, were equally certain of a constant source of recruits and their brevity of stay was partially compensated by a certain regularity of flow and by the alternation of arrivals and departures. Present-day mobility only substitutes new routes for those formerly used for exchange. Contemporary journeys are still oriented along certain specific axes; it is but a secondary form of social control.

Access to power provided by the journey

Circular mobility, moreover, regenerates some of the behaviour and attitudes typical of the traditional socio-cultural order. In the northern islands, the privilege of movement outside local frontiers was mainly held by 'big men' and those of high grade, being as it were a privilege of power as well as a means of acquiring it. Those who travelled returned rich with new experiences gained from a knowledge of other worlds and, more pragmatically, with various goods acquired from external transactions which were then redistributed.

At the beginning of this century, some plantation owners on Santo adapted their methods of payment to the expectations of their employees. At the end of their labour contract, the latter received not only cash but also the gift of a 'bokis': a wooden trunk, painted red and with brass fittings, which contained standardized gifts − cloth, axes, bush knives (machetes), tobacco, pipes, blankets − with the most valuable objects placed at the bottom of the trunk. Upon return, the plantation worker triumphantly set foot on his island bearing his 'bokis' full of gifts for the family.

Thus was revived a traditional practice of central Pentecost, whereby a person leaving on 'bisnis' should not return empty handed. His family expected him to bring back gifts and *rutrut* (a basket filled with red or white mats, whose value increased with basket depth and whose price equalled a pig with curved tusks). The *rutrut* was used as currency in customary transactions and was shared among family members. The 'bokis' thus evoked the role of the traditional *rutrut* and was particularly prevalent among labourers originating from Pentecost and Aoba. Although planters no longer use the 'bokis' as a form of payment, many wage workers today purchase similar trunks from chinese shops or plantation stores before returning to the village.

The distribution of 'bokis' goods and money conferred new prestige on returning labourers, reintegrated them more effectively into traditional life, and improved their personal status. By purchasing pigs and making various loans, for example, returnees could attain new grades and higher position. From about 1900, as a result, circular mobility very quickly assumed a role identical to that played by the customary journey of earlier times. Prior to marriage, any young man of some ambition had to work on several occasions at external plantations. Such a journey brought him face to face with the outside world, both testing and enriching him. Upon return he became a man of the world, got married, and raised himself through the traditional hierarchy.

Journeys 'in company'

Circular mobility is usually carried out within a group or, in Bislama, in 'company'. Formerly, when it was unwise to travel alone to the outside world, the traveller would go as one of a group for security. Today is no different: villagers leave with friends or relatives and try to obtain joint employment on the same plantation or with the same firm in town. Through this group, more appropriately called a fraternity as it is based on a common language, close social ties, and mutual help, links are maintained with the home territory. The communal nature of mobility, based upon a shared origin and kinship ties forged in the home island, is widespread. Groups forming at the place of destination (plantation, co-operative enterprise, settlement) very quickly evolve a new social structure. Some old people, who have attained a supervisory post or professional qualification, form a stable core and assume the role of leaders, rather like the old hands on plantations in Queensland. Around them recurs the fluctuating body of circular movers who come for shorter periods.

Journeys of youth

The demographic composition of circular movers echoes that of traditional models. Children, women, whether married or not, and old people stay at home. Those who leave the local community are the young men, warriors of former times, accompanied by some middle aged and more mature men. This explains all the difficulties encountered by planters who tried to establish married couples on their plantations. Such journeys are restricted to a particular time in the life of males. Later, when a man marries and becomes a father, he progressively acquires authority and becomes involved in the absorption and transmission of his customary heritage. It is no longer appropriate for him to travel.

Today, the demographic structure of plantation workers or urban migrants is characterised by a high proportion of the young and a marked degree of masculinity. In 1967, three quarters of the migrant population of Vila (the capital) was aged less than 30 and 54 per cent were between 15 and 30 years old (Bonnemaison 1974). Similarly in 1972, the sex ratio in Vila was very unbalanced: 173 men for every 100 women. The preponderance of young males is particularly evident among migrants originating from islands, such as Tanna and central Pentecost, which have retained a certain cultural cohesion (Bonnemaison 1977b).

Circular movement and cultural permanency

Initially, in Vanuatu, movement towards plantations or in connection with recruitment of boat crews took place within a society that conceived mobility as being geographically fragmented, socially formalized, and occurring in circular form. A return to one's territory of origin was not simply frequent, it was obligatory. Even today a villager cannot live for too long outside his home place without alienating his identity and his territorial rights, or without disregarding the ancestral laws that signify his territorial group.

None the less, circular movements bear the seeds of a more general transformation. Through contact with the outside world, horizons are broadened and periods of absence are gradually lengthened. The initial effects of formal schooling, the emergence of new tastes and patterns of living, and the acquisition of professional qualifications all contribute to the gradual detachment of some people from their original territory.

The extent of these changes varies considerably from island to island. In the northern islands, where christianity and the plantation economy were the most rapidly adopted, mobility soon became more intensive. These societies, accustomed to higher levels of interaction through the commercial exchanges of the grade system, adapted very quickly to the geographical fluidity of the new age. Islanders of Aoba were the first to leave in groups and work on the plantations of Espiritu Santo and were followed by others from Pentecost, the Banks, Paama, Ambrym, and Malekula (Fig. 4.1).

By contrast, the southern islands and particularly Tanna, the most densely populated, entered this movement cycle after some delay. The first participants were marginal men who, like the outcasts of traditional times, were politically or personally severed from their own groups. It was only during the second world war that considerable numbers of Tannese began to leave their island, when the American forces built military bases in Vila and Santo (Fig. 4.1) and appealed for local labour. Even today, the transformation of movement is slower in Tanna than elsewhere in Vanuatu. Circular mobility in the form of 'custom journey' remains the predominant model of movement and the attachment to one's territory of origin still is one of the essential poles of political and cultural life. Throughout the archipelago, the demographic characteristics and behaviour of wage labourers consequently vary according to the degree of acculturation of their society and specifically to the extent of commitment to Christianity and a plantation-type economy. Circular mobility analogous to the 'custom journey' currently persists in the more traditional island societies and is to be found among unskilled workers.

Upheavals created by external contacts, notably the introduction of plantation agriculture, caused greater disruption within the restricted space of small islands than in mainland areas. Particularly in these small islands, the loss of coherence in both the traditional world and its spatial equilibrium became expressed through departure and flight – initially to plantations and later to towns. This phenomenon is most noteworthy in the Shepherds and Paama (Fig. 4.1), 25 and 34 per cent of whose *de jure* population was resident in Vila and Santo in 1972 (Bonnemaison 1977a).

In the coastal regions of some larger islands, where population density is at least forty persons per square kilometre, movement is changing towards a classic, more permanent form explicable by economic factors. Here, and in certain small densely populated islands, the invasion by the coconut of usable agricultural land is gradually reducing the area available for subsistence gardening and shortening the period of fallow. The resultant decline in self-sufficiency is not offset by the returns from copra making, which fluctuate from year to year and are insufficient for those without adequate plantings. The coconut, furthermore, has reduced the flexibility of the traditional land tenure system: land-use rights are becoming more restrictive and the tenure of plantations is changing towards a permanent subdivision of land among a few families. Outmigration becomes increasingly imperative for those experiencing land shortages, especially the young.

The change from circular mobility towards a more permanent form of migration is indicative of a transition: from a traditional society with formalized mobility to one of peasants and planters, in which migration becomes either a means of advancement or an economic necessity. This transformation, however, is far from uniform. With a greater territorial mobility throughout the islands, the horizons of everyday life – the 'lifeworld' – have been enlarged. Space is now endless and uniform with alliance networks spreading much farther than before. Yet such changes in the scale and intensity of movement have not destroyed the traditional relationship between people and their territory.

Although niVanuatu are now much more mobile the land remains, and this continues to be the vital factor in contemporary social life. Today, as in the past, each man must have his own place in the territory of his group and must hold on to his resultant rights. The strength of this bond impedes the progression towards what Pierre George (1959:200) calls a 'turbulence', or total fluidity of population movements. Even when they travel far from their territory, islanders cannot sever the ties with their home place. By birth and heritage, they belong to this place as much as it belongs to them. As a niVanuatu has written: 'Land is the centre of the New Hebridean's social system. His whole life's activities revolve around land . . . The land that they live on nurtures and supports them and in turn they are responsible for the care of the land and its products' (Sope 1975:6).

Mobility in Vanuatu is thus dominantly circular and shall remain so as long as the actual relationship between people and territory endures. Circular mobility continues to be structured by the existence of routes, old or new, which link local groups to some plantation areas, urban enterprises, or suburban locations. Contemporary society also remains imbued with the spirit or *Weltanschauung* of the customary worldview. To attempt to account for social and cultural transformations in purely economic terms is clearly inadequate. Present patterns of mobility show both a series of rapid changes and some remarkably enduring features that can only be explained by a continuing cultural heritage; by, in particular, a profound sense of land and places. In this way, circular mobility appears a compromise between two sets of influences. There is, on the one hand, a cultural – if not religious – view of the world held by traditional society whose territorial links, principles of implantation, and forms of cultural fixity provide the general framework and, on the other, the pressures and attractions of an external economic and social space whose main components are the fluidity of people and goods.

Acknowledgments

I wish to express my gratitude to the people of Aoba, Pentecost, Tanna and Tongoa for their hospitality, trust, and co-operation. Thanks are also due to C.A.E. Pierce, Vila, for help in the English translation of the French original; and for editorial suggestions, to Dr Jean-Marc Philibert (University of Western Ontario, Canada), and Drs William and Margaret Rodman (McMaster University, Canada). The earlier and French version of this paper appeared in *L'Espace Géographique* (Bonnemaison 1979).

Part II
Transformations in circulation

5 Movement processes from precontact to contemporary times: the Ndi-Nggai, west Guadalcanal, Solomon Islands

Murray A. Bathgate

Most Melanesian societies, apart from those in the central highlands of Papua New Guinea, have been in contact with Europeans for at least a century. They have experienced many changes in economy and social organization during this time and their precontact patterns of movement have been progressively modified. To reconstruct behaviour before European contact and to establish similarities and dissimilarities between traditional and contemporary mobility requires a detailed knowledge of particular peoples. This paper is concerned with the Ndi-Nggai of west Guadalcanal, Solomon Islands (Fig. 5.1), among whom sixteen months' fieldwork was conducted from January 1971 to May 1972 (Bathgate 1973, 1975, 1978b). Its central argument is that, from decade to decade and from one generation to another, the dynamics of population movement reflect the complex interplay between intrinsic traditional and extrinsic modern forces; and that, at any point in time, neither set of forces is completely determinant.

Certain features of traditional society among the Ndi-Nggai are essential to an understanding of its mobility behaviours. Prior to contact the Ndi-Nggai numbered 2,500 and most lived in the rugged interior, a location that was occasioned by warfare. The littoral areas of west Guadalcanal were continually raided by parties from Savo (near Nggela), the Russell Islands, and New Georgia (Fig. 5.1) seeking human heads for use in rituals. Since local warfare also was endemic, most settlements were sited on hilltops and difficult to detect. Most were small and inhabited by one or two families.

Descent amongst the Ndi-Nggai was (and still is) matrilineal. The smallest unit was the *tina* (maternal grandmother, brothers and sisters, maternal cousins of the first order), several of which constituted a *puku*: all persons with a matrilineal blood tie to a common ancestor at least four generations removed. In turn, a number of *puku* formed a *duli*, which in effect was a clan whose members had worshipped the

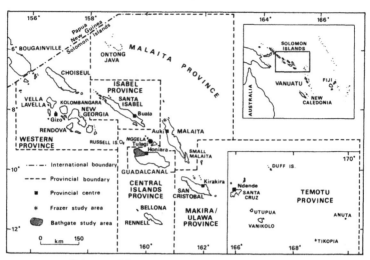

FIGURE 5.1 *Solomon Islands*

same totem in the past and acknowledged a common apical ancestor. Marriage was exogamous and there was preference for a son to marry his father's sister's daughter (patrilateral cross-cousin marriage), because of the close ties established between two *puku* over several generations.

Shifts in place of residence after marriage meant that members of any one *tina* and *puku* could be found in several hamlets. Even so, each *puku* owned at least one area of land, with inmarried women and their children being subordinate to the social group that exercised control and whose leader was responsible for collective activities. *Puku* acquired full rights to unclaimed land by holding a feast and sacrificing pigs to their ancestors. Then the district was divided into gardening areas and any land that remained unallocated as far as some prominent physical feature (a hill or river) was incorporated into the domain, but only after neighbouring groups had been consulted. Although inmarried residents were allocated working areas, only *puku* members held full rights to the land and could place a large flat stone, or *kokomu*, on it to signify the place where propitiations were made to ancestors. Thus the *kokomu* was both the spiritual and the spatial focal point of a *puku* and only shifted when that group vacated the region or sold the land to another *puku*.

The regular incorporation of persons after marriage meant that social organization was loosely structured. Each *tina* had a 'big man' (*taovia*) and the system paralleled that described for other Melanesian societies (e.g. Hogbin 1938; Sahlins 1963; Burns, Cooper, and Wild 1972). Amongst the Ndi-Nggai, the 'big man' obtained his position through primogeniture and ability in oratory; organized people in daily affairs;

and could call upon pigs, shell valuables, and garden produce that belonged to others for use in competitive feasts with rivals to maintain or improve the status of both himself and supporters. Hamlets often coalesced as a 'big man's' following widened within and beyond his lineage.

Competition at feasting and other forms of social activity, plus the demand for status, was reflected in a high degree of lineage factionalism and intergroup rivalry. *Malagai*, renowned warriors and in fact professional assassins, often were employed by 'big men' to eliminate rivals. *Vele*, sorcerers who practised death magic, could also be called into service. Rather than engage in open battle, a group that felt disparaging remarks had been made about them would deploy a small band to make a quick raid, a few people would be killed, and the warriors would retreat with some captives.

Precontact movement

Amongst the Ndi-Nggai, movement occurred within and beyond the local territory, with the latter being both temporary and long-term. Most moves were of short duration, occurred within the vicinity of the hamlet of residence, and reflected the need to collect forest materials to build and repair homes, hunt game, trap fish, and go to the food gardens. All families had a garden shelter and at certain times — as when land was being cleared or during the yam-planting season — would reside there for a few days. Whenever the social group was at war they invariably dispersed to these shelters, returning to the hamlet when it was safe and staying away for perhaps several months. Feeding and capturing pigs for large feasts also entailed absence. Wild pigs roamed the secondary growth and to domesticate them families constructed shelters (*verana bo*) in their midst. Several times a week food would be taken and thrown to these pigs from the safety of the shelter.

Fear of attack by a raiding party from Savo or New Georgia or by a known sorcerer meant that individuals seldom moved beyond the local territory. Even within their own domain, people usually went in small groups, were well armed, and travelled during the day — for a *vele* struck at night and notably when there was no moon. When leaving on a journey most carried *ria*, ginger handed down from ancestors, in the belief that this would warn of impending danger and provide protection from the magic of others.

Nonetheless, some extra-territorial movement was necessary to acquire resources not locally available. Some *puku* of the interior had a *tsatsapa* (corridor) that led to the coast. Whenever salt water was required for cooking or fish and coconuts were needed to vary the diet, well-armed parties would travel the river courses and camp immediately behind the beach. Shelters were constructed if such trips lasted more than a few days and were reoccupied on later visits. Inland and coast communities also engaged in intermittent food exchange by arrangement and, wherever possible, on neutral ground midway between their places of residence. Economic necessity also demanded that trade (*tsabiri*) be conducted without interference, since this was the only way for the Ndi-Nggai to obtain canoes. The *tomoko* and *bibina* used for carrying war parties and the *roko* needed for fishing came from the Nggela group or Santa Isabel (Fig. 5.1) and were exchanged for pigs during reciprocal visits. Ndi-Nggai men from coastal villages stayed with close maternal kin, in the absence of whom members of their *duli* were responsible for providing food and shelter.

Payment of brideprice and attendance at feasts also generated movement between hamlets provided, however, that *puku* from different *duli* were not at war. In such an instance, only inmarried persons from the other *duli* could pass safely between places controlled by opposing groups and then on condition that they were not politically aligned. In contrast, reciprocal exchanges between 'big men' accounted for by far the most journeys beyond the local territory. A *taovia* would inform the people of his hamlet or settlement cluster that he intended to visit a counterpart, who would be notified and then pigs and produce assembled. The party would set out on the appointed day, pigs and food would be formally presented at a ceremony, and dancing and singing would follow. The visiting group might stay several days and would be hosted by members of their own *puku* or *duli*. Future marriages would be arranged and if brideprice were paid then some women would return with them. Since the receipt of live pigs created a debt, later the hosts would make a reciprocal visit and ensure further exchanges if a great number of pigs were presented.

How often and how far people moved was thus an index of their social status (cf. Bonnemaison this volume). *Vele, malagai*, and *taovia* were far more mobile than commoners. Whilst those who practised *vele* were rarely known, sorcerers were said to catch on the same night selected victims who lived 160 kilometres apart. 'Big men' and renowned warriors were forced constantly to travel from place to place, seldom sleeping for more than a few nights at one retreat or hamlet within their settlement area lest their whereabouts become known to those employed to kill them.

Marriage was the most common of many factors that led to shifts of residence beyond the local territory. Most unions were between persons from different areas and entailed that the wife transfer to the husband's hamlet (patrivirilocal residence). Place of postnuptial residence was determined by brideprice which, if paid, gave a man the right to take his wife to his hamlet. Lack of brideprice usually occurred at the request of the bride's *tina* and meant that the husband resided at his wife's place. *Tina* with few females living on their own land were often reluctant to have women leave after marriage, because there was no guarantee that their children would later return to assume control of the *tina* estates, especially if the wife went to a very distant hamlet. As a compromise, some couples followed a system of dual residence (*vera ruka*), alternating between the hamlet of husband and wife and spending from a few weeks to several months at either place.

Residential shifts also occurred late in married life. The eldest or most capable son of the senior woman of the *tina* would leave his father's hamlet to live with a maternal uncle, the *taovia* whom he would succeed as controller of the *tina* estates. Other sons of outmarried women quite often returned to their *tina*, for a core of men was essential to ensure the group's fighting strength. Both divorce and death of a male spouse also found women and children returning to their natal group. Yet the flexibility of Ndi-Nggai social organization meant that some sons remained in their father's hamlet to inherit property, and wives and children did not always depart on the death of their husband and father.

Some changes in place of residence were forced and not voluntary, as when individuals captured during a raid were taken back to the hamlet to become slaves (*tseka*) and do menial work. Sometimes *tseka* were sold, perhaps three or four times over a lifespan, and thus lived in several districts. Raiding parties from other islands also took men and women as captives to the Russells, Savo, and New Georgia, from where they never returned. If a man and woman of the same *puku* or *duli* committed line incest, the ultimate crime, they had to retreat to and live in an uninhabited area to prevent being killed by the local group. Men who broke more minor laws were also obliged to leave their district once the swift spread of gossip led them to be called a *chinoho*, or trouble maker. A *chinoho* endeavoured to move as far away as possible and to seek the assistance and protection of a *taovia*, who would permit him to stay, provide land for gardening, and let a house be constructed. Later the *chinoho* might show his gratitude by giving shell valuables to the *taovia* and perhaps even marry a local woman.

The most permanent transfer of residence occurred with the abandonment of old and the construction of new hamlets (*pipiverana*), partly as a result of warfare or *vele* sorcery but mainly because of the system of shifting cultivation. Once the land was cleared of forest, thin soils and rapid leaching meant that garden production declined sharply after the first year of cultivation. Families had continually to prepare new plots, and once these became too distant the entire hamlet relocated to a more convenient site (cf. Chapman 1970:72-3). Genealogies indicate that inland groups would establish as many as three or four settlements within a single generation. In some cases, members of one generation would construct a series of hamlets leading to the interior and their descendants of the next would reverse the sequence, often reworking garden areas and using the same settlement sites. Such long-term circuits of relocation did not prevail in other instances, with local groups and their descendants penetrating far into the interior as unclaimed land was successively used and abandoned. On the other hand, some coastal villages located in favourable areas were occupied for more than five generations. Although a core of men might remain on their *tina* land, augmented by the occasional return of widows and sons, 106 genealogies collected over three generations demonstrate that people were forever moving to new areas, a process that ensured the residential dispersion of the matrilineage.

Movement following contact: 1870–1921

From 1870, the number of visits made by overseas vessels to west Guadalcanal and the contact of the Ndi-Nggai with Europeans rose quite sharply. Labour recruiters came first, followed by missionaries and colonial officials (once the Solomon Islands had been declared a British protectorate in 1893), and finally by commercial planters of cash crops. The use of Melanesian labour to work the sugar plantations of Queensland and Fiji beginning in the 1860s has been well described (e.g. Churchman 1888; Belshaw 1954; Corris 1970, 1973; cf. Bonnemaison, Connell this volume). Recruiting vessels were visiting Malaita (Fig. 5.1) by 1864 and west Guadalcanal soon after. Since planters offered from £Australian 10 to £A 12 for each recruit, there was considerable incentive for shipmasters to obtain them. Certainly kidnapping often occurred, yet the Ndi-Nggai also had reasons to offer their services or cajole others to do so. Curiosity, desire to see the white man's world, need for a reprieve from local warfare, and escape from detection of a crime were all important. Far greater, however, was the desire to obtain trade goods: calico print, tobacco, clay pipes, fish hooks, tomahawks, and rifles. Traders gave those goods to Ndi-Nggai who persuaded young men to make the long trip to Queensland, the growing demand for which items could in turn be satisfied through wage work at two shillings and sixpence a week over a three-year term.

Gradually the labour trade both increased mobility and widened the movement field of the Ndi-Nggai across the Pacific. The regular flow of recruits from the 1880s to the sugar plantations reflected a growing awareness that long-term participation in wage employment could alter the local political situation. Wicker shields and wooden spears were inferior to tomahawks and rifles; many 'big men' and *malagai* quickly understood that securing more of the latter weapons would not only eliminate rivals but also achieve control over a wider territory. Thus many 'big men' sent younger relatives overseas. By 1890, several had small armies equipped with Snider rifles, with which old scores were settled, defeated groups sometimes dispersed to remote districts of the interior, and control sometimes assumed over parts of the coastal margin. For such conquests to be sustained and for other groups to avoid subjugation, dispersal, or even extinction, more arms had to be acquired so the departure of able-bodied males to Queensland became a vigorous, self-perpetuating process.

This trade ended with the passage of the Pacific Island Labourers Act in 1901, under which terms all sugar workers were to be repatriated. Until 1907, vessels periodically off-loaded men at the main anchorages in west Guadalcanal, some of whom did not belong to these districts or even to that island. Many, fearing they might be killed by enemies, decided to remain and at the cost of trade goods brought with them were absorbed into local communities. An unknown number of Ndi-Nggai did not return and took up permanent residence in Queensland, because they had married, preferred a more material lifestyle, or wanted to avoid retaliation for past crimes.

Both the Wesleyan (Anglican) and Roman Catholic missions established coastal stations at the turn of the century, the former at Maravovo and the latter at Leosa (Vissale: Fig. 5.2). Initially, both missions distributed trade goods to coerce people from nearby hamlets to visit their stations and hear about christianity. Next followed an attempt to demonstrate that christians could not be harmed by *vele* and had no need to make offerings to spirits, carry weapons, or take a sprig of ginger to guarantee safety of movement. The final emphasis was on persuading the 'big men' to attend church services, knowing full well

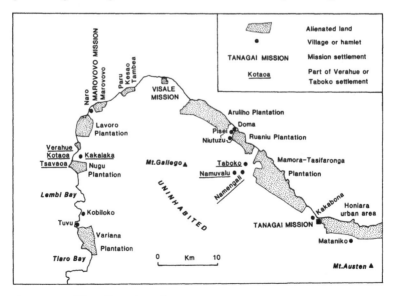

FIGURE 5.2 *West Guadalcanal*

that once they became regular 'hearers' and later converts, then others under their jurisdiction would follow.

The claim in 1910 by the Anglican and Roman Catholic missions that most of the Ndi-Nggai population had been converted to christianity was without foundation. Nonetheless some major changes had occurred. *Taovia* no longer deployed parties to raid each other, collect heads, and capture slaves, while the harassment from other islands (New Georgia, Savo, the Russells) also had ceased. With the enforcement of peace, the coastal margin and inland valleys of west Guadalcanal became safe places for all Ndi-Nggai to visit. In 1910, the official journal of the Melanesian/ Anglican mission reported:

> The influence of the line of [those hamlets which have accepted christianity] along some one hundred and fifty miles of coast extends in most places so far inland that a brother of many bush-chiefs told me that anyone can now travel safely throughout the bush (Southern Cross Log, 8 January 1910: 128).

Such an assessment was rather too general. Although the Ndi-Nggai no longer feared attack when travelling, some 'big men' remained bitter rivals and people both avoided trespassing into the domains of others and continued to take precautions against *vele* when going to their food gardens. Overall, movement did increase over longer distances and many 'big men' and renowned warriors began to stay longer at their central hamlet.

Another change was the abandonment of interior hamlets for coastal locations. As early as 1901 the Anglicans noted: 'People have come [to Maravovo] from all parts to see us, and have made friends, and people are coming out of the bush and building their houses on the shore close to us' (Southern Cross Log, 15 February 1901). Inland dwellers often moved spontaneously, but littoral settlement was fostered mainly by the missions because they found the many hamlets, often kilometres apart, difficult to reach. These small bands of people were directed to 'heap up': relocate to villages already established but still few in number, or construct their own. Some new villages and their surrounding gardens were quite extensive, incorporating up to 150 people. As before, outlying garden shelters were erected for periodic occupation and hunting trips made into the interior. However, *verana bo* were forbidden and pig pens had to be erected within the village precincts, since the two missions felt that absences to raise livestock too greatly reduced the level of church attendance.

Coastal villages were more permanent fixtures than the hamlets of precontact times, in the sense that people's mobility became bounded by their locations, while further temporal and spatial ordering occurred with the emergence of discrete religious territories. By 1915, Roman Catholicism was dominant in west Guadalcanal, but stretches of catholic villages from the far west in Tiaro Bay to Kakabona in the east (Fig. 5.2) were interrupted by smaller numbers of anglican ones. Several factors combined to ensure that gradually much mobility was circumscribed in terms of religious affiliation. Catholic missionaries forbade marriage between their own and Protestant converts, so that wives of the same denomination were drawn from one's own or neighbouring villages. Therefore close ties evolved between sets of villages, rather than *puku* as before, with kinship links and common religious affiliation merging very strongly in succeeding generations to define clear limits for movement of a marital or social nature. For instance, an annual round of feasts by Anglicans reflected the day on which each village church was dedicated.

Despite loss of access to the Queensland sugar plantations, between 1907 and 1921 there was a dramatic increase in the movement of men to wage work within the Solomons. Government offered jobs at Tulagi (Fig. 5.1), the protectorate's capital; missions provided employment at their stations; private individuals and limited commercial companies sought labourers for coconut plantations then being established along the coastlands of larger islands. At first, the Ndi-Nggai were reluctant to enter into employment; wages offered were lower than had been earned in Queensland and until 1914 both money and trade goods were available from the sale of land to Europeans. Once government forbade such sales, cash could only be obtained by making copra to sell to visiting traders or by engaging in temporary employment. Goods that were now seen as basic necessities − 'calico' (cloth), matches, tobacco, kerosene, rice, sugar, tea, tinned meats − could only be purchased with money.

The missions at Maravovo and Visale first provided regular employment in 1907 and hired Ndi-Nggai men as boatcrew, construction assistants, guides, and cooks. Most worked on a casual basis and returned to their village in about a month. Land for a coconut plantation was purchased in 1909 by the Maravovo mission to provide not only future income but also local people the opportunity to work under christian auspices through a 'civilizing' activity (Southern Cross Log, 8 January 1910:30). Apart from Maravovo, eight other expatriate-owned plantations were established in west Guadalcanal (Fig. 5.2) but drew most of their labour from the Kwaio and To'ambaita areas of Malaita (Fig. 11.1) on one- or two-year contracts (cf. Frazer this volume). Because of deep distrust of these Malaitans, few Ndi-Nggai men offered their services to the planters, and then only for brief periods of three to six months.

About one third of males taking up employment during 1907-21 went to plantations on Santa Isabel and islands of the western Solomons. This practise of interisland wage labour reflected the institutional context of hiring workers far more than any foreboding about Malaitans. All men in the Solomons who desired employment for one or two years had to be indentured and local planters were reluctant to accept Ndi-Nggai, because mandatory payments made to recruiters and government would be lost if workers became dissatisfied, broke their contract, and left for the village. Consequently it became standard to obtain plantation labour from islands other than that of residence so as to guarantee a captive work force for the period of contract (cf. Connell this volume).

Changes in mobility between 1921 and 1940

Intervillage movement increased during the 1920s for several reasons: more settled conditions, relaxation of personal fears about visiting the places of former enemies or their descendants, reduced distance between settlements because of their predominantly coastal location, and the greater number of potential movers as a consequence of consolidation. The development of better communications greatly facilitated this process. In 1917, on his inaugural tour, the first district officer for Guadalcanal instructed villagers to clear and maintain paths and within four years a trail linked all settlements along the north coast. In addition, transport by canoe became more prevalent. The Ndi-Nggai possessed few sea-going craft in 1910, but wage labourers saw the dugouts used in the Western Solomons and upon return began to manufacture canoes of the same design. Even former bush dwellers learned this art and soon dugouts became indispensable for fishing and for transporting families to neighbouring villages, to visit kin and attend feasts.

As socioeconomic change accelerated during the 1930s, certain types of mobility between villages and hamlets became less common. Whilst some large feasts occurred to denote status, to honour a dead person, and to celebrate past events, the number held and the crowd of participants was not as great as before contact. Fewer pigs were now raised and the missions worked energetically towards less lavishness in feasting, partly because of its pagan overtones and partly because deaths sometimes occurred from consuming poorly cooked or decaying pork. Local trade in canoes (*tsabiri*) also waned as Ndi-Nggai learned to make their own. Ecological differences between coast and interior were no longer important and food exchanges far less necessary now that former inland peoples had easy access to resources of the marine environment.

Despite fewer feasts and less trade, both the incidence of and reasons for local movement increased. From the 1920s men went more regularly to plant coconuts on *puku* land or to make copra from the groves of relatives. Some villages had trade stores by the mid 1930s and people from neighbouring settlements visited to make purchases. Schools were established at Maravovo and Visale, at which children spent several years, returning to their village when each term ended, while parents also travelled to the missions with food for their children or to participate in the more important religious services.

A head tax of ten shillings per year was levied in 1921 on all able-bodied men aged between sixteen and sixty. Two quite opposite changes were set in motion, as the Ndi-Nggai increased the size of their coastal coconut groves at the same time that more men left for wage employment. Soon a distinct pattern emerged: young men went to take up jobs but after marriage settled down to become cash croppers. Thus a period of wage labour became associated with a particular stage of the lifecycle; never represented a permanent commitment; was integrated with the changing extent of local obligations; and, under such circumstances, was a feature of the life histories of most Ndi-Nggai men (Bathgate 1975:941–54). Wage income provides a summary measure of this change. The amount earned from plantation work in 1914 was £A164 and in 1931 £635 – including short-term service as domestics and government labourers (Bathgate 1975:65, 75). This indicates a fourfold increase in movement to the places of employment, given that wage rates did not alter between 1914 and 1931, and also that the population decreased slightly as a result of introduced diseases.

By 1931, eleven per cent of adult males were working for wages and occupational histories collected from twelve elders in 1971 indicate the range of jobs available (Bathgate 1975:425–8). Apart from plantation work, there was employment at the mission printery, as deck hands on government ships and chinese trading vessels, as assistants to expatriate businessmen in Tulagi, and as catechists (many of whom moved regularly from one district to another). In the mid 1930s, a few men assisted prospectors after gold was discovered in the Guadalcanal interior. The case of Thomas of Taboko (Fig. 5.2), aged sixty-three in 1971, illustrates the circulation of Ndi-Nggai men to wage work. During his lifetime Thomas held seven jobs over a period of twenty-four years, but his work history is traced only to 1940.

Born at Mataniko village (Fig. 5.2), Thomas was forced into wage employment in 1923 when aged sixteen because neither his father nor *tina* had sufficient coconut palms for both subsistence and cash. Thomas travelled to nearby Ruaniu plantation, terminating his employment after two months once he had sufficient money to pay the head tax and buy some clothing. Following a short time at his village, he became an indentured labourer at Tasifaronga plantation. As a result of previous experience, Thomas received a monthly wage of one Australian pound and after two years returned to Mataniko with a new set of clothes and some tools. In 1926, Thomas accepted employment on a Chinese trading ship and, after his contract expired, became a butcher in an expatriate's shop at Tulagi. This job was held for five years and, late in 1932 after a sojourn at Mataniko, Thomas was hired as a labourer at Gold Ridge, in the interior of Guadalcanal. He remained nine years with this employer, but because of the proximity of Gold Ridge to Mataniko village was able to return regularly and assist occasionally in making copra or food gardens.

Despite much increase in movement between 1920 and 1940, for a time the protectorate administration tried to restrict any that entailed long-term absence from the village. Kings Regulation number 4, 1933 (Native Passes), stipulated that government servants, indentured labourers, and seamen were the only Solomon Islanders who could travel from one district to another without first obtaining a pass from the administrative officer (Pacific Islands Monthly, September 1933:7; December 1933:21). Costing one shilling, these passes were valid for one year, specified the district or place of proposed travel or residence, and required a legitimate reason for moving. Although the reasons for this regulation were never specified, there appears to have been official concern that higher levels of mobility were making it difficult to check the spread of disease and to undertake the annual village census, in order to calculate both the amount of head tax due and the rate of demographic change. District officers also found it troublesome to persuade people to remain in large coastal settlements, as well as to ensure that men fulfilled their local responsibilities.

The Native Passes Regulation had little impact when implemented in west Guadalcanal, especially since the district office was at Aola, more than 160 kilometres away. Men continued to take up casual employment on plantations, women to leave for their husband's village after marriage, and small groups to establish hamlets on their own coastal land — all without reference to the protectorate administration. Eventually, in 1937, the regulation was revoked and the High Commissioner for the Western Pacific High Commission was congratulated by several planters and traders, who had repeatedly criticized this interference with the personal liberty of Solomon Islanders.

Ndi-Nggai mobility, 1940-70

It was not until about 1946 that village life in west Guadalcanal returned to normal following the invasion of the Japanese five years previously. Patterns of mobility were radically altered by several major developments, most important of which was the establishment of a town — Honiara — on the fringe of Ndi-Nggai territory (Fig. 5.2). Originally an American supply and transit base, the British government purchased the extensive installations for a new capital of the Solomons and very shortly Chinese traders took up residence. By stages, a road was constructed from Honiara into west Guadalcanal and transport services slowly established. Until 1950, however, only villagers within about thirty kilometres of town could visit regularly to sell garden produce and buy trade goods, since trips from such peripheral settlements as Visale took a full day by canoe and at least two on foot.

Gradually, throughout the 1950s, the Ndi-Nggai established a transport system based on vehicular transport and the launches that eventually superceded the local canoe. The first vehicle, an American jeep owned by a Chinese trader, was acquired in 1950 by a man of Kakabona (Fig. 5.2). This initiated a ripple-like sequence of such purchases that concluded at the end of the coastal road, thirty-five kilometres west of Honiara, and was facilitated by increased incomes from selling copra (343 tons were sold in 1956 compared with 40 tons in 1931). Even so, jeep owners tended to operate close to their communities and to carry only local produce for sale in Honiara. Beyond the end of the road, five men operated launches and collected copra from a wider area, ferrying it to Yandina in the Russell Islands. Particularly significant was the fact that these launches and jeeps were not simply purchased to increase cash earned from copra but also carried kin and relatives to neighbouring settlements for feasts or surprise weekend visits. The launch operated by a Verahue villager took people as far as the Guadalcanal south coast, Savo, and Nggela (Fig. 5.1); for most Ndi-Nggai, it was the first time they had seen these places.

Between 1955 and 1960, the 'American highway' had been extended to Paru, fifty kilometres from town, rivers had been bridged, travelling time reduced, and six vehicles were licensed for the first time (Bathgate 1975:120). More villagers thus had improved access to Honiara; those from Taboko (Fig. 5.2), who operated six jeeps and a utility between 1958 and 1960, regularly sold local produce and many households went to market every two weeks. The road was constructed a further six kilometres to Naro between 1960 and 1970 and by the end of 1971 had reached Lambi Bay, sixty-four kilometres west of Honiara. All Ndi-Nggai communities were now connected to the capital; during this eleven-year period twenty-nine vehicles were purchased and the amount of rural-urban interaction rose dramatically. Between 5 November and 6 December 1971, the Ndi-Nggai made 601 produce trips to market and a further 200 to sell copra and cacao, for an average of three trips per person (Bathgate 1975:156–62). Another index of this transformation is that, whereas in 1964 ninety per cent of all copra was sold to visiting Chinese traders, in 1970 eighty-two per cent was personally delivered to the Copra Board in Honiara (Bathgate 1975:658).

Paralleling improved accessibility was an expansion of jobs available in both government and private enterprise. Employment histories collected in 1971 from forty-five men at Verahue and Taboko document that twenty-five different occupations were taken up between 1950 and 1970, of which only nine existed prior to 1940 (Table 5.1). Equally important was the fact that a slight rise in level of formal education over two decades allowed some men to obtain semiskilled

TABLE 5.1 *Occupations of forty-five men from Verahue and Taboko,*
1950-70

	No. of jobs	Employment time (years)
Within west Guadalcanal		
Plantation labourer	18	25
Catechist	8	39
School teacher (mission)	4	14
Mission carpenter	3	5
Government headman	2	6
Member of Visale native court	1	5
Elsewhere on Guadalcanal		
Deck hand on Chinese ship	8	17
Labourer for scrap metal company	2	8
Attendant at leprosarium	2	3
Other islands		
Plantation labourer	6	17
Mission printer	6	14
In Honiara		
Deck hand on government ship	8	29
Labourer/driver at public works	7	16
Police constable	5	16
Domestic servant	4	5
Government clerk	3	3
Agricultural labourer	3	3
Butcher	2	8
Assistant in geological survey	2	5
Hospital orderly	2	4
Restaurant waiter	2	3
Carpenter	2	1
Electrician's assistant	1	2
Sextant	1	.25
Plumber's assistant	1	.08

Source: Bathgate 1975:425-8.

positions: as clerk, carpenter, and electrician's assistant (cf. Frazer this volume).

The enlargement of job opportunities also saw changes in places and nature of wage employment. Numbers of men engaged on coconut plantations declined at rates similar to the rise in urban workers, for Honiara offered the prospect of not only a job but also sports events, movie theatres, and hotels. Between 1920 and 1954, forty per cent of all wage labourers from Taboko went to plantations as against eight per cent during 1955-70. Conversely, the percentage of men holding jobs in Honiara in 1950-54 was twelve (Verahue) and thirty-three (Taboko); by 1965-70 those proportions had risen to forty-six and eight-five per cent respectively (Bathgate 1975:425-30). Length of time spent at each job also declined, since the one- and two-year contracts of the early 1960s had given way to shorter ones (three to six months) or direct arrangements with the employer (Tedder 1966:37; Bellam 1970; cf. Frazer this volume). Consequently wage-work movement between village and town became more frequent.

A few young women also entered wage employment about the late 1960s, thereby breaking from their traditional roles of remaining in the village, making food gardens, doing housework, marrying as soon as possible, and raising a family. Some more progressive parents permitted their daughters to work for short periods at missions or local schools, but only after arrangements had been made for their supervision to reduce the danger of an unwanted pregnancy, which would lower both their status and their brideprice. This reversal of custom signalled other, equally dramatic shifts away from previous practice.

The war in the Solomons dislocated the interisland exchange of shell valuables with people from Malaita and it was never resumed. Trade contacts with Santa Isabel and Nggela, weak by the 1930s, had virtually ceased by the fifties, while the only surviving remnants of a once viable system of regional exchange were visits to neighbouring villages to buy pigs, local tobacco, or coral lime to mix with betel nut. From 1950, the area from which spouses were drawn widened beyond west Guadalcanal, as men returned from wage work with wives from such islands as Malaita and New Georgia. Within Ndi-Nggai territory, the greater availability of road transport encouraged personal visiting to more distant communities of the same religious denomination, while the attendance at Maravovo and Visale mission schools of children from throughout west Guadalcanal often resulted in friendships that led much later to marriage. Women continued to move predominantly to their husband's village after marriage, but liaisons between members of the same lineage (*duli*) became more frequent and no longer resulted in banishment to a remote district.

Contemporary movement

If the point of reference becomes the two villages of Verahue and Taboko (Fig. 5.2), then it is possible to examine contemporary Ndi-Nggai mobility within the context of socioeconomic organization and to assess what degree and kind of modifications have occurred to the traditional situation. Nowadays, as before, residential fluidity means that a high proportion of persons in any community are born outside it.

For fifty-five of the 122 *de jure* inhabitants of Verahue in 1971 and for 122 out of 192 at Taboko, this was their natal place. The others began life in twenty-four different villages in Ndi-Nggai, south Guadalcanal, and six other islands ranging from Choiseul to Tikopia (Fig. 5.1).

The continuity of this process over time is best illustrated by two case studies, for Paulino of Kakalaka hamlet, Verahue, and Daniel of Tavavao hamlet, Taboko.

> Paulino's mother, who was born in the interior, moved at the turn of the century to Verahue, where her *tina* had land (Fig. 5.3a). About 1910 she married, the couple remaining at Verahue rather than transfer to the husband's village. Paulino was born in 1913 and ten years later the family shifted to Tuvu (Fig. 5.2), so that his father could inherit land and coconut palms from a *tina* elder. From 1923 Paulino regularly visited Tsavagao hamlet, part of the Verahue complex, to see members of his *tina* and make copra from both their and his family's palms. He married a woman of Kobiloko (Fig. 5.2) in 1939, remaining at Tuvu to plant coconuts on her land. After 1946, he engaged in wage employment for six years, first on a government ship and then as a teacher at Tangarare mission. This experience concluded, Paulino relocated back to Tsavagao with his wife and constructed a new house, but they subsequently left for Naro to tend the coconuts planted by a deceased maternal uncle. Later still, they moved again to Tuvu. About 1955, Paulino was called back to Tsavagao hamlet by a *tina* elder, who wanted him to inherit his palms and be responsible for the *tina* land. Once more he built a house and a few years later constructed his own hamlet at Kakalaka, where he continued to live fifteen years later.
>
> Daniel was born in 1938 at Bogu, the antecedent village of Taboko (Fig. 5.3b). His *tina* had land in the immediate vicinity of both Bogu and Pisei (Fig. 5.2), with individual members also residing at Savo island and Tambea, near Visale mission. In his youth Daniel lived for several years at Kesao (Visale), Taboko, and Savo. Between 1955 and 1961 he worked on plantations in both west Guadalcanal and on other islands, returning first to Taboko to stay with his father and assist him in planting coconuts, thence to Tambea to help extend the groves of a *tina* elder. He arrived back from Tambea in 1964, living for three years at his father's new hamlet (Tavavao) until departing again for Savo. In 1969, having returned yet again, he married a woman of Popo, Visale, where he lived until the mid 1970s. Following another sojourn on Savo, this time for six months, Daniel returned to Tavavao to make food gardens and plant coconuts, leaving that hamlet towards the end of 1971 to establish his own eighty metres away.

As illustrated by the examples of Paulino and Daniel, the extent to

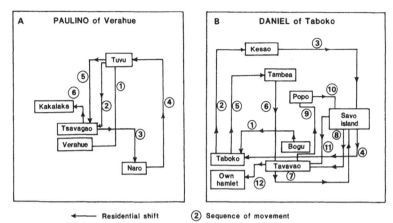

FIGURE 5.3 *Residential histories of two Ndi-Nggai males*

which men and women shift from one village to another varies according to social and economic circumstances. Most moves led to the establishment of food gardens, the planting of cash crops, the construction of houses and, in some cases, even the foundation of new hamlets. Most have also been of a circulatory nature. This characteristic is apparent from Figure 5.3, but becomes even more emphatic when it is noted that Bogu, Tavavao, and Daniel's own hamlet are very close to Taboko and that, in Paulino's case, Tsavagao and Kakalaka hamlets are part of Verahue village.

Residence and focality

People in any Ndi-Nggai village distinguish between *tinoni vera* (persons of the village) and *tinoni tavosi verana* (people of other places) to identify those inhabitants who, irrespective of place of birth and of current residence, hold primary rights to land as members of particular *tina* or *puku*. In 1971, there were twenty-nine *puku* among the *de jure* population of Verahue but only five (15%) had such primary rights; at Taboko, the number was five out of twenty-two *puku* (23%). Yet no less important is the contrary fact: lineage membership guarantees that most *puku* hold primary rights to land in other places. For Verahue and Taboko, these locations are concentrated in north, northwest, and south Guadalcanal, but if residents born on other islands are included then they extend to peoples as far away as Malaita and Tikopia (Table 5.2).

That so few people have land rights in their village of residence reflects both traditional and contemporary forces. Given a matrilineage system, the movement of women at marriage to their husband's place

TABLE 5.2 Duli *and* puku *of Verahue and Taboko, 1971*

Duli (clan)	De jure population	No. of puku	Location of puku land
Verahue			
Lokwili	44	12	Guadalcanal: south, northwest; Nggela islands
Kindipale	25	8	Guadalcanal: northwest
Haubata	23	6	Guadalcanal: northwest; Nggela islands
Kakau	22	3	Guadalcanal: northwest, east
Other[a]	8	3	Malaita; Russell islands; Tikopia
Taboko			
Haubata	55	8	Guadalcanal: northwest, north
Kakau	55	5	Guadalcanal: northwest, interior, north
Lokwili	36	5	Guadalcanal: northwest, south, southwest, interior; Savo island
Kindipale	32	3	Guadalcanal: northwest
Simbo	8	1	None
Other[a]	6	1	Guadalcanal: northwest; Choiseul

[a]Mainly persons born on other islands.
Source: Bathgate 1975: 182, 285.

is basic, but there are other reasons: at the turn of the century *puku* relocated from the interior, where they owned land, to the coast, where they did not; many more *puku* became represented in local communities through the combined impact of less attention to patrilateral cross-cousin marriage, which guaranteed the reciprocal flow of women between two areas, and an enlargement in the sociospatial range of marriage partners; plus an increasing tendency for women and children to decide not to return to the natal village after the death of a husband or father. This last trend, evident since the mid 1950s, is associated with the rise in planting coconuts and the inheritance or transfer of larger *tina* estates. A man and his sons plant the palms, his household makes the most of the resultant copra, and sons do receive a sizeable part of that estate, even though a man's *tina* also inherits some of it. If, as is occurring more and more, sons and mothers buy palms to seal their rights to them and thus acquire economic interests in the place of their father and husband, then eventual return to their *tina* land becomes decidely unattractive.

Whilst children may gain greater economic interests in their village of residence, in fact quite often neither they nor their fathers own the land on which the palms are planted. Nowadays, land-owning *puku* in both Verahue and Taboko seldom give plots to others for coconut groves, partly because children without primary rights are inheriting palms and partly because population increase is reducing the amount of land available per capita for future planting. Consequently disputes over primary rights have become common and two changes are emerging. First, the possibility of utilizing *puku* land in other locations is being considered, especially by residents of Taboko. People are visiting kin to *tutunga* ('story all about'), trace their *puku* origins, and determine how they secured primary land rights in particular areas. Many are travelling to the interior to locate former hamlets, abandoned *kokomu*, and other traces of historic settlements: literally engaging in *tsahorim-bulu* (following the road of the ancestors) so that they can return and occupy *puku* land should they be pressured to leave the coast. Yet others, whose mothers or grandmothers ignored customary practice and never returned to their hearth area, are presenting pigs to resident kin to offset costs not met when that relative was buried and, by this strategy, reestablishing rights to lineage land (Bathgate 1975:773–9).

The contrary development and preferred solution is that, whenever possible, land is being bought to ensure both primary rights and security, as well as to capitalize on previous effort and investment, rather than relocating to begin again at another location (Bathgate 1975:783–810). Conceptually, for the Ndi-Nggai, the solution of purchasing land is different from returning to it. Whereas return would reverse the expansion of a lineage through time-space, purchase represents consolidation at places where expansion has already occurred. Through common and combined purchase, the whole *puku* acquires additional focal points (*kokomu*) while individual members for whom such purchase is critical can transfer their economic, but not their social and sentimental interests, away from the hearth area to the village of residence.

In precontact times, dispersion from the main settlement resulted from social animosity and the need to live closer to distant food gardens; the contemporary parallels are friction over land problems, especially tenure, and the wish to reside near one's coconut groves. Viewed in this light, the division of villages like Verahue and Taboko underscores the nucleation encouraged by the missions and a colonial administration to be an alien imposition that even nowadays is difficult to sustain.

Movement within and beyond the local area

People spend more hours within the village and its immediate vicinity

TABLE 5.3 Average weekly use of time, in hours[a]: Verahue and Taboko, 1971–2

	Verahue			Taboko		
	Married men	Married women	Single men	Married men	Married women	Single men
Within village						
Store keeping	–	–	–	2.7	1.1	3.5
House construction	1.7	0.2	0.5	1.5	–	2.9
Council work	1.5	1.6	1.7	–	–	–
Church service	2.8	2.8	2.0	2.5	3.0	2.1
Church work	0.6	0.3	–	0.6	0.8	–
Household chores, leisure, sickness	56.4	62.9	65.5	42.3	43.6	58.7
Subtotal	63.0	67.8	69.7	49.6	48.5	67.2
Within immediate area						
Gardening	8.1	12.9	2.0	21.8	23.5	1.6
Fishing	3.5	–	2.0	0.3	–	0.9
Hunting pig, birds	–	–	–	1.2	–	0.8
Feast preparation	0.1	0.2	0.1	–	–	–
Preparing for market	0.2	0.2	0.1	2.0	3.0	–
Making copra, cutting cacao	4.5	0.4	1.3	1.2	0.6	–
Cutting firewood	0.5	–	0.5	0.2	–	–
Subtotal	16.7	13.7	6.0	26.7	27.1	3.3

Other villages						
Visits, *lela*	3.3	2.3	7.8	0.8	0.3	8.5
Subtotal	3.3	2.3	7.8	0.8	0.3	8.5
In Honiara						
Selling copra, cacao beans	0.6	–	0.4	0.6	–	3.2
Selling food	0.4	0.2	0.1	5.0	8.0	–
Social visits	–	–	–	1.3	0.1	1.8
Subtotal	1.0	0.2	0.5	6.9	8.1	5.0
Total	84.0	84.0	84.0	84.0	84.0	84.0

aBased on the hours 06.00–1800.
Source: Bathgate 1975:530–1.

than they do outside. Household chores and food gardening command most attention among the married, especially women; by virtue of hunting, fishing, and cash cropping, the ebb and flow of married men across the settlement periphery is more regular than for others; and males visit nearby communities far more often than women (Table 5.3). At any particular time, there are consequently many persons absent from either Verahue or Taboko (cf. Connell this volume) and their movements can be grouped into those completed within a day, those lasting two to seven days, and those entailing absence for periods from a week to at least six months (cf. Chapman 1969:132).

Short-term mobility, accomplished within the same day, is prompted by the need to buy pigs for reciprocal feasts, discuss land matters, visit stores, participate in court cases, observe meetings of the local council, and attend prayer sessions at the missions (cf. Chapman 1975:139–41). Feasts may be traditional in origin – for example, to mourn the death of kin – or reflect an introduced religion, as with the church-day feasts held by anglicans. The web of reciprocity involved is summarized in the fact that, on 1 July 1971, Verahue sponsored a church-day feast and attracted 500 people from seventy-eight villages and hamlets.

Beyond all these there is *lela*: to go 'walkabout' for a brief respite from one's community, to hear the latest gossip and, in the words of one Ndi-Nggai, 'see what there is to be seen' (cf. Chapman and Prothero this volume). Single men most often visit other villages (Table 5.3: row panel 3), usually through *lela*, which has caused elders to complain about too much 'walkabout' as a result of improvements in transport services along the road between the village, Honiara, and Lambi Bay (Fig. 5.2). Yet *lela* has its own sociospatial boundedness, for young men generally visit only the places of *tina* members or former school friends of the same denomination.

Noteworthy among absences that last several days is *toturahe*: 'stopping all about', practised primarily by young single males. Although like *lela* and more prevalent since the 1960s, *toturahe* is different in being socially unacceptable. Spending a few nights in one village, then another, and yet another is the lifestyle of a *mane hato* (lazy man) who, by continually shifting, avoids helping in the food gardens but always obtains sustenance through the courtesy of hosts. Wherever he goes, a *mane hato* literally 'stops nothing' and is 'fed for nothing'.

Copra production also generates a continual flow of people between settlements (Table 5.3: row panel 2). Such visits average three days, depend on economic circumstances, and vary across the lifecycle. Young men with few palms sometimes visit maternal uncles to cut copra, after which they produce a few bags that amount to payment in kind. Once married, the need for cash becomes greater and a man must rely heavily on the groves of *tina* members, his father, and father-in-law until his own plantings mature after five to seven years. By the age of forty, few visits are made to other villages to cut copra and gradually younger kinsmen are permitted to use his groves, with those men aged fifty or more the most important allocators (Bathgate 1975:465-98). The prevailing reasons for long-term absence from Verahue or Taboko are schooling at mission centres, wage employment, and *vera ruka* residence, which in total can account for a substantial proportion of the *de jure* population. Thus, on 27 February 1971, twenty-eight per cent of Verahue people and eighteen per cent of those of Taboko were away, all of whom had returned within ten months (Bathgate, 1975:936). Three men engaging in a *vera ruka* lifestyle maintained houses at Taboko and came regularly from Savo island and Niutuzu, while the two at Verahue often visited from Tiaro Bay. None contemplated single-village residence, because their dispersed resources were too large and *vera ruka* was the sole means by which they could be utilized.

When Ndi-Nggai have interests in land beyond their usual place of residence, visit relatives in other areas for lengthy periods, make copra from distant palms, occasionally leave to help plant coconuts or make new food gardens, and engage in *vera ruka* residence, it is clear that the concept of village or hamlet as a discrete unit and as having a single focal point runs counter to reality. The web of kinship is the most important factor that initiates movement, controls its direction, and influences its duration. Considered from the standpoint of a single community, since contact this web has been spun over a very large area. In Verahue in 1971, parents of twenty-two couples lived in ten other places and *tina* relatives were distributed throughout thirty-two settlements. Consequently, the movement spheres of households defined by social visits seldom coincide and their dimensions vary according to the number of places in which kin reside and their distance from the village of reference. However, kin in neighbouring settlements are visited most and for brief periods whereas those in most distant places are seen intermittently during sojourns that often span several months.

Changes in a person's kinship situation over the lifecycle also manifest themselves in the range and dimensions of movement. This may be shown schematically with reference to land belonging to four different clans (*duli*): Haubata, Kakau, Kindipale, and Lokwili. A single man of Kakau is born in the village of his father, who is a Haubata (Fig. 5.4a). This fact bounds subsequent movement, as do the places of residence of maternal uncles and aunts, married brothers and sisters. The bride-to-be, a Lokwili, most often visits the villages of matrilineal kin (Fig. 5.4b). When the couple marry, they may remain with the wife's parents, go to the village of the husband's father, or transfer to where the husband's *tina* has land. In this example, the third option was chosen (Fig. 5.4c), the social territory of both spouses widens, and occasional visits are made to both sets of parents as well as to members of their respective *tina*. A son born to them moves within an area (Fig. 5.4d) similar to that of his mother before marriage (Fig. 5.4b) but quite different from his father's (Fig. 5.4a), to complete a lifecycle sequence that has occurred for generations.

A. SINGLE MAN

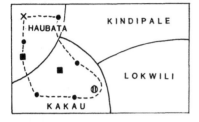

X Place of residence (father's village and perhaps place of birth).
⨀ Tina land where a maternal uncle lives.
● Places where tina aunts and uncles live.
■ Places where own married brothers and sisters have moved to.

B. BRIDE-TO-BE, BELONGING TO TINA OF ANOTHER PUKU

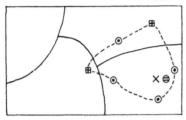

X Place of residence (father's village).
⊖ Tina land.
⊙ Places where tina aunts and uncles live.
⊞ Places where own married brothers and sisters have move to.

C. COUPLE AFTER MARRIAGE

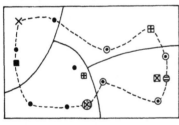

X Place of residence, males village.
⊗ Place where male's parents live.
⊠ Place where female's parents live.
⓪ Tina land of the male.
⊖ Tina land of the female.
● Places male's maternal uncles and aunts reside.
⊙ Places female's maternal uncles and aunts reside.
■ Places male's brothers and sisters live at.
⊞ Places female's brothers and sisters live at.

D. ADULT, UNMARRIED SON

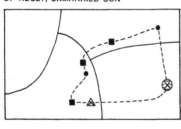

X Place of residence.
△ Place where parents live.
⊖ Tina land.
● Places tina uncles and aunts live.
■ Places married brothers and sisters live.

- - - - - - Boundary of movement area   KAKAU   Duli (clan)

FIGURE 5.4 *Context of kinship for mobility*

Rural-urban movement

Verahue and Taboko villagers visit Honiara for periods of 10–15 hours to sell garden produce, copra and cacao, to purchase goods at Chinese trade stores, to visit the agricultural department for technical advice, to attend the hospital and health clinics and, as an urban extension of a rural practise, to *lela*. Sale of subsistence and cash crops generates most movement: during 1971, 126 trips were made from Verahue (six per household) and 1,350 from Taboko (fifty-four per household), a differential that summarizes their distance from and travel time to Honiara (Fig. 5.2). Accessibility to the main town similarly influences the level of sales, not only for Verahue and Taboko but also for all Ndi-Nggai communities. A market survey of west Guadalcanal for the period 5 November to 6 December 1971 revealed that 0.47 return journeys per capita were made from villages 1.6–19.3 kilometres from Honiara (including Taboko); 0.21 from those in the 24–45 kilometre zone; 0.10 from those 48–64 kilometres distant (including Verahue); and 0.09 from those beyond 72 kilometres (Bathgate 1978a).

Copra and cacao are sold by men, whereas marketing root crops and other produce is an extension of gardening and the prerogative of women. From 5 November to 6 December 1971, Taboko women were part of eighty-nine per cent of all market trips but the males only forty-seven per cent. Consequently, Taboko women spend more time in Honiara than men; the reverse of the pattern for Verahue women (Table 5.3: row panel 4). Single men from Taboko seldom are involved in trading but ride on the market truck to go 'walkabout', which in turn accounts for the greater time they devote to *lela* compared with their Verahue counterparts (Table 5.3: row panel 3). In short, different levels of involvement in the cash economy are summarized in the ratio of rural-urban to intervillage movement: for Taboko people the former is higher but the latter overwhelmingly dominates among the Verahue (Table 5.3: row panels 3 and 4). This does not mean that contacts with kin have declined, only that the scene has changed. By design, not coincidence, several Taboko householders in 1971 always marketed on the same day as relatives from Mataniko and Kakabona (Fig. 5.2) and by this tactic were able to keep abreast of *tina* and *puku* business.

Movement to wage employment

To participate in wage work still involves circulating between the village and places beyond, the historical continuity of which is documented for four men born in 1927, 1935, 1942, and 1945 respectively (Fig. 5.5). Taking up and terminating paid employment is influenced by many factors but in general these fit Mitchell's classic characterization of labour circulation in south central Africa: 'the economic drives as a

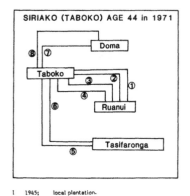

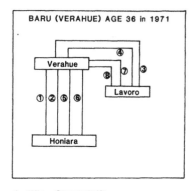

SIRIAKO (TABOKO) AGE 44 in 1971

1	1945;	local plantation.
2	1946:	village.
3	1948;	local plantation.
4	1948;	village.
5	1951;	local plantation.
6	1952;	village.
7	1972;	local plantation.
8	1972;	village.

BARU (VERAHUE) AGE 36 in 1971

1	1954;	Chinese trade ship.
2	1954;	village.
3	1955;	local plantation.
4	1958;	village.
5	1961;	Chinese trade ship.
6	1961;	village.
7	1966;	local plantation.
8	1969;	village.

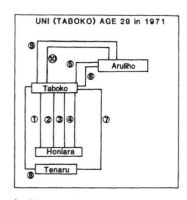

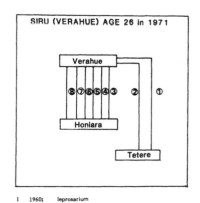

UNI (TABOKO) AGE 29 in 1971

1	1960;	Honiara,	Govt. orderly.
2	1960;	village.	
3	1961;	Honiara, policeforce.	
4	1962;	village.	
5	1964;	mission, carpentry.	
6	1965;	village.	
7	1965;	mission, carpentry.	
8	1969;	village.	
9	1971;	mission, carpentry.	
10	1971;	village.	

SIRU (VERAHUE) AGE 26 in 1971

1	1960;	leprosarium worker.
2	1960;	village.
3	1960;	Chinese trade ship.
4	1965;	village.
5	1968;	Chinese trade ship.
6.	1970;	village.
7	1971;	Chinese trade ship.
8	1971;	village.

FIGURE 5.5 *Histories of outside employment of four Ndi-Nggai males*

rule operate centrifugally to force men . . . to distant labour centres [and the] social system, operating particularly through the network of social relationships, tends to hold a man within its hold and to resist the influences pulling him away' (Mitchell 1961a:275-6). This tension is most obvious in the relationship between cash cropping and wage labour.

Since the early 1920s, lack of village-based resources has continually forced young men into paid employment to satisfy their need for cash, whereas concern for planting coconuts has led others to remain in the local community. Parents or *tina* had very few palms during the 1930s, hence wage labour was an important means to acquire money. Since 1960, the latter has become less significant, as most young men have access to the greater number of palms owned by fathers and maternal relatives. Yet since access to the groves of kinsmen cannot meet all cash needs of the married, both copra production and outside employment are pursued in sequence during the course of any particular year.

Money earned on the job in Honiara or at other workplaces may help plant coconut groves, establish trade stores, erect cattle fences and stockyards. Such local projects are usually initiated and controlled by *tina* or *puku* elders but single men must assist when asked, most often with their labour but at times with cash to hire workers from other districts to clear land, tend immature palms, or purchase fence posts. The need to contribute money may initiate the search for an outside job just as the lack of local labour may terminate one, but the link generally is more indirect, as men already in paid employment remit earnings because that is expected.

Since most Verahue and Taboko residents have only four or five years of formal education, most jobs held are unskilled or semiskilled (Table 5.1), for which they are poorly paid. Low wages continually frustrate villagers and are the reason many leave outside employment. As one Taboko man put it:

> We people of the Solomons live from day to day, we do not worry about tomorrow but think of our stomachs instead. However, this way, on a small wage, means that you never have enough money to buy some expensive item that you might like. Working really means working to buy food. This is why wage employment is not a good thing. It is not like living in the village, where you can decide when and if you want to work. If you want to sleep you can sleep, if you want money you can get it quickly. This is why we people say that working for the white man is *rotale*, it is hard work for nothing. That is why I left my job. (Translated from pidgin English: Bathgate 1975:436)

Monthly wages vary from $Australian 8-20 ($A 2.16 to £1.00), about two-thirds of which is spent on food, so that short absences are not only preferred but also sufficient to relieve boredom and satisfy individual curiosity. The economic circumstances faced during long-term employment compare very unfavourably with those in the village, where fewer hours are required to satisfy subsistence needs, leisure time is much greater, and the cash for a particular item can be obtained much sooner by making copra or selling a pig. This rationalization conforms with Berg's (1966:117) hypothesis: the decision to move and the time spent in wage employment depends on the level of income available from other (village-based) pursuits and the effort they require, although among the Ndi-Nggai these factors operate differentially with respect to age and marital status.

At first sight, the fact that men remain in Verahue and Taboko after marriage suggests they can ultimately gain higher incomes from establishing produce gardens and planting coconuts than from going away for wage labour. In Taboko for example, the cash earned at any given time from selling copra and market crops is six times higher than would be received from working on a plantation or in Honiara. Yet this differential, whilst real enough, is compounded by social factors. With marriage is initiated a whole range of commitments: food gardens must be established to ensure the independence of the nuclear family, prestations must be shared, pigs raised, and coconuts planted to provide for younger relatives. In short, the ability to choose between village domicile and outside wage employment applies more particularly to single males, who alternate between one and the other. The desire to see something of Honiara town and enjoy its sights and entertainment is a far greater incentive for seeking work than the wages offered, but in the end the long hours, low wages, and high cost of subsistence makes a village sojourn all the more attractive (cf. Frazer this volume).

The ways in which marriage markedly influences participation in paid employment is documented by life histories from forty men (Table 5.4). The average age at which wage work ceased (row 2) approximates that for marriage (row 1), with subsequent participation dramatically less (row 4). Because of constant returns to the village, the time actually spent in wage labour between first and last job (row 5) is depressed − a characteristic that can be expressed more precisely by relating years spent in outside employment to those between the age of sixteen and when the last job was terminated. The resultant percentages are thirty (Verahue) and forty-four (Taboko), far lower than those recorded in row 3. Finally, when marriage occurs earlier than the average (row 1), there is further reduction over the lifetime in total time spent in paid employment (rows 5–8).

For both single and married males, jobs on local plantations, at mission centres, and in Honiara dominate. On the one hand, these enable the unmarried to visit Verahue and Taboko at least during weekends; on the other, plantation work permits the married to remain in their community and make gardens, produce copra, and visit kin. By circulating between two points, both married and single men can balance the need for cash derived from outside employment with household and communal responsibilities. Extended employment in Honiara has decided social disadvantages for husbands, since not only is accommodation difficult to find but also wages are too low and living costs too high for an unskilled worker to support dependants.

Furthermore, the number and range of jobs open to the Ndi-Nggai is

*TABLE 5.4 Years spent in wage employment by married males,
Verahue and Taboko*

	Verahue (18 men)	Taboko (22 men)
Sample average		
Age at marriage	26.8	26.4
Range of years when employment undertaken	21.2-29.4	19.1-26.1
Portion of time in employment between first and last job	72.3	67.2
Employment before marriage proportional to all employment	72.3	88.1
Average for men marrying earlier than average age		
Total years in employment	4.8	3.5
Years in employment after marriage	1.8	0.4
Average for men marrying later than average age		
Total years in employment	7.9	10.1
Years in employment after marriage	1.7	1.0

Source: Bathgate 1975:445-9.

much limited by an unwillingness to move beyond the language area
and actively seek employment. Such hesitancy reflects the insularity
felt when amongst aliens in another area, the problems of finding food
and shelter without kin being nearby, and the fact that men can never
be sure of obtaining a job when local information about employment
opportunities is seldom available. Youths who have held jobs on islands
beyond Guadalcanal were invariably offered them by planters and
Chinese traders when visiting Honiara or places throughout west Guadalcanal.

When temporarily present in an area without 'wantoks' (people
speaking the same language), a Ndi-Nggai must rely on social relationships established with other employees on the basis of common interest:
hunting and fishing activities, island identity (being from Guadalcanal),
common religion. How well such adaptations occur very much determines whether a man remains on the job or returns to the village.
Relationships with employers are similarly critical. If a boss gives additional pay when asked or permits occasional village trips to fulfil social
obligations, then a longer term in paid employment is more likely. Such

employers are rare, however, and their general lack of generosity and maintenance of social distance both reinforces the feelings of insularity amongst Ndi-Nggai and compounds their rapid turnover. Between January 1971 and April 1972, the average time on the job for twenty-four men and two women was 2.8 (Taboko) and 3.5 (Verahue) months – quite similar to the three- and six-month averages reported for an inland and coastal settlement on south Guadalcanal (Chapman 1969:128).

Circulation for wage labour balances the need for some money and the desire to see the outside world against the commitments and attachment to a village lifestyle. Being undertaken initially by single men and a few women and being largely discontinued after marriage, it neither affects the local output of subsistence or cash crops nor places an extra burden on adults who remain behind. Increases in amount of formal education may result in more skilled and better jobs, longer periods in paid employment, and increased attractiveness of Honiara for the more highly educated (Frazer 1973:7-12; Bellam 1963:44-53; Bellam 1970: 82-83; cf. Frazer this volume). Amongst the Ndi-Nggai, for many reasons, involvement in urban wage employment will probably continue to be intermittent and punctuated bv longer stays in the village: cash incomes from local production are higher than for most parts of Guadalcanal; settlements are within range of Honiara and young men can move easily back and forth; past patterns of mobility behaviour are strongly ingrained; the capital of Honiara is considered an alien place where even subsistence involves money and hard work; and the social pull of the village remains strong.

Giving lie perhaps to the notion that higher education alienates youngsters from rural places, a survey in 1970 of secondary school pupils at Aruliho (Fig. 5.2) revealed them to prefer jobs closely connected with local development. Reasons given ranged from an emphasis on the negative aspects of urban wage employment; the desire to be close to parents, relatives, and friends; and the wish to preserve and later transmit one's cultural heritage to the hope of utilizing formal education to assist and promote social and economic progress within villages (Aruliho School 1970). The pupil's reasoning revealed the pull of the natal place, that wellspring of Melanesian identity, as well as the feeling that socioeconomic progress at the village level was a foremost priority impeded by life-long involvement in paid employment. Outside wage work was also considered to delay the local establishment of cash-earning activities – a view similarly held by several elders at Verahue and Taboko who, having circulated for many years as labourers, now regret that fact. If these school pupils translate their ideals into practice then they, like their forebears, will spend early adulthood in wage employment and become local cash croppers once married.

Sequence and continuity in the movement process

This examination of Ndi-Nggai mobility from precontact times until 1971 has revealed four long-term trends. Whereas movement in pre-contact society was restricted, it began to increase from 1870 with the recruitment of labour for Queensland, mission encouragement, and cessation of local and interisland warfare. Thereafter it expanded along with the development of wage employment throughout the Solomons, establishment of road links with Honiara, the gradual appearance of transport services, and the planting and sale of cash crops. Second, the territorial domain over which movement took place progressively widened from the local territory in precontact times to west Guadalcanal as a whole and ultimately to other islands. Third, group movement (for reasons of safety) gradually gave way to more individual behaviour; the freedom to journey and visit was no longer restricted to certain classes (warriors, sorcerers, 'big men'); and involuntary moves (of slaves, troublemakers) became rare. Finally, once a produce market was established in Honiara, women left the village more frequently for economic purposes and from the mid 1960s some even began to enter outside wage employment. Not only are these trends similar to those observed in Africa (Gugler 1968) but also they conform to one of the propositions advanced by Zelinsky (1971): that physical and social mobility in traditional society was limited but that rates of movement increase during a community's experience of the modernization process.

At another level, some of these changes in mobility behaviour are characteristic of matrilineal societies that possess patrilocal systems of residence. As Belshaw (1954) has noted for Nggela of the central Solomons, the web of kinship in such societies is territorially extensive and

a man may inherit land in widely separated areas, since his female ancestors and their brothers would have lived in many different places. Before the establishment of law and order men could work effectively only those areas which were close to their place of residence. Today, however, men actually work land in many widely separated parts and it is common for such wealthy people to regard one village as their home, but to spend much time living elsewhere cultivating other tracts of land (Belshaw 1954:102).

The same sequence occurred in west Guadalcanal, as the weakening of patrilateral cross-cousin marriage after contact and the enlarged socio-spatial arena from which spouses were drawn led to the dispersion of the matrilineage. This fact, coupled with developments in cash cropping, emerging problems over access to land, and improvements in transport services, means that nowadays the Ndi-Nggai not only regularly visit kin in distant places but also utilize their land resources and may even plant coconut groves.

Notwithstanding the emphasis in this paper on mobility change through time, it has been established that for the Ndi-Nggai many similarities exist from one decade to the next. Increased movement at any given period arose not simply from local socioeconomic transformations but also because additional reasons were added to preexisting ones. Some contextual factors have persisted from traditional times, so that nowadays movement reflects social and economic forces with both customary and modern origins. This interpretation can be underscored by listing the main reasons for movement over six periods, precontact to 1971, in terms of 'oscillation', 'circulation', and 'migration' (Table 5.5). Thus is revealed the decline in some traditional mobility (raids, visits to *verana bo* and *kokomu*, trading expeditions to sell canoes, journeys of *chinoho* and slaves) along with the continued importance of others associated with locally-defined economic activities, marriage, visits to kin, acquisition of pigs, two-village residence, reciprocal feasting, and the occasional relocation of entire settlements. The period 1870 to 1920 emerges as a transition, within which the evolution of wage-labour employment accompanied a realignment of much of the traditional structure of movement; whilst for the years 1950 to 1971 a cumulative increase in socioeconomic influences led to yet higher levels of mobility.

The coexistence of customary and modern forces in mobility behaviour is common to all Melanesian societies; it is only the ratio of one set to the other that differs. Amongst the Ndi-Nggai, the wide range of local economic activities pursued, the possession of vehicular transport, and modifications in traditional social organization mean that the ratio of modern to traditional influences is higher than for peoples along the south coast of Guadalcanal (Chapman 1976) or on Malaita (cf. Frazer this volume). Despite all this, Ndi-Nggai mobility still has a precontact character to it that parallels exactly the situation along the south coast, where movement continues to be controlled by the place of origin (hamlet, village, lineage land) and is predominantly circular whether undertaken to visit kin, make copra, sell local produce in town, or earn money from wage labour (Chapman 1970). Almost all Ndi-Nggai movement occurs within the language area; adult men and women return to their district after terminating outside employment; and the village has not been forsaken for long-term urban residence. In fact, permanent absence from their region would be tantamount to committing *sarosahi* (revoking kinship ties and land rights) and to denying one's identity as a Ndi-Nggai.

According to Chapman and Prothero (this volume), 'circulation as a form of mobility has not evolved from the impact of alien or western influences upon indigenous circumstances. . . . Rather, . . . externally generated changes have reinforced customary circuits of mobility and added new ones'. Of the several concepts proposed to describe this

TABLE 5.5 *Main reasons for Ndi-Nggai movement, precontact to 1971*

Reasons	Period					
	Precontact	*1870–1910*	*1910–20*	*1920–40*	*1940–50*	*1950–71*
Oscillation within local area						
Gardening	x	x	x	x	x	x
Fishing	x	x	x	x	x	x
Hunt in bush	x	x	x	x	x	x
Cut firewood	x	x	x	x	x	x
Kokomu	x	x	x	(x)	[x]	[x]
Verana bo	x	[x]				
Make copra/cacao			(x)	x	x	x
Oscillation beyond local area						
Raid other groups	x	x				
Visit kin	x	x	x	x	x	x
Attend traditional feast	x	x	x	x	x	x
Trade in canoes	x	(x)	(x)			
Exchange food (*sinei*)	x	x	[x]			
Buy pigs	x	[x]	x	x	x	x
Pray at mission			(x)	x	x	x
Medical treatment at mission			[x]	(x)	(x)	x
Village church-day feast			x	x	x	x
Buy goods at trade store			[x]	(x)	(x)	x
Make copra at village				x	x	x
Plant coconuts locally				[x]	(x)	x

Reasons

	Period					
	Precontact	1870–1910	1910–20	1920–40	1940–50	1950–71
Sell copra/cacao in Honiara					[x]	x
Market produce in Honiara					[x]	x
Purchase trade goods in Honiara					[x]	x
Medical treatment in Honiara					[x]	x
Attend council meeting					[x]	x
Attend Native Court hearings					[x]	x
Lela to another village			(x)	(x)	[x]	x
Lela to Honiara					[x]	x
Frequent circulation beyond local area						
Go to other house (*vera ruka*)	x	x	x	x		x
Attend mission school		[x]	(x)	x	x	x
Work at expatriate plantation			(x)	x	x	x
Go to mission centre			(x)	x	x	x
Work in Honiara			(x)	x	(x)	x
Proselytise (*Retatasiu*)				x	x	x

Migration						
International wage labour					x	x
Relocate settlement	x	x	[x]	x	x	x
Banishment (*chinoho*)	x	x				
Capture as slave	x					
After marriage (within island)	x	x	x	x	x	x
Interisland marriage	[x]	[x]	[x]	[x]	(x)	(x)

x Common occurrence.   (x) Rare occurrence.   [x] Very rare occurrence.

process, 'amplification' is most appropriate for the Ndi-Nggai of west Guadalcanal. Both *lela* and *toturahe* featured in traditional society, for instance, but not until after European contact did these forms of movement become common. Similarly, *vera ruka* residence was an important customary way of adjusting to social obligations in different places but nowadays a two-village lifestyle is more common. Given the increasingly dispersed nature of local economic interests, *vera ruka* represents an effective solution to competing needs, rights, and obligations among varying locations in precisely the same way as does circulation between the village sector and outside places of wage employment.

6 Copper, cocoa, and cash: terminal, temporary, and circular mobility in Siwai, North Solomons

John Connell

The people of Siwai occupy the central part of the south Bougainville plain. Bougainville is the easternmost island in Papua New Guinea and constitutes the province of the North Solomons. Geographically it is part of the Solomon Islands and the population of Bougainville tends to share characteristics with Solomon islanders to the east rather than Papua New Guineans to the west (Fig. 6.1). Siwai society was studied in detail between 1938 and 1939 (Oliver 1955); social and economic changes since then were examined from November 1974 to April 1976, based around the villages of Siroi and Maisua (Connell 1977, 1978). In 1940, there was a population of 4,500, which doubled in the following thirty-five years. There was then very little trade and contact with the commercial economy was primarily through movement to work on plantations. The domestic economy remained essentially one of subsistence affluence based on root crops (cf. Connell 1978:62-7). Following the second world war, Siwai became more closely incorporated into the national economy through cash cropping, initially of rice but ultimately and most successfully of cocoa. In 1972 the trans-island road was completed from the Panguna copper mine through Siwai to Buin (Fig. 6.1), thus dramatically increasing the cash income available through improved marketing facilities (for cocoa, vegetables) and increased accessibility of mine employment. A number of Siwai businessmen profited from these opportunities as the local economy shifted rapidly from subsistence to affluence.

Among the Siwai, much early experience of population movement closely paralleled that of many groups in the northern and western Solomon islands. In more recent decades, primarily because of cocoa and copper, it has taken on more unique features similar only to a small number of Bougainvillean societies. Siwai population mobility is now characterized by intermittent, short-distance journeys for social reasons and by a pattern of migrant labour which, in contrast to much of

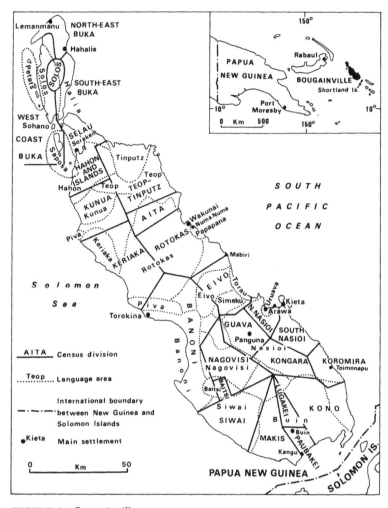

FIGURE 6.1 *Bougainville*

Melanesia, also occurs over short distance and involves high wage rates. Since 1900 the pattern of mobility has changed significantly and broadly reflects the incorporation of Siwai into a wider socio-economic and political arena.

The emergence of migrant labour

Although the history of population change in Siwai before the present century is almost unknown, historic separation and uniqueness were never absolute. In the nineteenth century marriages were contracted at

120

least as far as Nasioi in eastern Bougainville and with the Shortland islands of the Bougainville strait (Fig. 6.1), whilst the first accounts of European contact are derived from the oral histories of Siwais whose fathers or grandfathers worked during the same century on German plantations in Samoa. There, a small number of individuals gained their first experience of wage employment and from there a number of important plants and artefacts, especially steel tools, first came into Siwai.

In the first decade of the twentieth century and probably for some time before, a number of Alu (Shortland) islanders hired labourers from Buin, and probably from Siwai, to work their gardens and coconut plantations for periods of several years. On completing such employment they were given shell money, calico, axes, and boxes (Wheeler 1943:669-75). Thus by the end of the nineteenth century a small number of Siwais were directly familiar with wage employment, on either the European plantations of Samoa or the tiny plantations of Shortland islanders.

In 1886 Germany technically occupied the northern Solomons (Buka, Bougainville, Shortland islands, Choiseul, Santa Isabel: Fig. 5.1), although their first administration station in Bougainville was not established until September 1905, at Kieta on the east coast. When founded, its 'special task' was to increase labour recruitment from the German Solomons (Firth 1973:153). To some extent, this succeeded and more men went, especially from the Kieta area, but with little impact upon remote places like Siwai.

Plantations were begun on Bougainville at much the same time as attempts were being made to induce labour movement from the east coast. These seem little different from those elsewhere in New Guinea, combining a variety of tropical crops such as rubber, cocoa, and cotton. No plantations were very large; in 1921 the most sizeable, that of the Bismarck Archipelago Trading Company, was 4.3 hectares (New Guinea Gazette 12, 15 September 1921:63). In the late 1900s each plantation labourer received 18 grams of tobacco, a clay pipe, and 25 grams of soap once a week, a loincloth every month, a blanket and an eating bowl (the cost of which was subtracted from wages). Ten hours a day were worked, with three holidays a year besides Sundays (Firth 1973: 215 ff). The few surviving individuals employed on these early plantations recollect that the hours were very long, beginning before dawn; that they were often thrashed with 'whips like knives' or had their tobacco cancelled for any transgressions of plantation rules; and that the German money they received was later rejected by the Australian administration. Nevertheless it appears that the few labour migrants from Siwai went voluntarily. Working for Europeans was a novelty rather than a necessity and in general the Bougainvillean economy was scarcely changed.

When, in December 1914, the period of German administration abruptly ended during the first world war, few Siwai had travelled much beyond their own villages. Some had been as far as Samoa but perhaps no more than forty or fifty men had found wage employment outside; none are known to have worked in Queensland although it is probable that some Buins did. Siwais were little concerned with the world beyond, there was almost no trade of commercial products and they were not linked into a European-style economy for several more decades. In other parts of Bougainville, by contrast, especially on the east coast and Buka island, this had already occurred.

Pre-war migrant labour

The precontact distribution of population in Siwai apparently was much as it is today, even though people lived in small hamlets located on the land of a matrilineage elder or 'big man' (*mumi*). Population movements occurred between hamlets, either for marriage or because of the changing status and authority of individual 'big men', whilst hamlets themselves were occasionally relocated. The earliest years of Australian administration gave rise to one radical change: people of neighbouring hamlets were brought together and ordered to build a 'line village' consisting of straight rows of narrowly-spaced houses. Although it seems that by 1922 these villages were established in many parts of Siwai, for two decades only a minority of the population permanently lived there. Siwais therefore maintained dual residences, with houses in the line village generally used only at census times (cf. Hamnett, this volume).

Apart from this very little was changed in the initial years of Australian administration; plantation organization was modelled on that of the Germans and their nationals remained as owners, although no longer allowed to send profits to Germany. During the 1920s many new plantations were established in Buka and on the east coast of Bougainville. By 1923 there were 8,750 hectares of coconut plantations on the island, plus 185 hectares of rubber, 18 hectares of maize, .4 hectares of cotton, and 369 hectares of 'native foods' (Parliament of the Commonwealth of Australia 1924:93). Consequently plantation labour became much more common and in the same year there were 2,197 men from Bougainville working on the island, drawn mainly from the north and the east, together with a further 537 in east New Britain, 162 in New Ireland, and 33 in Manus district (Fig. 2.1). Both the Catholic and Methodist missions encouraged plantation work and Rev. Allan Cropp, the Methodist missionary at Soraken in north Bougainville, considered that labourers were being spoilt: 'The cutting out of the stick is doing quite a lot of harm and should be introduced again. . . . I have a great deal of sympathy with the plantation workers in the working of their labour which must be trying at times' (Parliament of the Commonwealth of Australia 1924:126). In principle, nevertheless, plantation work remained voluntary.

Following rapid expansion of the European plantation economy, an extensive framework of legislative provisions was elaborated to ensure that the plantation system operated efficiently. As Rowley (1972: 102-4) has written:

> . . . the long-term economics of the situation demanded that the village as the supplier of labour be maintained on a basis of welfare adequate for it to continue the supply. This required careful regulation of the movement and employment of the workers, out from the villages to the place of employment, and back again to the village; with maximum limits to the term of service, and a minimum period to be spent in the village between terms; careful setting of minimum standards of payment in cash and in kind, so that the incentive to go out for work would be maintained; the fixing of limits to punishments for breaches of labour discipline . . .; careful regulation of diet, health and accommodation standard. . . . [The migrant worker] was, as a matter of established practice, taken at least so far from his home [a minimum of 25 miles] that he had no chance of returning until the end of the contract term. . . . With a few minor exceptions, the story in New Guinea has been that of recruiting the man alone. The women should remain in the village and produce there the future generation of labourers. The village bore the costs of maintaining the wife and mother; and the wage was fixed in relation to the assumed needs of the 'unit of labour' only. . . . the return of the worker at the employer's expense was necessary to ensure that the women would produce future labourers; and that an essential minimum of social continuity be maintained.

Labourers were aged more than fourteen, employed for terms of three years and, after being returned to their home village, could not be recruited again for at least three months.

The administration imposed a head tax of ten shillings on all able-bodied males aged over fifteen, excluding contract labourers and other absent workers, village and mission officials, the sick, those aged over forty, and men with four living children by one wife. The head tax being exempt for those on labour contracts, there was both direct inducement to sign up and indirect pressure since few rural opportunities existed to earn money. This was especially true in Siwai where, in 1939, the only important source of cash came from working on plantations 'at about 70 to 120 shillings a year' (Oliver 1955:325). Among the adult male population of Siwai, only one in three were estimated to have to pay head tax, but Oliver (1955: 325) observed: 'It may not be officially intended . . . that taxation will force natives to work on plantations, but it does have that effect. . . .'

Oliver (1955:206) also noted that other factors influenced the decision to move to plantations, such as the desire 'to escape from an unhappy family situation and to spite their fathers'. Nevertheless, during the 'twenties and 'thirties, the acquisition of tax money was the sole economic reason for migrant labour since Siwai needs were few and insufficient stimulus by themselves. While under contract at coastal plantations, Siwai men glimpsed the white man's world, but from a distance.

A few may serve as houseboys in the homes of plantation managers, but most of them live out their contract periods between the coconut groves or copra sheds they work in and the barracks where they eat and sleep surrounded by other natives. . . . They receive little or no direct indoctrination in white men's values except for rules about punctuality, 'a full day's work', legally sanctioned contract obligations. . . . While on plantations, Siwai natives acquire few or no skills that have any relevance to their lives . . . They return home with a few Australian pounds in accrued earnings, a small wooden suitcase filled with calico, tobacco . . ., and a somewhat cynical disdain for the 'bush Kanaka' life of their elders (Oliver 1955:202).

Returning plantation labourers stimulated the extension of coconut plantations (Oliver 1955:480, 527), although the acquisition of metal tools were the only material benefit to the rural economy. Even so, Siwai men observed the organization and management of European plantations such that in post-war years they were able to adapt this to their own plans for communal cash cropping (Connell 1978:247). Oliver mentions no negative effects of migrant labour on the agricultural economy and absences from the village probably had minimal impact since it was rare for more than a third of all adult males to be involved, most of whom were young and unmarried. Plantation work was hard and few signed on for a second contract.

Although most migrant workers were employed on plantations, mainly in Bougainville, some became overseers or drivers, whilst a few served in the police force, became medical orderlies, or were hired for particular skills.

In the early 1920s Mapa, of Kaparo village (Fig. 6.2), was sent by a patrol officer ('kiap') to Rabaul, New Britain, to train as a policeman and subsequently posted to the new government station at Kangu, on the Buin coast (Fig. 6.1). He was the first Siwai to work in the police force, accompanying administration patrols through south Bougainville. Around 1926, Mapa was part of a patrol to an uncontracted part of Nagovisi (Fig. 6.1), which was ambushed and the

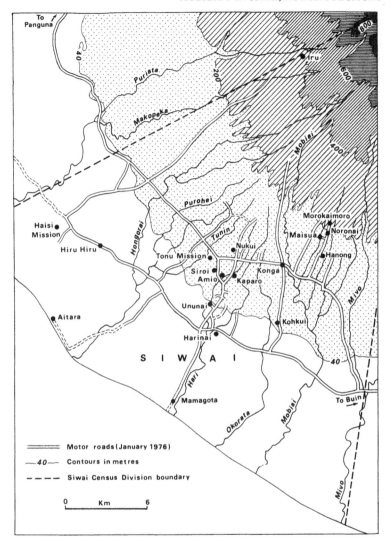

FIGURE 6.2 *Siwai*

'kiap' hit in the leg by an arrow. Mapa spent five years in the police force, including brief periods of duty in the Sohano and Kieta areas, before returning to Kaparo where the administration appointed him 'kukurai' (head man).

Although other Siwai became local teachers and medical aides, overall the pre-war years were characterized by a stream of plantation labourers, leaving and returning in groups, and learning something of the outside

125

world. Few migrants acquired skills and, except for a handful of mission catechists, returned to Siwai with little opportunity to practise them. The system of migrant labour essentially benefited the plantation economy, which originated and accentuated the dependence of the rural population.

Until 1940 plantation labourers, and all others working outside their villages, consequently existed in what Burawoy (1976) termed a situation of 'dual dependency'. His model describes migrant labour as a process involving dependency on both wage employment in one place and a basic lifestyle anchored in a different economy and place. As Rowley (1972) had earlier observed for Papua New Guinea in general, the physical separation of workers from their families was enforced by legal and political mechanisms which controlled spatial mobility; these mechanisms were made possible by the migrants' powerlessness in their work place, in the labour market, and under the legal and political systems within which they were employed.

Burawoy (1976:1052) argues that such a system of migrant labour contains elements of both capitalism and feudalism. In the former, 'the binding of production and reproduction is achieved through economic necessity: for the labouring population, work is necessary for survival'; in the latter, 'the unification is achieved through coercive regulation'. In Bougainville head tax, and to a much lesser extent a desire for certain commodities, created a need for money for which employment on European plantations was the only available source, whilst the migrant labourer required eventual support from his family engaged at home in the processes of renewal (essentially birth and child care, agricultural activities, and the maintenance of social and economic obligations). To maintain this dual dependency between two modes of production, the circulation of migrant labour was regulated by the colonial authority. As Burawoy (1976:1053) observed elsewhere: 'the state organizes the dependence of the productive worker on the reproductive worker, while the economy organizes the dependence of the reproductive worker on the productive worker'.

Post-war migrant labour

Battles were fought across south Bougainville during the war which disrupted village life and forced Siwais to move well beyond the traditional limits of settlement (Connell 1978:67-73). Large numbers of men, and also women, were used virtually as forced labour by the Japanese and to a lesser extent by the Australian and American armies. From this experience, many received a dramatic impression of alien wealth and organizational ability.

Immediately after the war there was no contract labour to plantations since villagers needed to reestablish their gardens and homes. An unprecedented labour shortage on New Guinea plantations prompted revision of the terms of agreement under which workers were employed: weekly hours were reduced and the minimum monthly wage was raised from five to fifteen shillings (Smith 1975:48). Within Bougainville a variety of socio-economic changes – the gradual emergence of cash-cropping, the extension of education, and the subsequent enlargement of job opportunities – steadily diversified work opportunities from a former dependence upon plantations. Contract labour gave way to more casual agreements between individual worker and employer. Consequently both components of the dual dependency of pre-war times were substantially weakened.

Although the plantation economy was essentially no different, the decline of the contract system meant that individual labourers were able to spend as much or as little time being employed as they wished. Increased expense meant that few Siwais travelled to plantations beyond Bougainville; even the more remote on their own island were less favoured and Siwai men preferred to walk over the mountains and seek employment in those nearby. While plantation employment became more casual and individualistic after the war, the way in which decisions often remained collective is reflected in the experience of Kamuai, a Siroi man aged thirty-eight.

Kamuai first went away to work in the early 1950s, as a boy of about fifteen. A Siwai recruiter for Arawa plantation had come to Siroi village (Fig. 6.2); three men went and, after they had gone, Kamuai ran and caught up with them, walking for three days to the plantation. Another three men from Siroi later joined them. After two years, Kamuai returned to Siroi with nothing to show for his experience. He married and soon afterwards travelled to the Shortland islands (Fig. 5.1), where wages were said to be higher, along with seven others from neighbouring villages. Work conditions in fact were worse than at Arawa and wages depended upon the level of copra production. When, after three months, another Siroi man lost his temper at the low pay, three of them returned by way of ship and car belonging to the Catholic mission. After his first child was born, following eighteen months in Siroi, Kamuai went with three other men to Buka (Fig. 6.1), again travelling on the mission ship. The others were employed on a native-owned plantation but, not wishing to work for 'a nothing man', Kamuai returned to Chinatown in Buka where a Siroi was storeman. There being few jobs available, after a month he travelled by ship to Arawa plantations. By now Kamuai had tired of hard plantation work and after three months he walked south along the coast some forty kilometres to Toimonapu plantation (Fig. 6.1) at which he found another Siwai man. Despite the efforts of a local plantation owner to recruit Kamuai, he remained at Toimonapu for two years and ended up as a tractor driver. He flew back to Buin (Fig. 6.1) with, this time, a transistor radio and a few Australian pounds to show for his efforts. Kamuai never worked outside his village again. Although more complicated than most, this example is typical of much post-war experience. In short, the regular, pre-war movements of plantation labourers had given way to a much more discontinuous and diversified pattern.

While few Siwai men went to work on plantations immediately after the war, within a few years the practice had resumed and even intensified. Compensation for war damage was not paid and cash cropping produced little revenue; in 1949 the 'kiap' (patrol officer) observed that 'outmigration has recently been causing concern'. The 1948 administrative census showed that about twelve per cent of the population were 'absent' from Siwai: this classification of plantation labourers and other migrant workers as 'absent' seems to reflect ambivalence in the view that population mobility was both crucial to the plantation economy and socially and economically disruptive to village life. By the early 1950s about thirty per cent of adult males were again employed outside Siwai and probably about a third of these beyond Bougainville. The change from contracts to individual agreements was reflected in one observation that most plantation labourers in Bougainville only stayed three or four months. To satisfy an increasing need for money, individuals were able to divide time between their cash crops, still producing a little revenue, and wage employment. In the 1950s administration reports noted that, in some parts of south Bougainville, high levels of outmigration had resulted in more village work being undertaken by older men and women, and in a neglect of communal tasks because of the absence of administrative-appointed local officials. Such complaints were less often recorded in Siwai yet, as elsewhere in south Bougainville, little of the financial rewards from migrant labour found their way to the villages.

The high rate of outmigration during the early 1950s, which coincided with a shortage of Bougainvillean labour on east coast plantations, initially declined in the Buka area but by the mid-fifties also began to occur within Siwai. This trend partly reflected high expectations of cash cropping, generally not to be met for another decade, and partly considerable diversification of available employment. Rabaul (Fig. 2.1) once again became an important focus, this time for a small stream of urban workers, and by 1962 two-thirds of the plantation labourers in Bougainville came from mainland New Guinea. This pattern continued, until nowadays only a small numbr of Siwai remain employed on plantations, but entirely in supervisory or technical capacities, and earning considerably more than labourers. Many have worked there for years, some are married locally, and a substantial number no longer consider the possibility of returning to Siwai in the immediate future.

A characteristic of the late 1950s and early 1960s was a balance between plantation labour and cash-cropping, with the former substituting in periods when crop sales were inadequate. Although one 'kiap' suggested that Bougainvilleans preferred to earn cash from plantation employment, most evidence suggests that this was true only for young men who had no cash crops and had not gone away previously. In the 1960s local growers were beginning to return from plantation work as soon as their cocoa and coffee holdings became established. One 'kiap' commented that in Buin: 'Development Bank loans have helped to stem the 'mass exodus' to Buin and Kieta around June each year. Young men still leave home to go, especially to Panguna, but this is analogous to Development Bank loans since money outside is sent back to maintain or extend present plantings' (Patrol Report Buin 5/69-70). Undeniably, the incidence of migrant labour was closely related to the rise of cash cropping and by 1967 a pattern of short-term circularity had been established.

This emergent affluence of rural Siwai is partly reflected in rural employment: until the 1960s, apart from occasional work on missions or in agricultural extension, no paid employment was locally available. When indigenous cocoa plantations began to extend throughout Siwai during the late 1950s, the earliest growers hired labourers from neighbouring Nagovisi (Fig. 6.1). These remained for up to two or three months at a time, earning about one Australian pound a month plus food and accommodation (Connell 1978:151). In general, Siwais were unwilling to take up such employment, preferring to invest time and money on their own plantations. Subsequently the Nagovisis returned to establish cocoa holdings at home, wage rates increased significantly in the 1960s – partly because the early growers were earning reasonable incomes, and Siwai employed each other. By the 1970s conditions for hiring local labour again had changed; mine employment at Panguna (Fig. 6.1) paid high wages and Siwais wishing to extend their holdings found it difficult to compete. Following a precedent already established in Buka, a few enterprising men introduced New Guinean labourers, some of whom had not been successful on east coast plantations or, in later years, at the copper mine. By 1975 there were about a dozen of these employed in Siwai cocoa plantations and trade stores, plus a few Bougainvilleans from the poor and remote west coast and some Shortland islanders.

Paralleling the local development of cocoa plantations in the 1960s was a systematic expansion of primary schools, hospitals, and agricultural extension throughout both Siwai and Bougainville. Each of these brought workers from other parts of Papua New Guinea to live and work in Siwai and was another measure of its incorporation into a regional and national economy. Although most labourers moved out of Siwai, there was always a thin trickle of outsiders in the opposite direction.

Changes in the distribution of Siwai migrant workers over the past

TABLE 6.1 Location of Siwai migrant workers, 1964-76

	1964-66	1967-69	1970	1974-76
Bougainville				
Panguna-Loloho	–	88	214	54
Arawa	–⎫	–⎫		16
Kieta	4⎬	41⎬	35	9
Rest of Kieta district	47⎭	15⎭		–
Buin	20	57	42	5
Baitsi/Nagovisi	9	16	12	6
Central Bougainville				
(Aita, Rotokas, Eivo)	49	25	17	3
North Bougainville	2	8	3	4
Buka	10	15	19	1
	141	265	342	98
Papua New Guinea Islands				
Rabaul & New Britain	35	50	46	8
West New Britain	2	2	1	–
New Ireland	2	1	7	–
	39	53	54	8
Mainland Papua New Guinea				
Port Moresby	6	7	20	8
Southern Highlands	3	3	4	1
Morobe	2	7	9	4
Madang	3	1	3	1
Chimbu	2	2	1	1
Other Provinces	1	4	7	–
	17	24	44	15
Solomon Islands	21	21	2	–
Unknown	2	9	5	–
Total	220	372	447	121

Sources of data:
1964-66: Siwai Tax Returns.
Includes the workplaces of all those who paid no taxes within Siwai.
In both this and the 1967-69 returns the villages of Haisi and Hiru-Hiru (Fig. 6.2) are excluded.

1967-69: Siwai Tax Returns.

1970: Patrol Report Konga No. 1/1970-71.
Includes all those who were absent; unlike other sources, it therefore lists those who were at tertiary educational institutions. Probably understates the number of workers in the Solomon Islands.

decade are indicated in Table 6.1, though the data are neither strictly comparable nor absolutely reliable and only broad conclusions are able to be drawn. The most substantial change is the growth of the employment complex at Panguna-Loloho-Arawa (Fig. 6.1), associated with the construction and operation of the Panguna copper mine which now employs more than half of all Siwai absentees. By contrast the traditional areas of plantation labour, especially the Kieta district and central Bougainville, have shown a continuous decline and by 1974–76 only five out of 121 workers were employed there. Towns are the sole centres of Siwai employment on the Papua New Guinea mainland and only recently has Rabaul begun to lose to Port Moresby (Fig. 2.1) its significance as a centre of skilled occupations. Overall there has been a substantial shift from rural to urban employment, coupled with a rising focus upon skilled work, in either the Panguna copper-mine complex or more distant urban centres.

In 1971, for the first time, a 'kiap' recorded the occupations of all Siwais with outside employment. These ranged from teacher to agricultural assistant; nun to radio announcer; medical assistant to mechanic; electrician to stevedore; and from storekeeper to labourer. From the central Siwai villages of Kaparo, Amio and Siroi (Fig. 6.2), for example, some sixty individuals were involved in full-time wage employment (of whom fourteen worked locally). Sixteen of these were labourers, but none on plantations, whereas there were eleven secretaries and clerks, ten teachers, and a variety of other occupations including one full-time Member of the National House of Assembly (Patrol Report Konga 1/70–71). These diverse occupations clearly indicate that by the 1970s a substantial number of migrant workers from Siwai, as elsewhere in south Bougainville, possessed trained skills. Incomes were also higher, especially for those employed at Panguna.

Female employment beyond Siwai did not begin until the 1960s, is still uncommon, and for eight villages in 1974–76 accounted for only fifteen out of 121 migrant workers. Almost all were nurses, teachers, or clerks; the first two occupations were the most acceptable since Siwais believed that unmarried girls were more likely to be in situations of firm social control. In addition, a growing number of Siwai women lived at the workplace of their migrant husbands.

Thus, in the three post-war decades, several crucial changes occurred in the system of migrant labour. Cash cropping in Siwai finally became successful whilst improved educational opportunities enabled greater diversity of employment. Plantation labour disappeared almost entirely,

1974–76: Survey Data (Connell).
Includes all those employed outside Siwai, November 1974–April 1976, in the eight villages of Siroi, Amio, Kaparo, Ununai, Maisua, Morokaimoro, Noronai, and Hanong (Fig. 6.2). Those few who worked in more than one place or for two periods are recorded twice. These villages are typical of Siwai.

to be briefly replaced by a range of jobs in varied locations. Subsequently, all but the most highly educated sought employment at a single location, Panguna (Fig. 6.1), where wage levels were unprecedented. The long-term labour contracts of pre-war years gave way to short-term circular movement to plantations and then to the copper mine whilst, towards the end of the period, women began for the first time to find work outside Siwai. Post-war there was a dramatic shift towards Siwai incorporation in a national economy, but paradoxically it was almost entirely voluntary compared with the 'dual dependence' of pre-war times.

Mine work

In 1960, copper ore deposits were confirmed around Kupei, in the Panguna valley, where there had previously been a tiny gold mine. When new technologies demonstrated the profitability of low-grade ores, the feasibility of mining was established. The construction phase began in 1969 and, apart from the mine itself, involved the building of two towns at Arawa and Panguna, a port at Loloho, a 26-kilometre pipeline to the port, and an access highway between mine and coast. At the peak of construction, 3,681 expatriates and 6,328 Papua New Guineans were on the payroll of the copper company and associated contractors. Much development was completed by 1972, when the first copper was mined, and since then the number of employees at Panguna has declined.

At the end of 1965, during the exploratory phase, more than 300 Bougainvilleans worked for CRA: Conzinc Riotinto (Australia) Proprietry Limited, the parent company of Bougainville Copper Limited and itself part of the massive transnational corporation, Rio Tinto Zinc. By early 1967, the first unskilled Papua New Guineans were recruited from outside the island (Bedford and Mamak 1976:170-1), even though the labour force came mainly from Siwai, Nagovisi, and Buka. Within Bougainville the Buin district, of which the Siwai census division (Fig. 6.1) is part, has consistently provided the greatest proportion of mine workers since 1970; during the four years 1970–3 Bougainville Copper Limited (BCL) employed about one tenth of the district's adult male population. Even more, especially from the Buin and Siwai divisions (Fig. 6.1), worked at the mine since completion of the trans-island road in 1972 (Bedford and Mamak 1976:172).

Employees whose home villages were in the Buin and Kieta districts tended to remain for shorter periods than those from more distant areas. The former were rarely accompanied by families, thus reducing their expenditure in the towns and reinforcing their commitment to the rural area, especially through intermittent visits. Bedford and Mamak concluded that in south Bougainville (Buin district) the response to wage employment was closely tied to participation in cash cropping, for when cocoa prices rose sharply in 1973 there was decline in the number who sought jobs. As early as 1971 the circulation of mine labour began to be recognized and one 'kiap' commented on western Buin: 'It seems that men spend three or four months in BCL and then return for three or four months to catch up on work required in the villages' (Patrol Report Buin/71-72). Movement histories for central Siwai do not corroborate this statement and, given poor communications at the time, such regular transfers into and out of employment seem unlikely. Even so, wages were high, access to jobs easy, and periods of wage-earning shorter, thereby indicating the emergence of a distinctively new pattern of movement to work.

In March 1976 a random sample of fifty Siwais employed at the mine, about thirty-seven percent of all Siwais there, revealed a mean gross income of $Australian 1,764 during the previous twelve months. When taxation, accommodation, and food charges were deducted, the net annual income averaged $A1,570 ($A1.54 to £1.00). Extrapolating this current income over a whole year suggests that, at current wage rates, annual net income would average $A2,004, almost all of which would represent potential savings.

Many of these mineworkers had no previous formal employment, a particular characteristic of Bougainvilleans, and even more had been longer with BCL than any previous employer. This was the first job for sixteen out of the sample of fifty; a further twenty-one had previously worked for BCL or with subsidiary contractors at or near the mine site. In part, this pattern reflects the youthfulness of the workforce. Data available for 87 of 135 Siwais employed at Panguna in February 1975 show seventeen (20%) were aged at least twenty-six but only three more than thirty-two, and fifty-four (62%) twenty-three or less. A year later, by comparison, thirteen out of forty-five Siwai employees were aged at least twenty-six (29%) and twenty-five (56%) at least twenty-three. Although inadequate by themselves, these data suggest a process of stabilization and ageing of the labour force. The longer unskilled workers remain at the mine, then the better their conditions: wages increase, particularly if skills are acquired, and housing conditions improve. These financial and social benefits of stability tend however to be welcomed more by distant mine workers than by Siwais who can circulate between town, mine, and village.

Outside the copper complex, a range of urban opportunities provide less skilled employment at lower wages. Some of these workers have been unable to secure a job at Panguna; others, especially in the past, have used their experience in town as a stepping stone to mine employment. Overall, the labour force outside the mine appears less stable and younger; consequently circulation also is greater. Despite increased education and a growing diversity of opportunities, one impact of more prestigious and highly-paid employment at the copper mine has been to dramatically reduce the range of locations hitherto chosen by Siwais. Wage work is therefore more spatially concentrated than at any time in the past half century.

For the Siwai sample of currently employed, the total time with BCL averaged 3.1 years. This figure contrasts strikingly with year-of-commencement data collected in 1974 by Bedford and Mamak (1976: 170), in which less than twenty-five per cent of all Bougainvilleans had been continuously employed for more than two-and-a-half years. Together, these data tentatively suggest the slow emergence of an urban-industrial proletariat, partly in response to a post-1975 decline in demand for mine workers; that employment at Panguna is increasingly the prerogative of fewer individuals; and that less turnover of labour has reduced the opportunity among Siwai for casual work interspersed with periods of village residence. In general, this recent decline in circularity means the pattern of migrant labour from Siwai has edged a little closer to that of many other Melanesian societies.

Circulation nevertheless remains integral to the expectation of most mineworkers, even if not always realized.

In 1968, after one year at high school, Henry of Kohkui village (Fig. 6.2) began work with CRA at the age of sixteen, doing what he described as a 'clerical assistant' and what BCL files record as a 'tea man and laundry hand'. He earned about $A.18.80 a month ($A2.16 to £1.00) but returned to Siwai after a year, claiming to be overdue for leave. About a year later he was back at the mine site to work as electrician for an independent contractor; after twelve months, frustration again set in, this time at lack of promotion, and he returned to Kohkui village. Two more urban jobs followed in the next two years, neither lasted more than six months, and in between he was back in Siwai. In August 1972, soon after the second job, Henry joined BCL as a Trainee Repairman Grade Three, earning $A22 every two weeks ($A2.05 to £1.00). Seven months later he was terminated for being too often absent; thirteen months elapsed in Siwai before another seven at Panguna, working for a construction company which then completed its project. A return to Kohkui lasted only two months before he again joined BCL, now as a Trainee Equipment Operator Grade Three at a wage of $A70 a month ($A1.76 to £1.00). Henry was still there fifteen months later – his longest period of seven in urban-industrial employment, between each of which he had returned to Siwai.

The more lengthy and diverse job histories are of those who have little if any education beyond primary school and less familiarity with life beyond Siwai; on both counts they experience difficulty in obtaining skilled and well-paid employment. Nor is this the limit of movements between town and village. Every five or six weeks, most mineworkers spend a weekend in Siwai. Such visits usually coincide with important public holidays, particular social events in home villages, or the decisions of other workers from the same locality. In addition, almost all Siwais take their one-month annual leave at home. This considerable flow of workers between mine and village is not reflected in crude employment statistics.

Despite little prior wage employment, the labour force of the Panguna complex contains most of the highly-skilled and experienced in Bougainville, and a high proportion of those in Papua New Guinea. Consequently it is almost entirely within mine occupations that Bedford and Mamak (1976:182–4) identified a 'proletarian' strategy of employment, whereby Bougainvilleans view urban wage-earning as the sole source of income and have become committed to urban residence. This strategy is more generally found among mineworkers from Buka whereas those from south Bougainville can be characterized more as having adopted a 'peasant' strategy, in which involvement in urban work is peripheral but cash cropping and entrepreneurial activity in rural areas is of primary importance. Not only do Siwais follow this latter strategy but also, being favoured by ease of communications, they can participate simultaneously in both wage employment and cash cropping. Unlike people from Buka, they obtain their security from cash cropping rather than mine employment.

Few individuals have such sizeable cocoa plantations that a quick need for great amounts of cash does not involve recourse to wage employment at Panguna. When a group of five men associated with Maisua village decided to organize what in 1974 became the largest final mourning feast (*domokomi*) ever held in Siwai, the mine became an important source of income for three of them. Potential resources had to be capitalized, since this feast cost the organizers about $A18,000 ($A1.60 to £1.00) and involved the distribution of 266 pigs.

> Siham, a fairly successful cocoa-grower with his own fermentary, expanded his operation by travelling through part of Siwai and purchasing cocoa as a private trader — at this time, an increasingly familiar operation for several businessmen (Connell 1978:225–6). Sukina was headmaster of the Maisua primary school and hence earned about $A2,500 a year; moreover he had a large cocoa plantation. Maria was already employed at Panguna, as personnel manager, one of the most senior positions then achieved by a Bougainvillean. Both Nasua and Nompiku, who was then about fifty, went to work as mine labour and contributed their savings — Nompiku from remaining there as long as seventeen months.

With the right connections, it is consequently still possible for even elderly men to work at the mine and earn a desired amount of cash.

Younger males have more diffuse and less specific objectives, both economic and social, tend to remain longer, and thereby contribute to a lessening of labour turnover. Older men find it easier to resist the financial attractions of wage labour at Panguna, partly because their requirements are few. Others rationalize this attitude through their conservatism: a fear of dying away from Siwai, dislike of machinery and of New Guineans, even discomfort of the cold at higher elevations. Yet despite a slow decline in circulation and the emergence of a more stable and skilled proletariat, most Siwai mine workers are still 'urban peasants'.

Siwais in town

Few Siwais live in towns: the growing commitment to mine employment is not yet matched by commitment to urban residence. There are very few Siwai married couples in Kieta, Arawa, or Panguna (Fig. 6.1) and even fewer in Buin, the small town of south Bougainville. Siwai men reside in Rabaul and Port Moresby, the two largest urban centres in Papua New Guinea, but have been there for years; virtually none have a Siwai wife with them or appear interested in returning to their natal group. Consequently no town within Bougainville or on the Papua New Guinea mainland contains a distinctly urban and Siwai population.

In early 1976, there were no more than twenty-six households in Arawa whose head was either a Siwai or married to one. Their mean period of urban residence was slightly less than two years; only two had more than five. Of the twenty-two households for which data are available, half consisted of married couples of whom one partner was not Siwai and observation suggests this pattern is even more striking in Kieta and Panguna. The most extraordinary characteristic of urban households is that the proportion of inter-married households is vastly greater than found among rural Siwai. With the exception of one Siwai married to an Australian, each urban household included permanent or semi-permanent residents from outside the nuclear family; the average size of household was therefore no different from rural Siwai.

Arawa, like Panguna town, is not far distant from Siwai and even closer to the accommodation at Panguna for unmarried mineworkers. Consequently the few Siwai households in Arawa are significant foci and most experience a steady flow of visitors from rural Siwai and Panguna. Siwai households are scattered throughout the town and so residents tend to be socially oriented towards these visitors rather than to other Arawa households, even those that adjoin their own or contain members from other parts of Siwai. This constant flow of visitors, the scattered distribution of town households, and the substantial amount of intermarriage tend to militate against the emergence of a specifically Siwai form of urban life (Connell 1979).

The urban future of households currently in Arawa is uncertain and those of the inter-married group the most indeterminate. Men who have married Siwai wives can establish themselves in Siwai villages, but access to land is more difficult for Siwai men with spouses from other language groups. Outsiders, both men and women and especially from beyond Bougainville, find settling into Siwai rural life awkward; some marriages have distintegrated and this seems indicative of a future trend. Compared with entirely Siwai households in Arawa, such couples have greater commitment to urban life, where they have access to both sets of 'wantok' (language group). Of the former group, only one household head stated that the family would remain permanently in town although he was well aware that they could return to their village whenever they wished. Single heads of household, even those in skilled employment, intended to return to Siwai within a few years. Decisions about urban residence are not usually made on the basis of job satisfaction but according to the need to accumulate capital with which to establish a successful cash crop or local business venture. Few doubted, however, that they would at some future time return to urban-industrial employment.

Despite the fluidity and recency of town living among Siwai, all urban households share characteristics distinct from their rural counterparts. First, urban households are closely integrated into a monetary economy; almost all food is purchased and most transactions involve cash. Although rural visitors invariably bring food, as do town dwellers returning from local visits, this accounts for a small part of total requirements. Secondly, Siwais are effectively distinguished from other groups by language. Partly because of this groups are formed for various social activities, specifically for sport and drinking alcohol — semi-formal and informal male activities for which there is no female parallel. Female social life revolves around casual visiting with a very restricted number of fellow urban residents or rural Siwais; the boredom that follows from such constraints is mainly responsible for female disenchantment with urban life and a consequent lack of commitment to long-term residence.

Although urban Siwais are economically distinct from their rural counterparts, they are rarely socially distinct. In town, they differ little from the broader group of south Bougainvilleans described by Bedford and Mamak (1976:181-2) and,

> In contrast to Bougainvilleans from further away, . . . [they] interact more frequently and regularly with relatives from their place of origin. In this way contact with events in both the town and the village is maintained. Many appear to be clannish and exhibit a greater degree of unity and solidarity amongst themselves than people from other parts of the District. Most are intolerant of other District groups. The system of urban social relations at work and

137

leisure repeatedly reinforces this identification with the village and co-member of the home area.

Two facilities of considerable value to Siwai are found in the Arawa-Kieta urban area (Fig. 6.1): the most sophisticated hospital on the island and the most profitable outlets of cocoa. Marketing is steady throughout the year and cocoa transported to Kieta in landcruisers, with part of the revenue usually spent on the wholesale purchase of goods for rural stores. Arawa hospital treats mainly the more complicated illnesses and injuries, most of which involve hospitalization. Bringing a patient to town demands a special trip and long-term patients receive streams of visitors; for the duration of illness in most serious cases, whole families transfer themselves from village house to urban kin.

Accompanying a cocoa seller or a patient provides an excuse to visit town but less frequently explains its occurrence. Most trips to Arawa are essentially indeterminate or, in pidgin, constitute a 'lukluk raun' (looking around); less often, their length is also undecided and dependent on circumstances. Travelling to town may be to visit relatives and friends, sometimes with important messages and provisions, to escape the monotony of rural life, make retail purchases, look for work, avoid rural disputes, meet court summonses, and take up wage employment. Most visitors will have more than one such reason for going and will also have received requests and tasks from other villagers. No one can visit town without being involved in a combination of activities, the relative significance of which may vary during a single day, so that secondary reasons may become primary and the likely duration of absence only established once a journey is completed.

James of Siroi village, aged twenty-four, is married with an infant son. He worked in Arawa for almost three years but returned to the village after marriage, remaining for about eighteen months while his son was born, working his food garden, establishing his first cocoa plantation, and organizing the construction of a new 'haus kuk' (kitchen). He often spoke of reclaiming his previous job, in a bakery, or doing a training course in Port Moresby so as to establish a bakery in Siwai. Yet he never followed up his inclination during frequent day trips to town. Another Siwai in the bakery sent word of a vacancy and, knowing about a day trip I was making to Arawa, James packed clothes and other possessions and went with me. At 10 a.m. he was left at a central place in the Arawa town centre, where other Siwais can always be found. Six hours later, at the same spot, he requested a lift back to Siroi village. In the course of a day he had visited and eaten at a friend's house, had drinks with some mineworkers, purchased various goods in the stores, but gone nowhere near the bakery. He decided it was not yet time to resume work there, although he subsequently rationalized this by blaming me for not having taken him directly there. The next day he was at work in his garden and although one or two trips were made to Arawa in the next two months, he expressed no further interest in urban employment. Four months later, immediately following a dispute with his wife, James set off for Arawa to rejoin the bakery and was still employed six months later.

This sequence is not unusual for young men, most of whom at some point journeyed to find employment, only to reappear in the village a day or so later. A phenomenon referred to in pidgin as 'taim bilong hambak' (time of fooling around), it would sooner or later be replaced by real work. On the other hand, few movements are so clear-cut that they can be unequivocally classified as going away to wage employment. The completion of the trans-island road in November 1972 radically altered the potential for urban visiting since, until then, casual travel was deterred by the fact that the distance had to be covered on foot, by ship or plane. Although improved accessibility resulted in a rapid growth of vehicle ownership, between 1972 and 1975 only a small proportion of Siwais visited town and for most it remained alien. Almost all Siwais in town are in practice short-term transients or what Skeldon (1976:19) has called the 'floating population'. Whereas the largest and oldest centres in Papua New Guinea contain a small and growing urban proletariat, this is scarcely true of Bougainvillean towns. In the latter, wage employment is essentially a peasant adaptation to the expansion, rather than the basic transformation, of an introduced economic system. Only those who die in Arawa and Kieta are assuredly not returning to Siwai and the agricultural economy will continue to be of overwhelming importance to the emergent urban peasantry.

Remittances

A major consequence of population movement is the transfer of cash or other resources between absentees and those remaining in the village. Despite apparently high urban incomes, relatively little cash reaches rural Siwai. Among a group of urban households not confined to Bougainvilleans, Mamak and Bedford (1977:445) concluded that the poorest relied upon friends and kin for meeting emergencies and also tended to contain wage-earners employed outside the copper mine. Only a detailed survey of incomes and expenditures could reveal if this were also true of urban Siwai. Observation suggests that for most married couples there is a net rural-urban flow of incomes through gifts and food donations, but not because of dependence on rural areas. For single workers, especially those at Panguna, there is a significant urban-rural flow of income although its magnitude is unknown.

In December 1973 the University of Papua New Guinea undertook an urban household survey, during which a random sample of sixteen Siwai mineworkers were interviewed (cf. Conroy/Curtain and Young, this volume). Four of these were married and all resided in single accommodation at Panguna. Each said they customarily sent gifts to village relatives; during the previous month four estimated these to have been worth $A205 whilst another nine claimed to have remitted $A985 in cash ($A1.60 to £1.00). Only six, by contrast, customarily received gifts from Siwai. Although field observations suggest these estimates to be exaggerated, their magnitude is not impossible and provides evidence of the close links between mineworker and village.

Direct remittances constitute a small fraction of the urban-rural flows, since most cash is brought back by workers upon completion of employment and with the aim of using it in the village economy. In the 1973 survey, all Siwais at Panguna intended to return with some cash, which would be used for cocoa planting (six men), to establish a store or other kind of business enterprise (three), and for brideprice (one). Of the nine who estimated how much money these objectives would require, those hoping to establish cocoa plantations gave totals ranging from $A100 to $A500. The two wanting to become store owners estimated $A500 and $A3,000. Even so, only three out of these sixteen intended to return to their village once those amounts had been saved. While these objectives typify virtually all Siwais in urban-industrial employment, most achieve little more than a cocoa plantation and a small store, independently or in partnership. On the other hand, most returning mineworkers have more than the $A200 required to hire labour to clear the land for planting. Unquestionably there is an urban-rural flow of cash, even if but a fraction of total earnings.

What is apparent about the process of going away to work, especially among the labour force at the copper mine, is that absence from the village is not intended to be permanent. While the household survey reveals the desire to take some cash back to the village, the relative ease of obtaining not only urban-industrial employment but also high net wages means that Siwai mineworkers are not as single-mindedly committed to particular targets as some of their counterparts described in the African literature (for example, Elkan 1967). Despite some unrealistic aspirations, most wage-earners feel it is not particularly difficult to earn sufficient money to meet immediate requirements and that they can subsequently return to obtain more should there be some need.

Nevertheless, cash remittances that reach Siwai villages constitute a smaller proportion of total earnings than that with which plantation workers returned in pre-war years. There is, moreover, a shift towards individualism. The contemporary earnings of mineworkers are for their own families rather than for kin, whereas pre-war it was economically and socially essential to share the resources derived from plantation labour. Nowadays, most remittances to kin are in response to particular requests, of which the most common is payment of school fees. Some earnings are converted into material possessions, especially the more ephemeral trappings of urban civilization – clothing, radios, watches – but also some goods which can benefit the rural economy. As with much poorer areas (cf. Connell, Dasgupta, Laishley and Lipton 1976: 98-9), most remittances go into everyday household needs or conspicuous consumption. On the other hand, the absence of able-bodied Siwais has little negative impact upon the village economy, unlike many rural situations in the third world where people leave because the subsistence economy cannot meet their basic needs. The poorest Siwai villagers, generally the elderly, do however suffer from the absence of cooperative labour especially in house building, and are less likely to benefit from the income redistribution brought about by remittances. In general, migrant labour improves the position of the already favoured whilst remittances reflect the self-interest of the mobile (cf. Connell 1980a).

Contemporary Siwai mobility

During the post-war years, population distribution in Siwai changed in several ways. For a time Siwais were convinced, as never before, that living in large nucleated villages was preferable to dispersed hamlets and several of the line villages regrouped into sizeable settlements. In many parts of Melanesia, this big-village pattern of settlement often occurred in conjunction with cargo cult expectations, although this was more true in neighbouring Nagovisi (Fig. 6.1; Nash 1974:79-80) than in Siwai, where amalgamation reflected a growing desire to establish cash crops – especially rice – which it was believed should be produced communally. These large villages did not endure but the Siwai population was more concentrated than ever before.

Although begun after the war, cash cropping was not successful until the 1960s when cocoa provided reasonable incomes (Connell 1978) and in 1975 averaged perhaps $A200 per household. Rapid extension of cocoa plantations necessitated clearance of land increasingly distant from the line village, especially since security of tenure was not otherwise easily obtained. Several households thus shifted from the village lines to build separate dwellings or re-establish small hamlets in areas where they had adequate claim to land. After about 1960 a considerable dispersion consequently occurred and, although few households relocated more than two or three kilometres and none beyond Siwai, this represented a significant change. Coupled with increased wealth from migrant labour and from cocoa, such redistribution emphasized the difficulties of organizing communal work and of maintaining social control in a dispersed population now more or less incorporated into a national economic and social system.

Despite economic changes, household relocation is still uncommon.

Even a shift away from the line village, although affecting about a third of the total population, rarely happens more than once in an individual's lifetime. Marriage represents the one permanent move made by all Siwais since couples invariably establish their household apart from either set of parents, but the distance usually separating partners prior to marriage is extremely short. The mean distance for 418 marriages over a sixteen-year period is no more than 2.2 km, as low as any recorded (cf. Livingston 1973:62), and scarcely causes any population redistribution. Settlements (villages, hamlets) and the households within them are quite static, a surprising fact since residences must be rebuilt every ten to fifteen years. Although the decay of timber houses may provide the final catalyst to relocation, dwellings are often rebuilt on the same spot.

However, historic and contemporary patterns of population movement in response to formal employment represent only a small proportion of Siwai mobility. In 1975, all movement into and out of three villages was monitored for a maximum of fifteen weeks, based on a record of every individual who either left or stayed in that community overnight. While even this approach is selective, village society is revealed to be much more mobile than suggested by the simple analysis of migrant labour.

Mobility data for Siroi village (Fig. 6.2) permits a new perspective on local population statistics. The administrative census substantially overstates the number actually resident in Siwai, by excluding at most only those who have married outside and remained away. Three categories of population may be distinguished (cf. Connell 1977:12). The first are the permanently resident, almost invariably present in the village but who may occasionally be absent in other villages, in hospital, or in town. The second group are those usually resident: effectively the legal definition of official residence used in the administrative census, which incorporates all school children, migrant workers, and other absentees considered to belong to that village, even though some may visit it but rarely (Fig. 6.3). Finally there are those whose 'nem emi stap': whose name remains in the village but who are no longer part of it, especially if married outside Siwai. Updating the administrative census for Siroi village in June 1975 produced a resident population of 128 and one usually resident of 162, with eleven pupils at secondary or vocational school and twenty-three migrant workers constituting the difference between the two populations (cf. Fig. 6.3).

Not surprisingly, the mobility record indicates that these different populations are subject to substantial and continuous fluctuation rarely reflected in official counts or estimates. Over fifteen weeks, for example, the resident population of Siroi was identical on only two successive nights; on all others at least one person either entered or left the village for twenty-four hours or more. Effectively, the village population was in a perpetual state of flux. Only one household out of

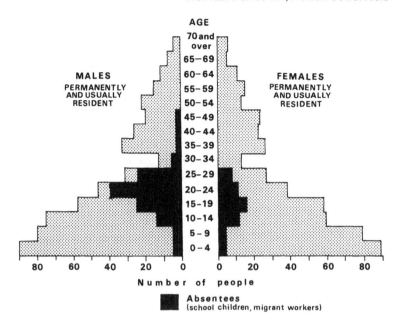

FIGURE 6.3 *Age-sex structure of eight Siwai villages, 1975*

twenty-seven had every member continuously present over the whole time: every other both lost members usually resident and gained some visitors. Over the 108-day period there was, however, a core residential population of forty-five individuals, mainly the very old or young and their mothers, who slept in the village every single night while the population fluctuated between extremes of 146 and 118. The lowest totals during fifteen weeks coincided with major feasts in other villages at which many Siroi men were present. Neither official census figures nor employment data on migrant workers even hint at this fluidity of village populations documented by the mobility record.

Those usually resident but nevertheless generally absent from Siroi village were receiving education and in formal employment elsewhere. All but two persons at high school or in tertiary institutions returned at some time during the fifteen weeks; those who did not were a university student in Port Moresby and an agricultural trainee at Mabiri on east coast Bougainville (Fig. 6.1). Of the twenty-one wage earners, only three were not recorded in Siroi during the survey period: one of these was a 'kiap' at Konua, north Bougainville; another, a girl, a secretary in Kieta (Fig. 6.1); the third was believed to be a carpenter on mainland Papua New Guinea. Most migrant workers therefore remained in close physical contact with the village.

Using the first and last days of the survey period, some assessment

143

can be made of the reasons for absence from Siroi (Table 6.2), with results surprisingly similar to those reported by Chapman (1975:133-6) for two communities on south Guadalcanal. Men are almost twice as likely to be away as women, especially because of wage labour but also because of formal education. Women leave for more obviously social reasons. Absence from the village is thus mainly for what Chapman (1975:134) calls 'economic and post-contact reasons', for the implication that mobility has substantially increased from pre-contact times is inescapable.

TABLE 6.2 Reasons for absence from Siroi village, 4 June-19 September 1975

	Male	Female	Total
Active migrants			
Education	7.5	2.5	10.0
Wage labour	16.0	4.0	20.0
Visit kin	3.0	5.5	8.5
Medical	0.5	0.5	1.0
	27.0	12.5	39.5
Passive migrants			
Accompany others	0.5	3.0	3.5
	27.5	15.5	43.0

If movement away from Siwai villages for education, employment, and medical treatment is considered to some extent self-explanatory, then those made to stay with kin are not. For Siroi, less than a quarter of the latter have easily identifiable explanations: the necessity to arrange kinship-based activities (including feasts and marriages), to maintain distant cash-crop plantations on matrilineage land, to look after children when parents are ill or temporarily absent, or even the inability through drunkenness to return home. Many others have a dual basis of maintaining matrilineage ties through personal contact on the one hand and of diversifying the occasional monotony of village life on the other. Thus individuals who contemplate organizing a feast which demands considerable cooperation from the matrilineage are more likely to circulate around Siwai villages, whilst a number of men have the reputation of a 'man bilong raun' (wanderer), in the sense of bludger or scrounger, because of an inability to remain content in a single place. Social reasons underlie most short-term, short-distance mobility, which parallels Chapman's (1975:143) statement that 'non-economic reasons are far more important than often appears from migration research', and yet – most crucially – these often have a strong economic content. Matrilineage ties are maintained because they enable access to land, labour, and capital, especially at times of crisis, rather than because they provide friendship, companionship, or relief from monotony.

Conclusion

Siwai is clearly exceptional; a rich rural area where cocoa provides high annual incomes and land is rarely in short supply. Moreover it is only eighty kilometres from the largest mine in Melanesia. Some facets of Siwai mobility are consequently unique yet many others have a shared relevance to dissimilar parts of Melanesia. Every phase of mobility reflects some adjustment to, and participation in, Siwai incorporation into a regional and national economy. Migrant labour began in the nineteenth century with a tentative interest in distant places; during the present century this interest became a commitment as wants became necessities and taxes proved inescapable. Absolute numbers of migrant wage-earners have grown steadily throughout the present century, interrupted by brief periods of decline.

Distances travelled have fluctuated from an initial fascination with Samoa, through a pre-war focus on east coast Bougainville and New Britain plantations, to a primary concern with Bougainville in recent years. The emergence of the Panguna complex since the late 1960s has concentrated all but a small, more highly-educated group of Siwai workers in the centre of the island. Especially in the 'seventies, the diversification of urban employment and changes in mores have resulted in women going away to work. More social movements, apparently unconnected with paid employment, have increased in range and duration and the 'man bilong raun' is a relatively new characteristic of Siwai society. Despite the possibility that the 'big man' was able to attract followers to his hamlet, mobility does not appear to have been a significant part of Siwai lifestyles before European contact. Gradual incorporation into a wider economy is reflected in greater participation in movement of all kinds and a lengthened range and duration. Mobility, if not permanent migration, continues to grow in importance.

Few village Siwais have suffered from the high frequency of migrant labour since the rural economy changed from one of subsistence to comparative affluence. Especially in the copper mining era, the proportion of able-bodied men away has accelerated the decline in social control and cooperative work, whilst the elderly share less in its monetary benefits. None the less, rural-urban income and welfare disparities are not evident for Siwai, although there are growing interprovincial divergences and hence the increasing movement of mainland labour into Siwai cocoa plantations. Young men have benefited especially from access to high mine earnings, enabling them to develop local cocoa plantations and business enterprises and to finance their own marriages, although this last fact has scarcely altered the principles of Siwai

145

nuptiality. Most men aged more than forty have ended their involvement in the wage-labour system, being satisfied with rural social life and their cash-crop earnings. In fact, the Siwai agricultural economy has never necessitated going away to work to meet basic subsistence requirements; only the imposition of local taxes forced participation in a system of migrant labour.

Compared with similar parts of Melanesia, both rural and urban-industrial incomes are high but what influence these rural incomes have had on local wage policy is unknown. High rates of saving are perfectly possible, if uncommon, in both rural Siwai and the mine, allowing structural changes in the rural mode of production. The unusual conditions of rural affluence and high urban wages have meant the disappearance of the 'dual dependency' of pre-war years. In economic terms migrant labour is no longer necessary, although young men are expected to finance the establishment of their own cocoa plantations, yet its underlying basis remains economic (cf. Bedford this volume). In future, as land becomes scarcer, movement is more likely to be in response to inequality.

Even for migrant labour, it is extremely difficult to identify the causes responsible for movements of particular durations and distances. Decisions invariably result from a number of inter-related factors; it is no longer possible to analyse the act of movement in a manner that is 'marginalist and individualist in spirit, focussing on factors affecting a single decision by a single individual at a single point in time . . . rather than as part of a process of structural change' (Godfrey 1975:9). The spatial mobility of people is a symptom of social and economic change: in this context the Todaro (1969) model, where the decision of labourers to move is a function of the gap in real income between village and city as well as the probability of finding urban employment, is no more than a simplistic description of observable facts. For Siwai, economic factors are both unnecessary and insufficient for migrant labour to occur, irrespective of whether the analytical focus is upon departure or return, and even though they underlie most Siwai mobility. There is invariably an economic incentive to labour mobility yet the actual stimulus to make a move, at a particular time and to a particular place, seems a social phenomenon (cf. Mitchell 1959:32; Bedford this volume). Among Siwais almost every adult male has been away to work, yet no one must. It is therefore impossible to agree with Skeldon (1977:411) who argues from the Peruvian case that 'the migration transition – the changing nature of the most important link between urban and rural areas – causes and is caused by transformations in the social structure of the areas of origin and areas of destination of migration'. Economic factors invariably underlie the social structure whilst in an economic environment that is increasingly uncertain, much movement behaviour is simply a process of experimentation.

These data on Siwai circulation and migration allow two further observations. First, there is essentially a continuum of population movements, so that differentiation according to reason or duration will vary according to whether the analysis is made before, after, or during the event. Secondly, the number and variety of moves made is so great that Siwai society is in a perpetual state of flux. While the first of these conclusions casts doubt on much historical data, most mobility behaviour has an underlying and identifiable logic to it. There is none the less a basic stability among the settlements and members of Siwai society which is striking, given that people are constantly on the move. Access to land by way of matrilineage affiliation is basic to much movement and becomes especially apparent in relocation away from the line village. The strength of matrilineage affiliation is noticeable even in the urban-industrial area, where members reappear in drinking groups and visiting patterns. Matrilineages, in short, are both social and economic entities.

In pre-war years, few migrant labourers circulated since not many cared to repeat or illegally break their two or three-year contract at a distant plantation. Post-war, circulation became more general: the end of contracts gave individual workers much greater freedom of decision-making; a growing diversity of employment opportunities encouraged movement between different kinds of jobs and places; and for two decades cash cropping seemed on the verge of rapid expansion, which was incentive to remain closely linked to a rural economy with apparent potential. The emergence of mine employment coincided with the success of cash cropping; since the demand for labour was great and earnings very high, thus attracting many skilled workers including teachers into the labour force, there was intense circulation during the construction phase. The proximity of the mine enabled workers to maintain their participation in Siwai economy and society: by maintaining some form of dual residence, a strategy with a long local history, movement from the village involved minimum risk.

Subsequently, and especially since 1975, circulation has declined as cash cropping and local business activities have become established and the demand for mine labour has lessened — notwithstanding that the direct cost of circulation is now quite small. The circularity in Bougainvillean urban-industrial employment is more apparent amongst the less skilled who are employed outside the copper complex, where abandoning one's job means the loss of less income and less prestige. Although the decline in circularity among Siwai workers hints at the emergence of a committed urban-industrial proletariat, this is more illusory than real. Some individuals seem permanent outmigrants and rather more believe they are, but rural Siwais consider them 'odd, not like us'. Nevertheless their numbers will grow as conditions of employment worsen. Siwai and Bougainville appear destined to follow the precedents set in much poorer countries.

Acknowledgments

Fieldwork on which this paper is based was undertaken from 1974 to 1976, while a Visiting Fellow in the Department of Economics, Research School of Pacific Studies, Australian National University. It was financed by the Council for Pacific Development Studies, to whom I am most grateful. I am indebted to Geoff Harris, formerly of the University of Papua New Guinea, for permitting access to questionnaires of Siwai mineworkers from the 1973 urban household survey.

7 Doctor in the Fijian islands: F. B. Vulaono 1924–76

Rejieli K. Racule

This paper is primarily an account by my father, Dr Filipe Bulivakatawa Vulaono, of his life and career in many parts of Fiji. He began his travels young, in 1924 at the age of six months, when he left his island – Lakeba, in the Lau group (Fig. 7.1; cf. Fig. 16.1) – with his parents. The family travelled to Nadi where his father, Mosese Vulaono, had been posted headmaster at the newly-established school: Provincial Western. Apart from brief revisits to Lakeba between 1924 and 1976, my father lived and worked there for only fourteen out of fifty-two years (1928-35, 1955-60). In 1976, he retired to return to his ancestral place but, such is the shortage of qualified doctors in Fiji, that he was immediately reemployed there. Thus most of Dr. Vulaono's lifetime has been spent in many other parts of the country, notably on the largest islands of Viti Levu and Vanua Levu, first to receive formal education and medical training, then as a doctor in government service. Since his own father was a school teacher, he is the second generation of my family to have lived in many places and among different peoples as a result of being a civil servant and being required to move to another location whenever directed by the employing authorities in the capital of Suva (Fig. 7.1).

My father belongs to a renowned sailor people, the Levuka, who are reputed to have transported the Melanesian ancestors to what is now Fiji from locations somewhere farther to the west. The Levuka people, collectively known as the *yavusa*, or tribe of Delaikorolevu, did not settle at a particular site for very long. As my father recounts:

> They sailed the seas in their great double canoes on voyages of discovery and carried on barter trading [as of yams and taro for forest products] amongst the populated islands. There seems little doubt that they were responsible for realizing migration of tribe and [Fijian] nation. They had much power; they were king-makers,

149

renowned warriors, ferocious cannibals, and great sailors. They were poets, composing music and *meke* (songs and action dances) where-ever they went. Like any other tribe who had to deal with the unknown, they developed their own religion. Because of the type of life they led, they were also known for their cleverness in surgery and in medicine. The Delaikorolevu had little interest in owning land except to establish shore stations as bases and workshops.

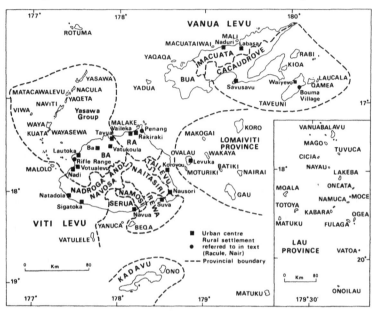

FIGURE 7.1 *Fiji*

The setting on Lakeba of the Levuka people occurred as recently as the 1860s at a place of the same name nearby Tubou (Fig. 16.1), today the largest village and administrative centre of Lau province (Fig. 7.1). Travelling over great distances, staying in strange places, and remaining away from one's relatives and friends for up to years at a time constitute the legends and oral history of their *yavusa*. The traditions of the Levuka people are based upon the practice of constant movement, which is in the blood of my father and his forebears. Perhaps this makes more understandable my father's acceptance of the peripatetic lifestyle of a government civil servant, that nevertheless was quite uncommon among Fijians of the generation born before the second world war.

As birthplace of many chiefs of high rank, Lakeba is known throughout Fiji as a 'chiefly island'. Most notably, it is the home island of the prime minister of Fiji since independence in 1970, Ratu Sir Kamisese Mara, who is also the *Tui* Nayau, paramount chief of all Lau. As Bedford (this volume) reports, Lakeba is located in a region that has become more and more peripheral from the socioeconomic mainstream of independent Fiji. Yet Lakeba still retains great historical importance and cultural position that outweighs these recent changes. On the island, as in my family's village of Levuka, the authority and leadership of the chiefs remain very important, many obligations and transactions between chiefs and people are conducted according to traditional practice, and such customs as the tribute of labour to the chief continue, these having lapsed in other parts of the country. The hold of the church on the social fabric of Lakeba is also very strong and complements that of the chiefly system.

Together, chiefs and church can command a great deal of the people's time but my father, as a traditional chief, knows that he must exercise leadership and counsel on behalf of his extended family (*i tokatoka*). This position comes to him through five generations. Each forebear captained a great double canoe and each, like my father, carried the title *koliloa*. But let Dr. Filipe Bulivakatawa Vulaono tell his own story (see editorial note).

Childhood and education, 1923–43

My father, Mosese Vulaono, was fourteen when he entered Lau provincial school in 1909, the year it was established. His teachers were Professor A. M. Hocart and Ratu Sir Lala Sukuna, the great Fijian statesman, scholar, and soldier. It was my father who helped Professor Hocart to learn Fijian and in much of the work that he wrote about Lakeba (Hocart 1929). Father left school when he was about seventeen or eighteen. Between 1914 and 1923, he spent some years on Bau, on Koro (Fig. 7.1), and also three years in Tonga (endpaper map), where he learnt the language and the music. When he returned he composed quite a number of songs about Lakeba that are still sung today. He was, of course, very fond of sailing. At one time a boat called in to Lakeba with the captain sick. Some young men of Levuka were detailed to sail the ship to Ono-i-lau. Father took over as captain, unloaded the cargo, and came back.

It was not until December 1923 that he accepted an official position, as head teacher of the newly-established school: Provincial Western, at Namaka, Nadi (Fig. 7.1). They say my father was a very clever man. He never had any formal training as a teacher but his first post was as a head. He became quite fluent in both the Nadi dialect and in Hindustani; he was also one of only two people in Lakeba about that time who could speak English.

I was just six months old when we left for Nadi, where we were for a little over three years (Fig. 7.2). Thus I was raised in an entirely different environment from here in Lakeba. I was brought up as a loving son

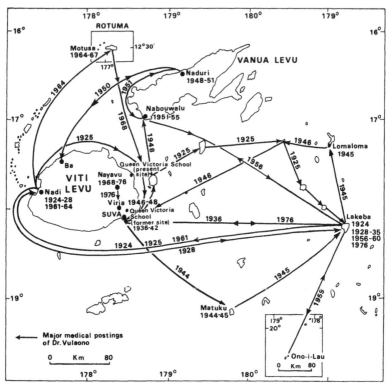

FIGURE 7.2 *Movement history of Dr. F. B. Vulaono, 1924–76*

of an official, of a teacher. My mother, Rejieli Koro, a beautiful woman from Ono-i-lau, I adored. In 1928 we moved back to Lakeba, where father was made first assistant master at Lau provincial school. I often accompanied him and sat under his desk. Among the things that I used to draw on the floor and which fascinated the other children were motor cars, trains, and big ships that I had seen in Viti Levu. They used to sit, watch, and ask questions while I would try to explain what the drivers of cars and horse-drawn carts did and where the passengers sat. Father had a horse-drawn cart in Nadi and we used it to take us to the township and various villages. Cars were very rare.

Sitting there under the table I learned quite a lot of what went on in the classroom. At about the age of five I could read and write but by 1931, when eight, I was still in the first class. My father kept me in mission and village schools for reasons of his own and I was free to do whatever I wanted. I used to go out spearing and bring back fish for my mother; go on long walks over the hills, across to other villages, and come back in the evening. I came to know a lot about Lakeba before I was ten. In the evenings I would listen to the

152

talk of grownups and to stories our grandmothers and aunts told us.

I had my own company of older cousins but I preferred to be alone most of the time. The games we children played included cricket, soccer, marbles, sprinting, and *veitiqa*, a stick shaped to bounce and skim along the ground and thrown like a javelin. We sailed toy canoes made from coconut shells, leaves, and sticks. We teased our grand-mothers, who rewarded us with sharp pinches from their claw-like fingers. We wrestled and had fist fights, with resultant black eyes and bruises. And of course we went swimming. I remember standing one day as an onlooker by the *yala*, the small channel separating Tubou and Levuku villages. The next minute I was drowning but managed to keep afloat, and fear made me swim for the banks. Although I cried like any other little boy, the next day I was swimming happily in the *yala*.

I used to follow father to his various gardens and learnt how to dig and turn the sods for *dalo* (taro) plots, how to channel water into these, how to use the *saupa* (digging stick), and how to distinguish the differ-ent varieties. During the yam planting season I learnt how to choose the seeds; how these were cut into pieces, cured, and buried; when the eyes sprouted how to plant them; and the care taken in the ensuing months to have a good yield. With my friends or by myself, I went out gather-ing coconuts and cut them for copra. A novelty was to carry a load of copra on my back.

One special occasion was when I sailed my own canoe. This was a small, double canoe, specifically made for the occasion. Father made me take this double canoe a considerable distance. He had prepared everything. There was no difficulty about steering because there were two ropes, one on each end. All I had to do was pull up one and, after tacking, push the other back into the water. My father probably knew he would not live long enough to teach me most of these things and so crammed them all into the few years that we had together before he died in 1935. It was probably why I was given so much freedom on Lakeba when I should have been in school like the other children.

My freedom finally came to an end in April 1931, when formal education started. At the age of eight I went into class two at Lau provincial school. In 1930, father had become the head of the mission school at Gaunavou, in Tubou (Fig. 16.1). Among my teachers at Lau provincial school were Ratu Edward Cakobau, later Ratu Sir Edward Cakobau, and Mr. W. J. E. Eesen, who was a district officer at one time, was manager to the firm of Carpenters at Rotuma (Fig. 7.1), and then served in the Fijian administration for a number of years before retiring. There were about fifty attending school. I was a day scholar from 1931 to 1933, when I became a boarder. By 1935 I had reached class eight. There were only five of us; I was twelve and the oldest was about eighteen.

Then one day, 15 May, our beloved father died. The shock was terrible. I kissed his forehead for the last time and, young as I was, promised that one day I would erect a memorial stone on his grave. The days after his death brought home to me a great sense of loss in that the man of our house was no more. Our poor grieving mother! She comforted us, saying that the living God would see us through our bereavement and that through his blessings we would achieve a better life. That, she said, could only be done by hard work, and that from then on I had to act and work like a man, thinking always that there was my younger brother and mother to look after.

My father's cousin's did not want me to return to school, but I did nevertheless. Later in 1935, I took the entrance examination to Queen Victoria school, then located in Suva (Fig. 7.2). I passed, but because my father had died and the price of copra had dropped – this was during the great depression – it was rather difficult to attend. However, since I had come second in the entrance examination, I was offered one of the six exhibition scholarships. The rest of the money and school equipment was provided by my godfather, Ratu Tevita Uluilakemba, father of the present *Tui* Nayau and Prime Minister of Fiji. Queen Victoria school was a government institution especially set up for the education of Fijian boys. I entered at the beginning of 1936 and worked very hard. I began in class four, three weeks later moved to class five, and within two years had completed the school course. I was forced to remain in class eight for three years, being too young to sit for further examinations. I took all the prizes that could be won. I was prefect for two years and then head prefect in 1939. I was a sergeant-major in the cadet corps at the beginning of the second world war, the highest position in those days that a Fijian boy could aspire to. In 1939 I took the qualifying examination for medical school, topped the list, and began my studies on 15 January 1940. I qualified as a doctor in 1943 after four years.

What of my brother and mother who were left struggling in the village? My mother was a beautiful woman and imbued with a lovely character. I still have a very vivid memory of her standing on the shore with my brother and crying one fine day in January 1939. I was leaving Lakeba by the ketch *Tui Tubou* to return to Queen Victoria school. Even when the ship was entering the passage in the reef, I could still see her waving a white handkerchief. It was the last I saw of her. Leaving my young brother Mika in the care of relatives, she went back to Ono-i-lau for a visit. She died there on 22 November 1939 at the age of thirty-six.

On December 1943, I entered the Royal Fiji Naval Volunteer Reserve and joined the HMNZ *Venture* as chief petty officer in the sick berth. Apart from Dr Macu Salato, recently Secretary-General of the South Pacific Commission, I was the only Fijian who was billeted and ate with the European commissioned and petty officers. I was not yet twenty-one. At first there was some racial feeling towards me on the part of some officers, but that soon wore off and we became good friends. As a result, my whole attitude towards Europeans changed. Those that I had previously come in contact with had always been people in authority: my teachers and tutors at medical school. Now I saw that Europeans also were ordinary folk, with all their human failings, weaknesses, and strengths.

The doctor, 1944–76

Matuku, Lau province, 1944–45

My first posting was to Matuku rural hospital in southern Lau (Fig. 7.2), where we travelled on the *Adi Moce* in February 1944. I had married Lisoa Raibe Tuwai, of Sorokoba, Ba (Fig. 7.1), on 15 January 1944. She comes from the chiefly Tio *mataqali*, the subtribe from which the paramount chief of Ba is chosen. I had met, fallen in love with, and courted her while she was at nursing school. My dear wife has been a source of strength throughout my career as a doctor. Matuku is a special place for us for it was here that our eldest daughter was born. We named her after my mother, Rejieli Koro Vulaono.

I found life on Matuku very slow after the capital of Suva. Many people attended outpatients, but very few had to be admitted to hospital. However, I learned to become a villager and started a yam garden. The people helped me with its preparation, but the actual planting and maintenance I did myself. I also had a *dalo* patch. There were very few places suitable for planting taro, mainly because of scarcity of water. The villagers had a communal garden, with two or three plots for each person. I also did a lot of fishing on Matuku. This was in the days before nylon lines and we used cotton ones.

During war time, we were on an isolated island, which was very hard. There were few things obtainable from the Chinese store and we had to live on my salary, at the time five pounds a month. When I was in the navy, I was getting £28 12s 6d. For many months I did not receive any salary at all and had to wait until the district commissioner came around on tour, when I could draw it. The people of Matuku were very good to us and even provided the money for church collection on Sundays, as well as for soap and other necessities. I found out later that I was related to the chiefs of Matuku, through a woman of Levuka who married into their family some four or five generations before.

Perhaps the greatest event during our tour of Matuku was when there was a difficult birth and I had to call Suva for help. A Catalina flying boat from the army brought some dried plasma for transfusion. We could not do blood transfusions in those days, but I had had experience with plasma transfusions at the Colonial War Memorial Hospital in Suva, so it was not new to me. The arrival of the Catalina caused great excitement and later another brought more plasma. The people, knowing my father and me, were very impressed about this.

Lomaloma, Vanuabalavu, Lau province: December 1945–June 1946

We had to leave Matuku because of some difficulty at Lomaloma hospital. We left by the ketch *Adi Moce*, went up to Lakeba, and arrived there on 24 December 1945 (Fig. 7.2). After the Christmas feast at Levuka, we left Tubou for Lomaloma, arriving there the next day, when I took over running the hospital. Lomaloma is on the beautiful island of Vanuabalavu (Fig. 7.1), with a scenic harbour opening to the southeast. Outside the barrier reef are the two islands of Munia and Cikobia. My family and I spent many evenings in its tranquil beauty.

We had very few nurses in those days. I had none in Matuku and in Lomaloma there was only an old nurse. At that time, we were introducing the new series of sulphanamide drugs, which had to be given every four hours to maintain the high concentration in the blood for them to be of any use. I could not just rely on the nurse to do all this and, whenever there was a patient, had to keep the drugs going myself day and night. I also had to do many deliveries myself. As a matter of fact, I had to attend my wife and deliver my first born. I have had quite a lot of experience of midwifery at both the Colonial War Memorial Hospital in Suva and Matuku, and of course in Lomaloma, where I established a new obstetric hospital. This was in the house that used to belong to clerks of the district administration, for by 1945 the district commissioner had moved from Lomaloma to Suva.

Dr Evans, who was then director of medical services, told me later that the reason I was posted to Lomaloma was because I was from Tubou and the people would respect me, Tubou being the seat of the paramount chief of Lau and my being related to the chiefs. There was no more trouble in the six months we were at Lomaloma.

Viria, Naitasiri province, Viti Levu: 1946–48

Then we were posted to Viria in Naitasiri province, Viti Levu (Fig. 7.1). We left Lomaloma on the ship *Adi Tui Lomaloma*, sailed two days up to Yacata, where we spent three days, then sailed another four to Suva (Fig. 7.2). We had to lie on copra sacks, for the boat was small. Luckily we did not have very much in the way of personal effects in those days and had only Rejieli, one year old, to cope with.

In July 1946 I started work at Viria. The station had been vacant for nine months and was overgrown. Whereas in Lomaloma we were living in a wooden building, I found in Viria there was only a small, kitchen-type house for us to occupy. The station was at the edge of both the village and the sugar-cane plantations that extended along the Rewa valley. The people spoke a different dialect from the ones I was accustomed to and were very unfriendly at the beginning, because of the rumour that I had been posted from Lomaloma as a result of having affairs with girls. They thought their new medical practitioner was not a good man for them and would not help me prepare the house or clear the ground with a cane knife. I had to attend to all this myself. Luckily I was still young: only twenty-three.

Two days after our arrival in Viria, I was called to my first obstetric case. I was told the woman was at a nearby village (Nawaqabena) and started walking there about five in the afternoon, eventually reaching the place at nine o'clock. It was absolutely dark. I had been told that after we ascended two hills we would come down to the village, but found out there were no less than six and I was a little angry when we arrived. The house was lit by only one, very small lamp. I walked right in, going to the room where the patient lay. I still had on my haversack, knelt down, and examined her abdomen. I stood up: 'Look here, you said you've been in labour for three days, but you only had an abdominal pain. Your people have heard about your being in labour and are coming to await birth. You were so ashamed to tell them you are not in labour that you had to send for me to come all the way in the dark to attend you. You can lie there and wait for three weeks and then have your baby. It will be a boy. I'm going back.' While talking to the patient, I realized that I had passed a crowd of people, whom I had not seen in the dark. Not only that, I heard someone say: 'He's an Indian', in reference to my impetuous entry. I came out of the room, started to shake hands all round, eventually sat down, clapped my hands as a mark of respect, and said I was leaving. I eventually reached Viria at five in the morning.

What I did not know was that the village people were counting the days and, three weeks after I had visited the woman, she gave birth to a boy. The event in itself was nothing uncommon but what made it news was that my forecasts were correct. The incident made people realize that this man, to whom they had shown enmity might not be the same person whose poor reputation had reached them. That one incident made them accept me as I was.

It was at Viria that my reputation as a good doctor was established. I had patients coming all the way from Namosi in the hills, from Tailevu, from the capital of Suva, and even as far away as Ono-i-lau (Fig. 7.1). I made regular health visits to all the villages up the Waidina river, as far as the border of Namosi province, and far up the Rewa river. When Dr Macu Salato returned from England to start the tuberculosis campaign, we initiated the pilot survey using the new tuberculin in one of the village areas.

I particularly remember Viria because it was here, in 1946, that my second daughter, Vikatoria Tabuisulu Sainimere Vulaono, was born;

and two years later, in September, our third daughter: Jesepeli Tauga. Then in November, we learned that we were to be posted to Naduri in Macuata province, Vanua Levu (Fig. 7.1).

The people of Viria went to the medical department and even presented a *tabua* to request that I stay. The Indian people sent a delegation to their member of the then legislative council, who wrote a letter also asking that I remain. However, we had to go. I found out later that there had been some difficulty in Naduri and the high chiefs of Macuata had approached Ratu Sir Lala Sukuna for help. Wherever we went, we established close relationships with the people. We grew to appreciate their way of life, their problems, and did what we could to help. Leaving a station was always a wrench, with us being farewelled with feasts, parties, gifts, and tears.

Naduri, Macuata province, Vanua Levu: 1948–51

From Viria, the family first went by launch for ten miles and then another eight by bus to Nausori township (Fig. 7.1). At Nausori post office I received a letter saying that my only brother, Mika Mateiwai Vulaono, had died at Lakeba. He had been to see us in June, bringing yam all the way from our home island where he had been teaching, and by September he was dead. It was a sad trip. We had belonged to a small, compact, loving family. We had lost our father when I was twelve and Mika eight; our mother when I was sixteen and Mika thirteen; and now I had lost my talented young brother. The trip was made still harder because we had three young children, one of them only a month old.

After a day in Ba on the other side of Viti Levu, to see my wife's people, we spent a week in Suva and then took saloon passages on the *Adi Rewa*, sailing on to Levuka, on Ovalau, the next day across to Nabouwalu, and then northwards along the coast to Naduri, reaching it two days later (Fig. 7.2; cf. Fig. 7.1). Parents travelling with several small children on interisland ships have my sympathy. Fortunately mine were good sailors, like all their tribe, but my poor wife was always seasick.

By now my salary had increased to twenty-five pounds a month and I had bought some carpentry tools while at Viria, which proved very useful. I had also begun the matriculation course to the University of New Zealand with Hemmingway's correspondence school and continued while at Naduri. My subjects were English, history, geography, book keeping, commercial practice, and Latin. Here, too, the children were growing up. There was no school nearby, but after a while the people started a new school next door to the dispensary and it was to this that Rejieli first went. Vikatoria more often than not followed her elder sister to school.

At Naduri, we slept in a small *bure* (house) and another of the same size was our kitchen and dining room. The people of Macuata are famed for house building; constructing one for a chief used to mean human sacrifice. Naduri chiefs had a notorious reputation for this. Posts left in three unused chiefly houses were sometimes wrapped with a piece of *masi* (a particular kind of tapa) and were a constant reminder of the atrocious practice of *ai vakasobu ni duru*: burying a man with the house post. It was 1948, but the people believed this was still being done and whenever a house in Naduri was under construction, they avoided going out alone. Thus, when a chiefly house was ordered built for the new doctor, people were wary, most especially the few Indians living on a nearby plateau: they had heard too many rumours! When the great logs were carried in, sometimes by as many as forty strong men, there would be one or two of the oldest sitting atop the log, beating time to the eerie chants and shouts of the carriers. Four logs of newly-cut *buabua* trees, a native hardwood, were transported in this manner.

We worked in Macuata province for two years, then in 1950 took our first vacation leave of three months. The previous year we had received word of the death of my wife's mother, so when we went on leave to her village (Sorokoba, Ba province: Fig. 7.2), we were prepared for the traditional requirement of *i reguregu* (formal condolence). This consisted of presenting several *tabua*, drums of kerosene, and some fine mats to my wife's relatives. Fortunately, the women of Macuata were famous for weaving this type of mat.

One day, in Sorokoba, I was approached by the then *Tui* Ba, Ratu Nale, to see if I could remove a tumour on his chiefly skull. This was surgically done at a nearby dispensary, where the Rev Dr. Waqabaca of Waciwaci, Lakeba, was in charge. Every soiled swab and towel was collected and taken away by the chief's son for special burial. On our return to the village, we found the people prepared for a special cere-mony. The surgeon was to be honoured and his blessings evoked for the quick healing of the scalp wound. I was made to enter the most chiefly *bure* from the side door and sit on the prepared pile of mats. A huge *tabua* was presented by the *matanivanua* (hereditary spokesman), saying that I had done a most noble job and that I had not only touched the head of their chief, but also shed his blood – both of which are strictly forbidden in Fijian custom. As this had been done at the chief's insistence and with his permission, I was expiated, exonerated, and exorcised all at once of any malevolent effect of his *mana* (power) on me, my wife, and children. At the same time I was requested to grant that my ancestors' and my own *mana* be bestowed in abundance on the chief, his family, and his people. I accepted the *tabua* and to this speech answered as required by tradition. The *magiti* (food for a feast) was then offered and received.

We returned to Macuata via Suva, travelling to Labasa (Fig. 7.1) on a flying boat of the Tasman Empire Airways Limited, and thence by launch to Naduri. We had come a long way from sailing vessels and auxiliary ketch to an aeroplane.

I had started being a lay preacher in Lomaloma in 1946 and have been one in the Methodist church ever since. There was a beautiful church in Naduri but very few preachers, and on some Sundays I took two services. It was here also that I met Dr Lindsay Verrier and, while visiting villages, collected material for his demographic work on the Fijian population (cf. Ward 1961:265, 270).

I had done some horse riding at Viria, but here in Macuata the only way to go quickly from one village to another was on horseback. The area to be served was quite large and included the islands of Macuata-i-wai and Kia, inhabited by famed fishermen. I used to ride once a month to Labasa (Fig. 7.1) and beyond, up to fifteen miles [twenty-five kilometres], and paid health visits to each village every three weeks. I had a very clever mare, Maqe. Sometimes, while twenty miles [thirty-two kilometres] from Naduri and wanting to give news to the family, I could send her alone with a letter. She answered only to my call, even if in hiding somewhere.

Included with my first revision of salary at Naduri were some arrears. It was enough to buy a new *yaka* (hardwood) bed for our eldest daughter, two armchairs, and good wrist watches for my wife and myself. We also bought a new portable typewriter, which proved most valuable. Later we purchased our first Telefunken radio set, which had valves the size of light globes and worked on a nine-plate battery. This had to be taken to Labasa every two weeks for recharging. People used to crowd into our house from neighbouring villages every Tuesday evening to listen to the Fijian programme.

Then one day, while painting the dispensary, a runner from Labasa arrived to say that I had to leave early next morning for Nabouwalu, Bua province, to take over from the doctor. Next day, the motor vessel *Tui Taveuni* arrived and there was a touching send off from the Macuata people I had come to know so well, all lining the jetty. My family stayed at Naduri for another two weeks.

Nabouwalu, Bua province, Vanua Levu: 1951–55

The morning after arriving at Nabouwalu I went over to the hospital. The doctor of course did not know I was there deputizing for the medical officer in charge of the district. I was kept very busy at Nabouwalu. Quarters for the doctor and nurses were not in the best condition and needed renovation. The hospital kitchen was a dilapidated, old, wooden building. By the time we left a concrete structure had replaced it, there were new quarters for the nurses, a kitchen for the doctor, and improvements in the toilet and bathroom facilities for patients. New equipment had also been provided for the hospital. I travelled on health visits to the villages both on horseback and by launch. Again, the area was quite large and included the island of Yadua.

In Nabouwalu, Rejieli and Mateyawa (an adopted son) were joined at school by Vikatoria (1952) and Tauga (1954). Maintaining four children in school was becoming costly. By then, ten years after qualifying as a doctor, my salary had risen to forty-two pounds a month. But we still could not make ends meet and slowly accumulated debt at the local Chinese store. Most vegetables came from our garden, but people often brought foodstuffs. This is a Fijian custom: all civil servants in rural areas receive gifts of food but this obliges them, by tradition, to feed those who bring them, buy tobacco or *yaqona*, give them a *tabua* or sometimes even a drum of kerosene. Besides that, there were collections for church and school, all of which meant money. On Viria, Naduri, and Nabouwalu, I received a horse allowance to buy saddles, riding outfits, and ropes. Money used on spear guns and fishing lines was well spent, because Nabouwalu is a very rich fishing ground.

Given the hospital's central position on southern Vanua Levu (Fig. 7.2), we received many official visitors. Once the chief justice of Tonga and his wife came to stay. There were district officers from Labasa and Savusavu (Fig. 7.1), medical officers from Labasa, catholic priests from nearby institutions and visiting Wesleyan missionaries, police officers, the chiefs of Macuata and Bua provinces, our friends, and our own people from Ba and Lau. Seldom a day passed without a visitor at home and much more money was spent on them than on ourselves. In December 1953, Her Majesty the Queen visited Fiji as part of her Commonwealth tour after coronation. This created great excitement amongst Fijians, who have great love and loyalty for the royal family. My wife and the children were amongst the thousands who flocked into Suva to see the Queen and the Duke of Edinburgh. They went with most of our savings.

In 1954, Rejieli entered Ballantine Memorial school, in Suva, as a boarder, returning to Nabouwalu at the end of every term. All she had to do in Suva was report on board any of the ships plying between Suva and Labasa, via Nabouwalu, whose officers and men were our regular visitors and knew all the children. The following year Mateyawa entered Ratu Kadavulevu school, Tailevu province, also as a boarder. My own studies continued and in 1952 I gained a diploma in book keeping from Hemmingway's correspondence school in New Zealand. For many years there were only two Fijians with this qualification. Later. in Lakeba, this knowledge proved very useful.

Vacation leave was due in 1955 and we were saving to return home to Lakeba for the first time since our one-day visit ten years earlier. In July, we left Nabouwalu, expecting to return after 180 days' absence (Fig. 7.2). We took only what we would need at Lakeba and the bulk of our luggage consisted of sandalwood logs, bundles of mats, a crate of *tabua*, added to in Suva by some drums of kerosene. All this was required for the formal presentation of my children to my people, for I was taking them home for the first time (cf. Tubuna, this volume). Since our leave coincided with the school holidays, in Suva we collected Rejieli and sailed to Lakeba on the *Adi Maopa*. At Naikawakawasinu, my ancestral residence, the presentation was offered and our relatives made theirs in return, to welcome us after eleven years. Most of the younger people did not know who we were. It was immediately apparent that we needed a bigger house; Naikawakawasinu had to be renewed and enlarged. It would take almost a year's salary, but we decided to do this straightaway before returning to Nabouwalu the next year.

Lakeba, Lau province: 1956-60

Moves were made by the *Tui* Nayau, *Sau ni Vanua* (paramount chief of all Lau), to have me posted to Lakeba hospital. This was done and the government administrative officer for the region, who happened to be his son, directed that our personal effects left at Nabouwalu be shipped to Lakeba. My wife went to Suva with Rejieli and Mateyawa, the former to enter Adi Cakobau school and the latter to return as boarder at Ratu Kadavulevu school in Tailevu province. Adi Cakobau is an exclusive government school for Fijian girls and much more expensive than Queen Victoria. Properly to equip a student initially required no less than one hundred pounds. What with new sets of tailor-made uniforms, bedding and other requirements, fares to and from Suva, and pocket money, I ran into debt. But my daughters needed only the best and they all went to Adi Cakobau: Rejieli from 1956-62, Vikatoria in 1958-64, and Tauga 1960-65.

Soon after taking over Lakeba hospital, I organized our *mataqali* cooperative store and transferred to the subtribe all the cash due from my cess-money – a levy on copra production to support Fijian development – which derived from a copra plantation at Vakano, another village in Lakeba. Within four years, we were able to build two modern houses and a store. Jimi, a boy from the *mataqali*, went to secondary school and had his fees paid by the cooperative. He is now a civil engineer. I had become the choir master in church, sports organizer and cricket coach, and had assumed my place as adviser and helper to the high chief. I relaxed in my garden, planting *dalo* and yams. When, in 1958, Ratu Sir Lala Sukuna died in the Indian Ocean and his chiefly body was brought back to Fiji and then to Lakeba, I was in charge of both preparing the tomb and the actual burial. This was my tribe's prerogative, for the first piece of land given to it in Lakeba was returned to be the chiefly burial site. Only the people of Levuka can touch the bodies and tombs of chiefs without any ill effects from their *mana*. The site is of course *tabu* and once I left the tomb, no one else was allowed inside. I had identified myself with my people and with their traditional obligations.

I was thirty seven and decided to return to postgraduate studies, having in 1959 attended a seminar in social and preventive medicine at the Fiji school of medicine. En route to Suva, my wife, Jone (a recently adopted son), and I visited Ono-i-lau, where we observed the traditional requirements for honouring my mother's grave and taking my wife to my mother's people for the first time. It was also the initial visit of one of my children and our relations gave us an overwhelming welcome. For three days, baked pigs and *magiti* were brought to us, mats and *tabua* presented, and yams dug up and taken on board the *Vuniwai*, the medical boat that was talking us to Suva. Dr. Vuli Mataitoga, of Ono-i-lau, was my colleague from Fiji to attend the newly-introduced course for the certificate of public health. While we lived at the school of medicine, our families were found lodgings in town.

Thus began a new chapter in my career as doctor. It was 1961 and for six months I attended lectures, doing assignments at various villages on Viti Levu; at Queen Victoria school at Tailevu (Fig. 7.2); the Emperor gold mines at Vatukoula (Fig. 7.1); the sugar mills of Rakiraki; at Makogai leprosarium, Lomaiviti province, where I spent two weeks relieving the medical officer in charge; and at Matuku, where I led a health survey for three weeks, before taking final examinations in September. In this class were a doctor from the Solomon Islands, two from Western Samoa, one from the Gilbert and Ellice Islands [now Kiribati and Tuvalu], one from Papua New Guinea, and one from Rotuma (see endpaper map).

There had been tremendous advances in all branches of medicine and surgery. Government had mounted national campaigns to eradicate yaws, tuberculosis, and filariasis. Family planning and immunization had been introduced; antibiotics and antihistamines were coming into general use. Mental stress and various health hazards, among them malnutrition, had become identified. Advances in medicine were closely related to, if not actually caused by, the introduction of new industries and the awakening danger of population explosion. Under Dr C. H. Gurd, the medical department was being thoroughly reorganized. Specialization was introduced and more workers had to be trained in public health. Incidentally, the political and economic situation in Fiji was being awakened, so presumably what was happening in the medical department was also found in other government departments. Maybe the generation of young people, returning to Fiji with new ideas from overseas universities, were the cause when they found the archaic order of things and the paternal outlook of the colonial power on a docile, multiracial society being more or less perpetuated. Changes meant reorganization, planning, training. My work during the next seventeen years was oriented to new ways of life being introduced into Fiji. We went home to Lakeba only five times, a total of five weeks, during that period.

Nadi hospital, Ba province: 1961-63

Life in Nadi was very different from the rural areas where we had previously lived. Most patients were Indian: businessmen and traders in town and farmers from the surrounding area. Fijian patients came from local villages and migrants from other parts of the country who had leased land from sugar-cane farms. Nadi was the only place where I did not have a garden, since there was not enough vacant land on the hospital grounds. We had to buy food from the shops and market. However, my wife had many relatives near town and when they visited, they invariably brought crabs and tilapia: a delicious freshwater fish abounding in the nearby streams.

At Nadi hospital, I worked very closely with the police. There were postmortems almost everyday; my daughters must have found it a great shock to realize that their father cut up dead bodies! I appeared in court as a crown witness on cases of murder, rape, and traffic injury. It was a hectic time, but then my profession was also undergoing drastic change. The hospital was understaffed and inadequately equipped to cope with the population pressure of the area. Working with me were another doctor and six nurses, while hospital administration required much paper work. By the time we left Nadi, three years later, a concrete block for the casualty department as well as a new kitchen were in use and the foundation of the present, double-storied, maternity unit also had been laid.

Rotuma hospital, Rotuma island: 1964-67

'Land ahoy! Land ahoy!'. We had been sailing for fifty-two hours on the *Aoniu*, a quite fast motor vessel, when I heard the call. I came up on deck and there, still shrouded in mist, was the island of Rotuma (Fig. 7.2). We were approaching from the south and steaming towards its northeastern end. My impression from the deck was one of complete isolation. Rotuma is a very small island, even when compared with other land masses in the Pacific. In the old days, sailing vessels took about three weeks to go from Viti Levu to Rotuma, with the right winds, and one could easily pass the island without sighting it. I do not believe that the people deliberately set out to settle Rotuma; they more likely drifted there, for it is 300 miles [482 kilometres] from the nearest land.

The Rotumans appear a mixture of various Pacific races, which is reflected in their physical appearance, culture, and language. From a medical point of view, the island was an interesting community. Although small, it had almost the whole range of tropical diseases and most other bodily and mental ailments common to human beings. Apart from Jone, who went to school there, we did not speak the language but this was not as great a barrier as one would have thought. The nurses or the dresser could interpret and there was always someone with a smattering of English or Fijian.

As the only medical officer on the island, I had to be surgeon, obstetrician, dentist, radiologist, and health inspector all at once. My achievement in public health at Rotuma was rated exceptional. Out of the three hundred tuberculosis cases under register when we arrived, only one remained at our departure and even he was under domiciliary treatment. The Rotuman house is built of crushed coral without any reinforcement, so constructing more permanent dwellings was encouraged, and water-sealed latrines were introduced together with the smokeless stove and family incinerator. Rotuma has no running water and in nearly all villages people depended on rainwater stored in huge concrete tanks. There were one or two wells, but at times the water turned brackish. In spite of this, mosquitoes and flies were a major problem and experiments were being conducted to find a suitable biological control to overcome that.

On Rotuma, there were four other Fijian families of government officials. Our wives, in particular, felt the isolation and formed themselves into a close unit, visiting each other, weaving mats, creating other handicrafts, exchanging recipes and news, getting to know the Rotuman women, and keeping themselves busy with church matters. Our services were held at the *bure* of the Great Council of Chiefs situated at Ahou, the government station, and the preacher, usually from one of the villages, had lunch with us every Sunday. As in all other rural areas of Fiji, in Rotuma the trained doctor is a very prominent figure and respected. Thus it was that all government officials (medical, educational, police) who visited from Suva, as well as church dignitaries, all came to our home.

In December 1965, the government officer in charge and I organized a special gathering for the Methodist Youth Fellowship at Ahou. We invited everyone: the chiefs and their people; the Roman Catholic priest, the sisters and members of their church. Rotuma had a history of very strong denominational rivalry resulting in interdistrict warfare and this meeting in 1965 was historic in that, for the first time, the different churches came together for a combined service. It was a great success. We grew to love the Rotumans. They paid me a great honour in 1967, during the celebration of the 87th anniversary of their cession to Great Britain. His Excellency, the Governor of Fiji, was unable to attend and the Great Council of Chiefs unanimously nominated me to be guest of honour, which had never previously been given to a non-Rotuman.

Nayavu health centre, Tailevu province: 1968-71, 1972-76

After far-flung Rotuma, it was back to another rural area in Viti Levu: this time at Nayavu health centre, Tailevu (Fig. 7.2), where we spent a total of eight years. Nayavu is about eighty kilometres north of Suva, along King's Road (Fig. 15.1; cf. Fig. 10.1). My visits to Fijian villages and Indian settlements were made by government landrover, certainly a long way from horseback. In Rotuma, there had been an ambulance and a motor bicycle.

I was now a senior medical officer with the accompanying salary, and the children were becoming self-supporting, so at least I was in a position to buy a car. That meant we became more mobile and numerous were the visits we made to see Jone at Queen Victoria school, where he was deputy head prefect (1974) and head prefect (1975); Vikatoria and her family at Ratu Kadavulevu school, Tailevu, where her husband was farm manager; and Tauga, Rejieli and their families in Suva, where we did most of our shopping. My wife often went to Ba and Nadi to visit her relatives.

Much of the work I did at Nayavu received only a line or two in my monthly and annual reports. How the land dispute in the Nakorowaiwai area was settled and why the new village of Naituvatuvavatu had to be registered as part of Wainibuka, Tailevu, and separated from another village of the neighbouring province of Ra (Fig. 7.1). How I spent two days alone in the upland forests surveying for a possible water catchment and how, a year later, I was invited to officially open the new water supply to several communities. How I led the survey team to Wainibuka (Fig. 10.1) after the devastating flood in 1964, having worked for a week with victims evacuated to as far as Nausori town (Fig. 7.1). Of my role in founding Wainibuka junior secondary school with a committee of villagers, organizing a feast to help raise school funds, and inviting the Governor General, Ratu Sir George Cakobau, to declare it officially open. Beyond that, there was the professional and personal help I was able to render people with whom I came in contact; often matters never noted in my reports but sources of great satisfaction, for they gave me a sense of achievement, of being useful.

It was now time that I retired, and I did in June 1976 with the intention of returning to Lakeba to renovate our permanent home, Naikawakawasinu. However, lack of doctors in Fiji meant that I was almost immediately reemployed by government and since September 1976 have been the senior medical officer stationed at Lakeba and responsible for the entire eastern region. Naikawakawasinu has been rebuilt. It still has the original Tongan shape but the walls are of concrete blocks. There is now electricity in the village, so we have electric lights and use an iron. These, with our kerosene refrigerator and gas stove, make things much easier than in the past. We have bought a carrier van, which is proving very useful for both our personal use and as an ambulance – since the hospital does not have one.

That then has been the pattern of my life. Since early childhood I have been constantly on the move, first with my parents because of their profession, then for formal education, and then as a doctor in government service, but always coming back 'home' to Lakeba whenever possible. My family is now scattered. My wife and I, with Vikatoria's elder child who has been with us since six weeks old, are here at Levuka. Tauga and her husband are now in Wellington, New Zealand, where he is a law student. Vikatoria and Jack, with their younger son, have migrated to New Zealand and are now living in Auckland. Jone is working in Suva and Mateyawa in Labasa, on Vanua Levu (Fig. 7.1), where he has been living with his young family for the last ten years. Rejieli and Amani are well established with their two children in Suva. In all likelihood, my grandchildren will never know the life of villagers, let alone the difficulties that my children and I have experienced. Their problems and their lifestyles will be very different.

On being a Lauan: Rejieli Racule

It was not until 1955, when he had vacation leave, that my father took us to our village of Levuka. I spent six months there before departing for education in early 1956. During the five years (1956-60) when father was stationed at Lakeba as doctor, I returned only for Christmas vacations, remaining five or six weeks each year. Not until seventeen years later, during Easter 1977, did I revisit Levuka but stayed only three days. Later the same year, I returned with my husband and children for a longer period (ten days). As is important in Fijian custom, and as my father has already described, we went to formally present our children to our relations (cf. Tubuna, this volume) as well as to celebrate new year with my parents. I have hardly lived in Lakeba, yet regard myself a Lauan. There are many reasons for this: parental influence; our social standing due to my father being a government official; the Fijian culture and system of inheritance; and the social, political, and economic situation peculiar to Fijians in their own country.

As is clear from this account, my father was steeped in love for Lakeba, for his tribe and its history, and for Lauan tradition and culture. He was proud of these and never too tired to talk about them. We children grew up to love the legends and stories of Lakeba he related to us at bedtime and we listened avidly to accounts of his childhood experiences. For father, Lakeba was 'home'. For us children, 'home' was wherever our parents happened to be, but we knew that one day we would return to what was really 'home': Levuka, Lakeba, Lau.

Father was never stationed sufficiently long in one place for us to feel completely attached to it. We grew up knowing that we would always move after a few years. We were the children of a government official, a man from another part of Fiji. We went to school with the local children; played, fished, and swam with them; almost everyday brought friends to our house; but were regarded as being in a different category. We dressed and spoke differently − although after awhile acquired the local dialect; our house was not the same as others and we had another lifestyle. This attitude of the local children helped make us conscious that we were something of outsiders. Such was made very plain whenever there were arguments and tussles, when we were called derisively *Kai tani*! *Kai Lau*! (Foreigner! Lauan!), or teased mercilessly: 'Go away − Swim back to your island!'.

The fact that my parents were stationed in Lakeba during the years when I was eleven to sixteen helped consolidate in my mind that I was a Lauan. Although I actually lived there for twelve months overall, this was a formative time when I was absorbing impressions and the very atmosphere of Lakeba. Thus my love and appreciation for life in this island world became cemented.

In addition, there were the particular social, political, and economic contexts that govern Fijians and reinforced our awareness of being Lauans. At that time, the Fijian administration required all adult Fijian males to pay a provincial tax to finance local projects (cf. Nair, this volume). In my father's case, this tax had to be paid annually to Lau, even though he was resident in other provinces. Apart from such levies, father often sent money to Lakeba, either to relatives who had asked or along with other contributions for the development of his island of birth, as distinct from the entire province.

All Fijians belong to subtribes (*mataqali*) and tribes (*yavusa*), members of which are descended from a common ancestor and usually live in a village (*koro*) or several neighbouring villages. Groups of *yavusa* may, for political reasons, form an alliance (*vanua*), several of which may in turn establish a confederacy (*matanitu*). Each *mataqali*, *yavusa*, and *vanua* has a traditional role to play in the maintenance of Fijian society. Chiefs, for instance, are chosen from the chiefly tribe; there is the *matanivanua* (face-of-the land) *mataqali* who are the spokesmen, the go-between, and the mediators between chief and people. These roles pass from one generation to the next. Although there can be only one chief, all members of the 'chiefly clan' are respected and should provide leadership whenever appropriate. This is why my father, in the absence of the chiefs of Tubou and Levuka (Fig. 16.1), deputized for them during customary ceremonies and social occasions.

Each *mataqali*, *yavusa*, or *vanua* has a name, or *yaca buli*, that refers to both people and chief. These titles are always mentioned in the formal speeches that accompany a customary ceremony or ritual, when *yaqona* (kava) is offered, or there is a presentation of *i yau*: whales' teeth, fine mats, tapa, rolls of manufactured cloth, and drums of kerosene (cf. Tubuna, this volume). In the same way, the tribal names of both husband and wife are referred to in any family ceremony. Thus wherever Fijians find themselves, the fact that they belong to a particular tribe is always acknowledged. This is and was an everyday event for my family as we moved around the country; we are labelled Lauans.

The Fijian administration has jurisdiction over all land belonging to the Fijian people, most of which is owned communally by the *mataqali* or *yavusa*. To ensure that all Fijians have secure rights to this land, they are registered a member of their father's *mataqali* and *yavusa*. Official documents such as birth certificates and medical records include this information. Since these records are required during one's school life, and for medical and legal purposes, a Fijian knows his village and tribe if not his genealogy.

The class of land termed *kovukovu*, most of which is acquired as dowries, belongs to individuals or family groups rather than to the tribe. Two pieces of such land came to my family in this way, while another was given my great-grandfather by the chief in whose family he was raised. Land is a fundamental part of Fijian society and culture and deep emotion attaches to it. Our attitude is not mercenary; rather, the land is held with reverence, since it represents the basis and sustenance of life. For this reason Fijians regard land as security, perhaps even more so nowadays than in the past by those who have access to it. Consequently it would be unthinkable to consider selling our *kovukovu* lands at Lakeba, for they constitute part of my family's history and its heritage (cf. Nair, this volume).

Finally, my feeling about being Lauan is much affected by the fact that I am the eldest child of a man who is the last male in a long lineage of distinguished ancestors. While a woman, one day I shall inherit all my father's responsibilities to our relations. It is to some extent true that Fijian society is patrilineal and male dominant but, as my father's account makes clear, its women are not down trodden, are respected as individuals, and have an important role to play. Hence I am now very conscious of this future duty that will be thrust upon me, although this realization has come very gradually. Wherever we lived, relatives were always coming to visit or for counsel, and by tradition brought small gifts of food, mats, or perfumed coconut oil. We, in turn, provided shop-bought goods and money. This was reciprocity, but to young eyes it appeared that our parents could nave better spent their money on us children and themselves.

Let my father continue:

> The children thought that *nana* (mother) and I spent our money on useless things, forgetting that we were committed to our people, our relatives, right from the very beginning. So even today, when I am retired and working again in Lakeba, our many obligations to our

169

relations and friends must still be met. It is insurance to the time when we are no more. Our daughters will still retain all their rights to their lands, their home, and their relatives, who will always help as much as they can. Perhaps this is the very essence of our way of life: the way of togetherness, never without a friend or relative to turn to in time of need, never a relative being turned away without giving consolation in words or kind.

If this is ever abolished, then we would be reduced to the same pattern of life as highly-industrialized countries, whose people had to move out to live in new communities of workers. In Fiji, this would be like the gold mining town of Vatukoula in western Viti Levu (Fig. 7.1) or perhaps the housing estates like Raiwaqa in Suva (see Nair, this volume). Perhaps the western influence, so powerful in urban areas, combines with integration with other races and other cultures to lead people to lose their culture and tradition, their dialects and languages. Perhaps, sooner than they realise, they lose all knowledge of their inheritance besides the immediate need of being alive and for money to maintain life. Such a situation should never be allowed into my family, nor for that matter into any Fijian family. I know of many families from the islands who have migrated to other parts of Fiji, have acquired lands and are settled there, but still maintain their ties with the old place. This is good as it offers them some security.

As my father says, in Lakeba we have our relations, our home, and our land. This represents security and something permanent; it constitutes our roots. For me, relations, home place, and land are part of my identity, hence my being a Lauan. But what of my plans and intentions, those of my children and my husband? These are very difficult to say, depending very much on the perception of our needs and, in turn, on the circumstances in which we find ourselves.

My husband, Ratu Amani Delana Racule, is the second son of the chief of his village: Matameivere, in Namara, Tailevu province (Fig. 7.1). This community is about thirty kilometres from the capital of Suva and easily accessible by road. The village land, totalling only 37.36 hectares, is communally owned by the *yavusa*. Population pressure is already high and more than half the tribe, about 100–150 persons, has moved to urban centres where they can find employment or to other parts of Fiji where they can acquire land. People of Matameivere, for instance, constitute about three quarters of the village of Navaga, in Koro, Lomaiviti province (Fig. 7.1). There is a piece of *kovukovu* land that came to the Racule family through Amani's grandmother, Adi Maca Marama, a lady of high rank from the neighbouring village of Nakorolevu, but the family of my husband's elder brother hold first rights to that land. As a result, Amani and I bought a house in Suva as security. Before this, we organized the building of a house at Matameivere village for the use of the Racule extended family. This was necessary because none of my husband's parents, four brothers, or one sister live there, but need somewhere to stay whenever they and their families visit – as we all do regularly.

Since my husband also teaches secondary school in Suva, our two children are being raised in an urban environment. Our eldest child is a daughter named after her paternal great grandmother, Adi Maca, and second child a son who carries my father's name. Because of our present lifestyle, Maca and Filipe will have a very different upbringing from those of their parents. Amani's father, Ratu Isireli U. Racule, was a teacher and like mine posted to different parts of Viti Levu. Yet Amani was essentially raised within a village environment, for his parents lived in Matameivere while Ratu Isireli taught at the local school and later in the community of Amani's mother where Ratu Isireli also was stationed. By the time his parents transferred to Suva, Amani was in boarding school; his background is thus quite different from mine. In Lakeba, the traditional way of life is still very strong, partly because the island is so far removed from urban influences and partly because the chiefs, most of whom individually own land, provide strong leadership. Matameivere, by contrast, is closer to urban centres (within thirty kilometres of Suva, Korovou, and Nausori: Fig. 7.1), while lack of land there and of strong leadership has led to a startling breakdown in customary adherence to and cohesion of the social unit.

Amani has a strong sense both of obligation to his relatives in Matameivere and of etiquette and thus cannot assume leadership, for this is the responsibility of his father, his elder brother, and his father's first cousins. To maintain contact with kin, he runs an association comprising Matameivere people living in Suva, who meet regularly and raise funds to help the village. In the same way, there is in town a Delai-korolevu association of Levuka people from Lakeba (cf. Nair, this volume). For both my husband and myself, cultural contacts are actively maintained with kin in both Suva and our respective villages of origin.

Whether Maca and Filipe will continue to sustain these connexions to the same extent is difficult to predict. Because of their urban upbringing, a growing shortage of land, and their genealogy on my husband's side, they might choose to make their lives and careers in urban centres. This decision would depend in turn on the level of formal education they reach and what they have been taught to need. On the other hand, because of my genealogy and the fact of being male, Filipe – while registered in his father's tribe – will legally inherit the land at Lakeba that will pass to me. Consequently he will have several choices of where to live. Thus whatever our children decide will depend very much on their own perceptions of their needs and the circumstances in which they find themselves. But what they do will be with the memory of their grandfather, Dr F. B. Vulaono, who died on 1 July 1980.

Acknowledgments

I wish to record my deep gratitude to my father, Filipe Bulivakatawa Vulaono, for his account is of great personal value to his family, apart from its academic interest. My thanks also to Professor Murray Chapman, who initiated the move to put in writing my father's experiences; and to my husband, Amani, for help with preparing this paper.

Editorial note

This chapter is excerpted from a much longer story dictated on tape by Dr F. B. Vulaono and transcribed by his daughter, Rejieli Racule. Sections on tribal mobility and genealogy have been omitted almost entirely. Apart from deletions, necessary through constraints of space, this narrative remains basically as told by Dr F. B. Vulaono. Thanks are due Richard Bedford and John Campbell for advice about treatment of the original manuscript.

Part III
Economy, education, and circulation

8 Weasisi mobility: a committed rural proletariat?

Ron Bastin

Movement for wage labour has been the subject of anthropological enquiry for the past forty-five years. First observed in central and east Africa, the pattern of movement alternated between rural areas and centres of wage opportunities (Elkan 1960; Garbett 1960; Richards 1954; Watson 1958). Originally, this was viewed as a temporary phenomenon which would diminish as commitment to wage labour increased and there emerged a residential urban proletariat totally reliant on cash for subsistence. Despite high rates of urban growth throughout Africa and increasing lengths of stay in town, villages remain a focus for investment and much intellectual effort has been devoted towards explaining the continuing importance of the rural sector (Berg 1965; Elkan 1967; Garbett 1975; Gugler 1969; Mitchell 1959, 1961b).

This process of partial involvement, termed both 'labour circulation' (Mitchell 1961b:232) and 'circular migration' (Elkan 1967:582), has also been documented for the island Pacific. A principle reason advanced to explain the persistent circulation of labour has been the local development of plantation agriculture oriented to export markets. Although certain areas of Africa have proven capable of cash-crop production, the growth of a peasantry has been inhibited by price fluctuations for export crops and increased urban wages. Under conditions of free labour movement this has resulted in rural-urban drift.

Favourable climatic conditions throughout coastal Melanesia have allowed the establishment of coconut plantations, while higher inland regions have proven suitable for growing coffee and cocoa. Studies of labour circulation in Papua New Guinea, for example, have shown that much wage employment has been undertaken to finance the extension of local plantations (Finney 1973; Ogan 1972), although recent research has indicated limitations to potential expansion in this and other sectors (Howlett 1973). The history of Melanesian plantation agriculture has led Howlett (1980) to argue that the recruitment of workers from disadvantaged areas is giving rise to a class of landless labourers, or 'rural proletarians' (Mintz 1974), who have chosen to permanently relocate in their places of employment.

By contrast, the position taken in this paper is that resettlement is not a necessary condition for the emergence of a rural proletariat and that village reliance on wage incomes may be more crucial than is realised. Using evidence from a community in Vanuatu, formerly the New Hebrides (Fig. 4.1), it will be shown that, even among local groups who have cash crops, population increase is outstripping natural resources and basic needs must be met partially from wages earned outside the village. Although the people appear at first sight to be peasants deriving a livelihood from their own land, the hypothesis may be advanced that reliance on outside employment opportunities is such that we can speak of a growing class of committed rural proletarians who retain their village base only through periodic wage labour in town.

This perspective may appear to conflict with recent evidence. In an historical study of niVanuatu labour mobility, Bedford (1973a) documents a transition from long-term contracts to short-term circular mobility. He also notes what appeared to be a growing commitment to urban residence among skilled and semi-skilled workers, although his final comment was that 'New Hebridean migrants are not a wage-dependent proletariat yet; they have vested interests in property and security in the rural areas' (Bedford 1973a:117). In a later investigation of urban residence, Bonnemaison (1976) suggests that the process of movement to the town of Vila (Fig. 4.1) has changed from 'organised' circulation linked to and controlled from the rural areas, to a situation of 'uncontrolled' or spontaneous migration in which individuals are 'cut off from the backing of their community structures [and] show an increasing tendency to form a predominantly young and male urban proletariat, going through a rapid process of acculturation' (Bonnemaison 1976:11).

Bonnemaison attributes most of this movement to groups who are firmly involved in the plantation economy and in the forefront of modernization. Such groups are characteristically found on small or medium-sized islands, where population densities are high and the extension of plantation agriculture has occurred on land customarily used for food gardens. The critical level of population density is set at thirty persons per square kilometre and the acreage under plantations at more than one per head of population (Bonnemaison 1977a:131). Under such conditions traditional bases of cultural cohesion, particularly the collective ownership of land, are increasingly weakened and if rural incomes fall, the prevailing mechanisms of social control are insufficiently strong to prevent large numbers of islanders moving to towns (cf. Bonnemaison, this volume).

High rates of rural outmigration need not be the only response to population pressure and increased need of cash, a conclusion which derives from data gathered in 1976–77 during anthropological field-work on Tanna, in southern Vanuatu (Fig. 4.1). In 1971 the population of this island, some twenty-two kilometres by thirty-two in extent, was estimated at 12,850. Research focused upon the village of Weasisi, which in December 1976 had a *de jure* population of 262. Descent and inheritance in this society is through the father and the rule of residence is patri-virilocal (Bastin 1980).

For Weasisi, despite increased population and a decline in the significance of copra income, circulation remains the rule. Absentees retain a strong rural focus and strategies are geared to creating economic opportunities in the local community such that wage-labour mobility is no longer necessary. The few individuals who have succeeded in this aim are able to employ fellow villagers and there is emerging a system of stratification based on cash-oriented relationships.

Changing economic relationships in Tannese society

The discovery in 1825 of sandalwood in Vanuatu (Fig. 4.1) created a demand for indigenous workers. The southern islands initially provided labour for sandalwood stations, although recruiting soon spread throughout the group. By 1849 the sandalwood market had collapsed and recruits were taken farther afield to meet the increasing requirements of plantation agriculture in Queensland, Fiji, Samoa, and Hawaii (cf. Bathgate and Connell, this volume).

Labour recruitment had a marked effect on Tannese life. Workers demanded muskets as payment and their introduction led to a high degree of internecine warfare, which continued unabated until the 1890s. By this time the disruption of everyday life had reached unacceptable levels and large numbers of christian converts rejected traditional practice in favour of the stability offered by the Presbyterian mission. Christian influence was spread through a system of indigenous government known as Tanna law. Persons infringing local regulations were punished by courts, which were given official recognition by the condominium government in 1909.

The first missionary came to Weasisi Bay in 1882 and under the safety offered by christianity a number of hamlets grouped together to form what is today Weasisi village (Fig. 4.1). This security was conditional on rejection of much traditional behaviour and missionaries attacked those aspects viewed as heathen. Drinking of kava, a mild narcotic root which facilitated communion with ancestors, was banned and participation in pig exchanges discouraged. Missionaries sought to destroy stones associated with the practice of sorcery but informants state that many other stones used in gardening magic were also inadvertently removed.

In addition to undermining the social fabric, christianity created a new system of material requirements. Penis sheaths and grass skirts were replaced by clothes and the desire to own European goods generated a demand for cash. Missionaries consequently promoted copra production as an alternative source of income to the three-year contracts offered by labour recruiters. This policy, designed to prevent the long-term absence of males and to arrest the population decline evident in coastal areas since European contact, was generally successful. By the early twentieth century 'Tanna provided very few recruits, despite the long traditions of Tannese labour migration in the nineteenth century' (Bedford 1973a:35).

The relationship between cash cropping and labour mobility for niVanuatu can be traced from contact (Table 8.1). During periods of high copra prices, as in the 1920s and late 1940s–early 1950s, the production of an indigenous market surplus generally satisfied income requirements and resulted in fewer men going away to work. Conversely, participation in wage labour increased during times of low copra prices: for instance, during the early 1930s. Although the establishment of indigenous cash cropping did not provide sufficient income to eliminate the need to participate in outside employment, the amount of cash locally available was such that long-term labour contracts became unacceptable and by the late 1930s casual employment was preferred.

Overall, the effect of these changing relationships between the local and external economy was that individuals, by engaging in short-term circulation, were able to consider far more the demands of village society. Yet these demands also change as different roles are assumed throughout one's lifetime. The changing nature of wage opportunities and varying demands on the individual throughout the developmental cycle are apparent in case histories, especially of older Weasisi males.

Case 1: Martin (a pseudonym)

Born in 1920, Martin first began wage labour when aged fifteen and worked on a plantation on Efate (Fig. 4.1) along with several older males from Weasisi. After a year he returned to the village for two months before undertaking further employment for six months on the same plantation. After three months back at home, he worked for six months on an inter-island trading ship. At age eighteen Martin returned to marry, at which point his father allocated him land and part of the family copra plantation. He remained for a year but left for wage employment after the death of both his wife and infant in childbirth.

Accompanied by a fellow villager, Martin worked on a ship for six months, followed by another three spent in Weasisi. He next found employment on an Efate plantation, for six months, before leaving for home to marry a second time. After six months with his wife, he again

TABLE 8.1 Wage-labour opportunities in Vanuatu, precontact to 1972

Pre-contact period	Limited mobility due to high degree of internecine warfare.
1840s	Arrival of labour recruiters. Destinations initially islands of Vanuatu, but changed to Queensland and other Pacific islands.
1880–1920s	Increased number of European settlers establishing plantations and greater demand for indigenous labour. Three years the basic length of contract favoured by European planters to ensure stability of labour force.
1920s	Steady decline in number of contracts signed. Contracts of one year or less favoured by niVanuatu and a marked reluctance to sign three-year contracts. High copra prices.
Early 1930s	Increase in labour offered as a result of low prices for cash crops.
Late 1930s	Growth of casual employment led to niVanuatu being able to gear periods of employment to village requirements.
1942	Establishment of military bases on Efate and Espiritu Santo. High wages paid to indigenous workers and considerable fringe benefits.
Late 1940s-early 1950s	High copra prices. Demand for labour led to 150% increase in wages offered but niVanuatu retain preference for six-month contract.
Late 1950s–1968	Rapid expansion of Vila and Santo offered opportunities in construction industry. Growth of educational and medical services throughout the group.
1968–1972	Expansion of nickel industry in New Caledonia offered high wages for short-term contracts (3–6 months).

Source: Bedford (1973a:21–56).

returned to plantation work on Efate which, after only ninety days, was interrupted by the outbreak of the Second World War.

Between 1942 and 1945 Martin was employed as one of a local crew of deck labourers on a supply ship for the Allied Forces. Wages were high compared with previous levels and Martin claims to have earned £Australian 10–14 a month. After the war ended he continued for three months as a deck hand at a monthly wage of £A16. Following a year and a half at Weasisi, Martin became a general labourer in Vila for £A2 10s a month. There was next a long stay of two years in Weasisi, a further six months of plantation work, and residence in the village for another year. Martin then went to Efate with two fellow villagers securing employment with a Vietnamese plantation owner whose monthly rate was £A10. This was far more than that paid by European employers, but Martin stayed only six months before returning to Weasisi for the express purpose of making new food gardens.

By now, Martin had five children and decided to remain awhile with his family. Periods in the village of two and a half and two years were interspersed by earning £A40 from four months' work on a ship. The need to raise school fees for his eldest son forced him back into wage labour and he again worked on board ship for three months at £A10 a month. Thus was established a pattern: periods of one year and six months spent in Weasisi both followed by six months at sea. On this particular return, in 1964, Martin secured work as the labourer responsible for maintaining the newly-bulldozed track which linked the bay with the main road. He held this position for ten years before deciding to retire and transfer it to Donald, a fellow villager.

Martin's employment history is representative of wage-labour mobility among older Weasisi men. The pattern of movement is circular, with the length of absence altering in response to different social roles and changing economic needs. While single, more time was spent in wage labour than residence in the village. Marriage represented a significant change in status to household head and Martin received his own land and coconut plantation in keeping with this. He stated that, with marriage, he accepted he must stay longer in Weasisi to look after his family and help with gardening activities. Following the death of his first wife, the pattern of circulation changed to one of six months in the local community and six in outside employment.

The outbreak of war in the Pacific led to an increase in indigenous incomes. Because of Weasisi's location, villagers were able to undertake well-paid employment with little disruption of social life. The vessel on which most men worked collected them while en route from Australia to Vila (Fig. 4.1). They laboured for two weeks at unloading and were able to disembark at Weasisi on the ship's return voyage.

After the war the elapsed time Martin spent in wage labour fell from half to just over one quarter. From a total of 9.3 years, only 2.3 were spent outside Weasisi, and he expressly stated that he wanted to spend more time with his growing family. Despite this desire, periodic excursions into outside wage labour document the inadequacy of local cash cropping. Marriage gave Martin rights to land which, although yielding income from copra, was only sufficient to meet family needs at times of high copra prices. Beyond this, cash had to be earned in the wider economy. After the war Martin undertook wage labour solely to meet such specific requirements as to raise school fees for his son. Finally, the opportunity to earn money locally allowed him to remain permanently in Weasisi.

Possessing no saleable skills, Martin had to take whatever opportunities arose. Although frequently working beyond Tanna, the continuing pattern of circulation shows that wage income played a subsidary role to rural interests. Between marriage and the Second World War, this alternating sequence of six months in wage labour and six in the village complemented the agricultural cycle, which requires high labour inputs for only half of each year. The importance of rural interests is further underlined by his return, within six months, to establish new food gardens when earning above-average wages from the Vietnamese planter.

Much social change has occurred throughout Vanuatu since 1945. Increased access to formal education and greater involvement in the external economic sphere have allowed the younger generation of Weasisi males a much wider opportunity structure, as the following two case studies will show.

Case 2: Donald

Born in 1943, Donald was initially educated at the village school and then was selected to attend the primary school run by the Presbyterian mission. When aged eighteen he ran away to sea with a friend and worked for three months as dishwasher, before being promoted to operate the ship's launch. Employed in this capacity for two years at a monthly rate of £A25, Donald replaced his departing friend in the engine room where he stayed for a further two years. This experience finally earned him promotion to captain at a wage of £A30 a month, a position held for twelve months before the owner sold the vessel.

Donald then obtained employment for a year on another trading ship, as quarter-master, after which he returned to Weasisi and married at the age of twenty-four. He went back to sea after six months, after only a short stay in the village, again working as a quarter-master. Word arrived that his mother was sick and he left for Weasisi at Christmas. Three months later she died, leaving him compelled to assume the responsibility for both his wife and father.

Donald estimates to have earned $A2,500 during seven years (1961-68) but about $A2,000 was spent on alcohol, luxuries, and generally enjoying himself. Returning to Weasisi, he invested $A400 ($A2.16 to £1.00) of his savings in a small dinghy and an outboard engine. This he used for fishing expeditions and to transport copra from outlying villages to the local co-operative store, in which he also invested $A20 to buy a share. For the past two years (1975-1976), Donald has held the position of road repairer in succession to Martin.

This case history shows how a wider opportunity structure amongst the younger generation in Weasisi results from increased length of absence. Unlike Martin, who gave precedence to rural interests, Donald was away for several years and, despite marriage, returned only when compelled by family circumstances. This continuous period in wage employment beyond Tanna allowed the acquisition of skills, which both yielded a good salary and provided a position of responsibility and prestige.

The accumulated capital enabled Donald to acquire income-producing assets and thus to improve his economic standing, so that he no longer had to sell his labour in the external economy. Hiring out his boat together with the job as road repairer, provides Donald a high income by Weasisi standards, as well as a position of influence. His multiple economic activities sometimes enable him to hire fellow villagers and such employment − although infrequent − is much sought and his resultant patronage valued. The growing importance of access to cash being based on patron/client relationships is further demonstrated in Donald's becoming road repairer. A number of Weasisi men, including Martin, were obligated to Donald because of employment given on his ship. When Martin retired, Donald exercised these obligations and obtained the position. Since these two men belong to different lineages and are not directly linked by marriage, this job transfer was not subject to established inheritance rules but was given in return for past employment opportunities.

Increased participation in the wider economic sphere can improve the wage labourer's position by both providing savings for personal enterprises and creating cash-based obligations, which in turn may be used to advance one's local status. Participation in labour circulation is not, however, the sole route to economic advancement. Formal education has brought an increasing awareness of European systems and offers the prospect of creating local economic opportunities without the need for outside wage employment.

Case 3: Harry

Born in 1939, Harry was the second son of a Presbyterian church elder. Three years were spent at the Presbyterian primary school on Tanna and another three at the Theological Training Institute on the island of Tongoa, near Espiritu Santo (Fig. 4.1). Upon completion, he was appointed village school teacher in Weasisi at an annual salary of $A60. Harry taught there for six years and for another year on the island of Futuna (Fig. 4.1). Married in 1962, three years later he was appointed teacher at the local primary school run by the British administration.

This position he resigned in 1968 and took his family to Vila, obtaining work as a cargo checker at a daily wage of $A2.60 ($A2.16 to £1.00), a considerable increase over his teaching salary. However, after only four months in town he acceded to a request to return to Weasisi and become church elder. In 1971 he initiated the Kasali Co-operative Society, which was launched under the guidance of the New Hebrides Co-operative Federation. The capital for a store was raised from forty-five shares of $A20 each ($A2.16 to £1.00) and represented the end of a long battle to establish co-operative enterprise locally following the failure of several earlier business ventures. Harry, originally chairman of the co-operative committee, became secretary in 1972 after several persons proved unable to cope with the bookkeeping and other procedures.

This personal history documents how formal education may create economic opportunities in the village. All of Harry's education was received from the Presbyterian mission and he was eventually offered a position in the church hierarchy. He attached sufficient importance to this to leave a well-paid job in town and was able, as religious leader, to institute collective enterprise. Despite little practical experience with trading, Harry's education was employed to explain to villagers the benefits of a co-operative store and convince sufficient of them to raise the necessary capital. This success not only increased his prestige and income but also contributed to the economic development of Weasisi (Bastin 1981).

The strategies adopted by Harry and Donald indicate the general trend of economic and social change in Weasisi. The existence of a cash nexus enables them to establish local enterprises and earn incomes sufficient to meet their needs. Yet Harry and Donald represent the exception rather than the rule. Most villagers are involved in wage-labour circulation, as was Martin, although with certain important differences. Social and economic pressures have caused an increase in both the incidence and duration of outside employment and to this taking precedence over rural interests. Moreover, these pressures change throughout adulthood.

The developmental cycle

Mitchell (1959) has argued that a migrant worker is subject to conflicting influences throughout a lifetime which both draw him away from the village (centrifugal tendencies) and back again (centripetal tendencies). This idea can be used to summarize the pressures which operate on the average villager. Before marriage, a Weasisi youth has few social responsibilities and during this period is able to see the outside world and indulge in the pleasures of town (Table 8.2). Since marriage involves no brideprice, a specific level of saving is not imperative and young men may spend their earnings as they wish. Cash needs increase with marriage and the birth of children but, although the social pressures are to remain in the village, most men continue as wage labourers.

TABLE 8.2 Paradigm of wage-labour circulation for Weasisi males

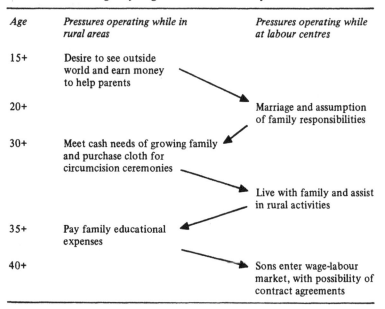

Age	Pressures operating while in rural areas	Pressures operating while at labour centres
15+	Desire to see outside world and earn money to help parents	
20+		Marriage and assumption of family responsibilities
30+	Meet cash needs of growing family and purchase cloth for circumcision ceremonies	
		Live with family and assist in rural activities
35+	Pay family educational expenses	
40+		Sons enter wage-labour market, with possibility of contract agreements

In addition to a growing family to feed, 'calico' (cloth) must be purchased for presentation at lineage exchange ceremonies. Boys, circumcised between the ages of six and nine, undergo a two or three week seclusion until the operation scars have healed. Male members of the mother's lineage are responsible for their supervision during seclusion and fathers make large presentations by way of payment. Traditionally, such an exchange was a minor component of the prestation cycle between groups who had exchanged women (*nierle*). Presentations were made at the birth of all children, being larger to mark that of the first male, and continued through such manifestations of physical development as growth of hair and walking until circumcision or first menses.

Circumcision presentations are no longer limited to the boy's father and matrikin but, because of greatly increased size, may involve all members of both lineages. Customarily, only baskets of root crops, pandanus baskets and mats were given, along with a large stump of kava. If the father felt especially generous, a pig would also be contributed; small puddings were cooked by both sets of relatives to feed spectators. Nowadays, by contrast, a pyramid is constructed of bags of root crops and bedecked with much 'calico' and one or two blankets. Four or five stumps of kava are given away and two pigs normally sacrificed, in addition to one or two cooked before the ceremony to feed guests.

The introduction of 'calico' into circumcision presentations occurred after the second world war and is indicative of the greater use of money. At the ceremony, the receiving side makes note of all items given other than root crops since repayment will occur at the next one. The enlarged size of presentations has increased the community's need of cash because one man can seldom meet all costs and male agnates must assist, otherwise the lineage loses prestige. Economic pressures involved in prestations also extend beyond the purchase of 'calico'. Rising demand for garden land means that the number of pigs kept has declined and they must sometimes be purchased outside Weasisi.

When these demands are related to the development cycle, the continuing pressure of social obligations becomes clear. Involvement of more and more paternal kin in ceremonial exchanges has given rise to greater demand for cash and the desire for a constant monetary income to meet increased obligations. For most men, the only means of obtaining such income is to become wage labourers until their sons are sufficiently old to undertake outside employment and subsequently send remittances.

Discussion thus far has been concerned with the notion of choice and indicated how the opportunity structure for individuals has altered, especially since 1945. The developmental paradigm has illustrated the changing economic pressures on a person throughout his lifetime, but only in general terms and without consideration of differing propensities to move. The focus now becomes the effects of labour circulation on village life. Having argued that changing social needs have resulted in a greater incidence of circular mobility, its degree and the current importance of the rural base remain to be evaluated.

Contemporary wage-labour movement

In the decade since the late 'sixties, Weasisi has witnessed an increase in both population and the need for cash. Comparison of the *de facto* base population in 1967 with that ten years later reveals an annual growth rate of slightly more than one per cent (Table 8.3). In contrast with females, the growth rate for males is far less, which provides a crude index of the number absent in wage employment (cf. Fig. 8.1).

Rising population has resulted in several changes in agricultural practices. Informants state that the fallow period has declined from approximately eight to three or four years, with the same plot being used for two yearly cycles instead of one. A cumulative loss in soil fertility has resulted and there has been a shift from growing yams to the greater-yielding crops of sweet potato and cassava. Increased use of land has also changed labour inputs: the absence of secondary growth means it is less arduous to clear gardens, so that women can now do much of the work.

TABLE 8.3 De facto *population growth, Weasisi, 1967–77*

	1967	*1977*	*Annual growth 1967–77 (%)*
Total population[a]	162	183	1.2
Index	100.0	113.0	
Male population	86	91	0.6
Index	100.0	105.8	
Female population	76	92	1.9
Index	100.0	121.1	

[a]The 1967 census of Vanuatu (McArthur and Yaxley 1968) was a *de facto* count. For this reason, the total population reported for Weasisi excluded those who were usually resident but had temporarily relocated at census date for ideological reasons. To permit comparison, the 1977 field data (*de jure* count) thus exclude both immigrants (1967–77) and families who relocated in 1967 but subsequently returned to Weasisi together with their descendants. Thanks are due Dr. Norma McArthur, Australian National University, for naking the 1967 data available.

A similar situation has occurred with copra production. Fears of land shortage have led to only one small plantation of 125 trees being planted since 1945 and existing groves are left to regenerate. Nuts are collected when sufficient numbers have fallen to provide a viable return, their flesh gouged out and smoked on a raised bamboo platform to prevent deterioration. The copra is then bagged and carried to the co-operative store where an on-site price is paid. Any further return from overseas sales becomes part of an annual dividend to shareholders.

Nowadays most tasks in processing copra have been assumed by women even though, apart from nut collection, this was traditionally a male occupation. Men may still produce copra but, except for one man being needed to split the nuts with an axe and maintain the bamboo platform, women may shell, smoke, and carry the harvest. This increasing involvement of women does not, however, represent a significant change in their social position. Increasing population and diminished agricultural productivity have meant growing reliance on European foodstuffs, with tinned fish and rice now an accepted part of the household diet. Yet copra does not provide the regular income necessary to meet such requirements: between August 1976 and September 1977, for example, the return on copra fluctuated between nine and twenty cents a kilogram. As a result, men increasingly seek wage labour opportunities in town.

From September 1976 until August 1977, a record was kept of all wage-labour movement between Weasisi and Vila, the major urban centre (Fig. 4.1), data from which indicate the diminished importance of males in agricultural tasks. Two thirds of all movement occurred between December and February, the time of Christmas festivities (Table 8.4). Many absentees schedule their visits to coincide with these celebrations before leaving again for town, although their stay in Weasisi may extend into early March if sea transport is not available.

An important feature is that labour movement is not geared to agricultural requirements (cf. Chapman 1975). Christmas occurs during the rainy season, November to April, the months of harvest and the slack period. The time of intense, often arduous garden work begins in late April/early May with the preparation of sweet potato gardens and ends after the planting of yam tubers in October. Rather than a net return of villagers before the season of planting, there is in fact a net outflow with two men and a woman coming back and seven men and four women leaving between the months of December and May.

The age-sex structure of the *de jure* population in 1977 also reveals the extent of labour mobility (Fig. 8.1). Absent males account for slightly more than thirty per cent of the total male workforce (those aged between fifteen and sixty), while absent females comprise about fifteen per cent of those of working age. Many wage earners are less than thirty years old and therefore likely to be single or recently married. A person's marital status can be indicative when examining reasons for movement. Earlier discussion has revealed how important are remittances in relieving older men of having to take up outside employment and most single men cite the desire to help parents as their reason for entering wage labour. Wanting

TABLE 8.4 Timing of labour circulation for Weasisi village, September 1976–August 1977

Reason for move	Sep–Nov 1976	Dec 1976–Feb 1977	Mar–May 1977	Jun–Aug 1977
Male				
Visit home	–	4	–	–
Return from wage labour	–	2	–	–
Visit Vila	1	–	3	–
Undertake wage labour	–	5	2	1
Female				
Visit home	–	2	–	–
Return from wage labour	–	1	–	–
Visit Vila	–	3	–	–
Undertake wage labour	–	2	2	–
Total moves	1	19	7	1

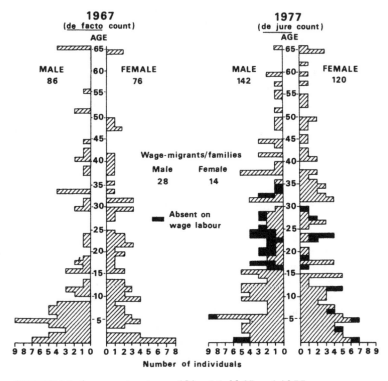

FIGURE 8.1 *Age-sex structure of Weasisi, 1967 and 1977*

to see the town and experience a different lifestyle are also influential.

Married men face a rather different situation. Gaining a wife no longer causes a distinct break in the pattern of labour circulation and young husbands may still be away for several years at a time. The proposed duration of absence affects, however, the arrangements made for dependants. If the husband intends to be away for less than a year, then his family can be left in care of immediate kin and this service will be repaid with gifts upon return. If the period absent is likely to be longer, then his family will probably join him in town and a 'contract' (agreement) will be made with someone to look after his village interests.

Use of the term 'contract' denotes a conceptual distinction between traditional, kin-defined obligations and cash-based transactions. The contractor agrees to pay a pre-arranged sum for a specified task and is under no further obligation after this particular contract has been fulfilled. This latter condition is especially important in safeguarding a family's rights to land. The entire nuclear unit may be away for several years, a prolonged period which may result in land rights being disputed. By making 'contract' payments, the absent villager keeps active his rural interests, as well as having no obligation to grant usufruct rights upon return — as would have been the case with traditional principles of reciprocity.

'Contract' agreements also operate within the village and cover a multitude of tasks, such as working in the co-operative store, cutting or carrying copra, and assisting in the clearing of land for new food gardens. Villagers are reluctant to discuss 'contract' transactions, because they are seen in opposition to established values of reciprocity, and thus no systematic account of their extent is possible. None the less, wealthier villagers like Donald and Harry figured significantly as 'contract' employers, generally of older men. The importance of such employment opportunities becomes apparent when related to the developmental cycle. As a man ages he is increasingly unable to compete with younger, fitter individuals entering the urban labour market. His rural income may also decline if he has married sons, because the family estate will have been subdivided and his copra holdings reduced accordingly. Yet cash needs continue and, although such men may receive remittances and other assistance from their children, the 'contract' can provide an important local source of income.

Conclusion

Wage-labour circulation may be seen as a process of movement subject to social and economic change. Population growth has placed constraints on the expansion of coconut plantations and led to the reduced importance of copra, while intensified use of land has seen women assume greater responsibility for gardening and copra production. This has both resulted in and permitted higher incidences of labour circulation by males and in their spending greater lengths of time outside the village to compensate for declining rural incomes.

Despite increased participation in wage employment beyond Tanna, villagers retain a rural focus. The existence of the 'contract' shows that long-term movers make periodic payments to protect their local interests and that these now constitute an important source of income. Within Weasisi, 'contract' payments indicate emergent patron/client relationships because richer villagers control access to employment opportunities and their patronage is actively sought by older men no longer able to compete in the urban market.

Although the rural economy has proven increasingly unable to meet subsistence requirements, there has been no rural exodus of the kind Bonnemaison (this volume) outlines for several northern islands of Vanuatu. Instead, adult males seek to establish business ventures in the village which relieve them of the necessity to circulate for wage employment. That certain have succeeded in this objective does not alter the overall economic pressures. No longer can villagers be described as peasants, because cash cropping does not meet their income needs, while changing social relationships document that men must increasingly sell their labour locally in order to ensure a basic minimum income.

The strategies adopted by Weasisi villagers indicate a high degree of commitment to wage labour. Increasing lengths of absence, the growth of monetary payments for tasks formerly undertaken as mutual assistance, and the desire to initiate business enterprises all underline their reliance upon cash incomes. Yet the people do not seek permanent urban residence. Instead they wish to retain the security of the local community even though this entails longer periods of absence from it. In many ways this situation indicates a greater degree of commitment than found amongst rural proletarians in other areas of Melanesia. Among the Waghi of the New Guinea highlands, for example, persons who have migrated see benefits to be gained from voluntary relocation (Heaney 1977:5). Weasisi villagers, by contrast, believe there is no advantage in urban settlement and aim to compensate for population growth by purchasing imported foodstuffs. But they are able to do this only at the expense of traditional modes of assistance and by becoming, in Mintz's term, rural proletarians.

Acknowledgment

This paper is based on doctoral research conducted between August 1976 and September 1977 under the sponsorship of the Social Science Research Council, Great Britain. I thank all my friends in Weasisi who made this study possible.

9 Circularity within migration: the experience of Simbu and New Irelanders, Papua New Guinea

Elspeth Young

Movement is the main reason for the redistribution of people. In terms of population policy, it is important to distinguish between transfers which involve a change of residence (migration) and those whereby individuals eventually return to their place of origin (circulation). This distinction, however, is not always clear. Persons who have been away from their origin community sufficiently long to be considered migrants may eventually return, while others who left intending to go back may find unexpected opportunities and become permanently absent. In practice, it is only with the death of the mover that the existence of circulation or migration can be proven. Residence histories can reveal to what extent movers have circulated before the time of interview and thus indicate varying degrees of permanency in their absence.

Descriptive typologies recognise the problem of distinguishing between circulation and migration, by incorporating both within a total system of population mobility. Such an approach, while useful, may be misleading because each type is presented as a separate entity. In fact, people may participate in different forms of movement during their lifetimes and even combine them: as, for example, establish permanent residence elsewhere and make short-term visits to their previous domicile. In this paper, based upon village studies undertaken during 1974–76 in the province of Simbu (formerly Chimbu) and New Ireland, Papua New Guinea (Fig. 2.1), the focus is on moves that were perceived by the person involved to be a change of residence.

Migration research in Papua New Guinea

Existing studies suggest that internal migration, essentially from rural origins, has been strongly circular. Whether this circularity persists from precontact days is unclear, because most research excludes traditional forms of mobility (cf. Watson this volume). Customarily, migration resulted from marriage, warfare, famine, or natural disasters. The significance of migration for marriage, the main type for which there is direct evidence, varied according to the social organization of the particular group. Most Simbu women, for example, leave their birthplaces on marriage because the society is patrilineal and settlements consist of people from the same exogamous subclan. In New Ireland villages, by contrast, several subclans are found within one settlement and the social structure is matrilineal, so that men sometimes migrate upon marriage.

Different data sources suggest that more contemporary migration in Papua New Guinea has been mainly circular. In the studies by Ryan (1968), Oram (1968a, 1968b), Salisbury and Salisbury (1970), Baxter (1973), and Strathern (1972, 1975), interviews were conducted at places of both origin and current residence, thus permitting direct analysis of the amount of circularity. They also contain statements by respondents who had departed intending to return and who, while absent, maintained contact with kin through visits and exchange of gifts. Other research, as by Whiteman (1973) and Rew (1974), is based entirely on interviews at migrant destinations and provides only indirect evidence of circularity, despite the strong links which most absentees maintain with their natal communities. Conversely, my earlier fieldwork among the Agarabi and Gadsup, Eastern Highlands Province (Fig. 2.2), considers the migration experience of village residents so that, by definition, all previous movers had circulated (Young 1973, 1977a). The strong evidence these studies present for circulation does not mean that more permanent forms of movement did not also occur. Often the relative importance of circulation and migration was not investigated and such assessment would require a broader research design.

All the results discussed thus far are from small-scale enquiries among tribal groups of varied regional origin and socioeconomic characteristics. Consequently they are not strictly comparable: the migration experience of the Hula, a quite sophisticated group of coastal people (Oram 1968a, 1968b), is likely to differ from that of the Siane (Salisbury and Salisbury 1970) or Hageners (Strathern 1972, 1975), both highland societies with less than forty years' contact with Europeans. Problems of comparison with evidence derived from local research can be overcome by using national or regional data. Large-scale studies of migration in Papua New Guinea are still few and, because of problems of data collection, contain many inaccuracies. Prior to the first national census in 1966, Brookfield's (1960) study of a sample of contract labourers in 1957 was the only analysis available of migration streams. Ward (1971) subsequently examined a comparable sample of labourers in 1967, together with the major flows revealed by the 1966 census. Neither study provides direct evidence of the circularity within migration, being based on cross-sectional data and lacking historical depth.

Most contract workers, however, were assumed to have circulated through being repatriated under the terms of their work agreements. The 1971 population census also fails to indicate the degree of circularity in migration, despite the inclusion of an additional question on length of residence at place of enumeration. During the inter-censal period, 1966–71, the rate of population turnover was extremely high, which Skeldon (1977b) assumed was due mainly to circular movement, especially between rural villages, towns, and such rural non-village settlements as schools, missions, and plantations.

All national data available until the early seventies were collected by official government bodies. In 1973–74, a joint team from the Australian National University, Canberra, and the University of Papua New Guinea, Port Moresby, conducted a sample study, the Urban Household Survey, of the characteristics and experiences of people resident in seventeen urban centres (see Conroy/Curtain, this volume). Unlike the inhabitants of rural villages, most of those questioned were migrants and were born elsewhere. Many migrants had lived in the longer-established towns for ten or more years (Port Moresby 29 per cent, Daru 23 per cent, Rabaul 30 per cent, Madang 26 per cent (Fig. 2.1); Garnaut, Wright and Curtain 1977:60). In the seven major towns, more than half were accompanied by their wives. Some movement had therefore resulted in quite permanent urban settlement, with little intention to leave, even though many migrants said they had recently visited their previous village homes and had sent and received gifts, money, and letters. These results show that urban dwellers maintained close contact with their natal communities but do not establish the degree of circularity in migration and constitute an unsatisfactory basis from which to argue that urban dwellers are 'still integrated into the peasant economy' (Curtain 1980b:54).

Thus far government organizations have accepted results from both micro and macro studies as proof of the dominance of circular flows and have continued to plan mainly for an urban population which will not remain permanently in town (Papua New Guinea, Central Planning Office 1976:37; Papua New Guinea, National Planning Office 1977:1). In the process, indicators of long-term migration apparent from micro-studies undertaken by Dakeyne (1967:155), Ryan (1968:62–65), and Oram (1968a:265) have been ignored, along with warnings that the 'urban population is much more permanently committed to town than is commonly supposed' (Ward 1971:102). Problems of evidence and contradictory interpretation can be alleviated by the use of residence history data, analyzed to document whether circularity existed in the past and whether, in more recent times, it has become more or less intense. It is to such data that this paper now turns, of nine villages located in the provinces of Simbu and New Ireland.

Research methodology

Simbu and New Ireland were chosen for study because of the way their several contrasts might be expected to affect migration. The former province (Fig. 9.1), a rugged land centrally located within the highlands, has rural population densities five times the national average. Lack of cultivable land restricts the spread of cash cropping for coffee and, since the provincial towns of Kundiawa and Kerowagi are small, there are few local economic opportunities. Simbu is accessible to other parts of the central highlands and by highway to the coastal city of Lae

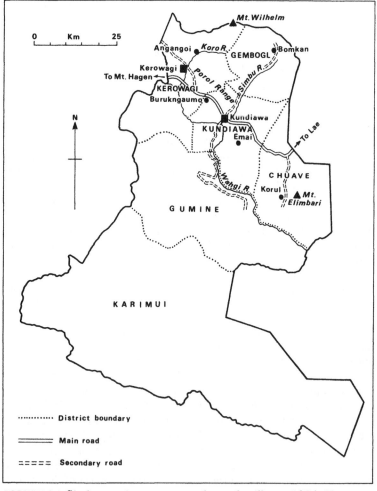

FIGURE 9.1 *Simbu province: towns and sample villages, 1974–75*

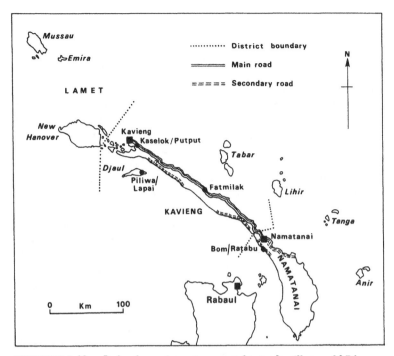

FIGURE 9.2 *New Ireland province: towns and sample villages, 1976*

(Fig. 2.1). Ease of access, coupled with high population density and a low level of economic growth, has encouraged outmigration and in 1971 more than twelve per cent of those born in Simbu were living in other provinces compared with the national average of 7.3 per cent. New Ireland, by contrast, is an archipelagic province consisting of two large and many small islands relatively isolated from the mainland of Papua New Guinea (Fig. 9.2). With population densities close to the national average, New Ireland theoretically has no land shortage, although the fertile coastal plains of the east are intensively used and large areas have been alienated for commercial plantations. Most New Irelanders have established substantial groves of coconut, with cocoa in favoured areas. Although they may earn cash by working on nearby plantations, few take up the opportunity. Rather, being semi-literate, they prefer to rely on cash cropping or, when qualified, to accept one of the limited number of skilled jobs available within their province. New Irelanders feel less need than Simbu to migrate and in 1971 only eight per cent were enumerated outside their province of birth.

While the 1971 census revealed the differential rate of outmigration from Simbu and New Ireland, previous studies contain few hints about the possibility of those flows being circular. Both Brookfield (1960) and Ward (1971) identify Simbu men as major participants in the Highlands Labour Scheme organized by the government between 1949 and 1974. An unskilled workforce of highlanders was channelled to commercial plantations on the coast and recruits signed contracts of between one and two years, upon completion of which they were repatriated to their home villages (West 1958). Since the early 1960s Simbu have preferred more casual and independent employment, as knowledge of such opportunities became more widely spread and official restrictions were gradually relaxed on the movement of indigenous workers to towns and between provinces (Rowley 1965: Chapter 8; Wolfers 1975). It is therefore possible that the degree of circularity has declined although, as Salisbury and Salisbury (1970) have shown, the links of Simbu with their tribal origins have remained strong (cf. Connell this volume).

Previous evidence on New Ireland migration is even more limited. In the latter part of the nineteenth and early twentieth centuries when New Ireland was part of German New Guinea, Firth (1973:48, 108) reports that many New Irelanders worked as plantation labourers in Samoa, New Britain, and the New Guinea mainland (Fig. 2.1). Although some never returned most were, as their Simbu counterparts at a later date, repatriated at the end of their contracts. Department of Labour records show that the participation of New Irelanders in work contracts declined from the first world war and had virtually ceased by 1940, even though the unrestricted flow of labour did not occur before the late sixties. Evidence of New Ireland migration since 1945 is slight, but increased availability of formal education and therefore marketable skills probably meant a marked rise in the amount of time individuals remained away from the local community.

Constraints of time and resources limited fieldwork to nine villages: five in Simbu and four in New Ireland. Chosen purposively, these varied in their accessibility to the main provincial town and those of other provinces (Figs. 9.1, 9.2). Access, measured by the cost of reaching town by public transport, provided a basis for examining intra-provinci. variations in the migration process and permitted assessment of the relationship between distance and migration, which some research had indicated to be rather weak (Lea and Weinand 1971:133; Strathern 1972:20; Young 1973:57-9).

The term village, suggesting a nucleated cluster of dwellings, is somewhat misleading since most settlements were partly dispersed. Three Simbu villages contained members of at least two exogamous subclans, whose men's houses were located in the ceremonial grounds (Bomkan, Emai, Angangoi); the remaining two belonged to single subclans (Korul, Burukngaumo: Fig. 9.1). Three New Ireland villages (Kaselok/Putput, Bom/Ratabu, Piliwa/Lapai) comprised a pair of adjacent settlements because one would have been too small for an adequate sample (Fig. 9.2).

Such structural variety was unimportant since residents of all nine villages were similarly influenced by the same factors – access, economic opportunity, social change – that governed their original selection.

During a month's stay in each village, attempts were made to interview all adults aged fifteen or more years, although some were missed because of the frequency of short-term absences. Sample villages ranged in size from 120 to 200 in densely populated Simbu and from 90 to 120 in less tightly settled New Ireland. Most of the interview involved construction of a detailed migration history. For each move perceived to be a residential shift, questions were asked about not only its timing, destination, and current age of the person involved, but also the mover's employment, education, accompanying dependants, and reasons for leaving the village. Other details recorded for each respondent included socio-demographic characteristics, information about close relatives who were absent, and the most recent visits made to selected towns. The last-named data, not discussed in this paper, demonstrated how villagers gain extensive experience of urban centres without becoming migrants as well as some of the local effects of more ephemeral mobility.

Inevitably, migration histories collected retrospectively are affected by recall, but for several reasons the amount of error in this study is quite low. First, interviews concentrated on eliciting a migration history by itself rather than as part of a lengthy questionnaire. Secondly, experiences were related as a story with little intervention from either interviewer or interpreter. In most cases, events were told in chronological order and their resultant details commented upon by interested bystanders. Such comments provided a valuable cross-check, especially in Simbu, where many movements in the 1950s had involved groups of men. Interviews, thirdly, took place when and where people chose, mainly in the evening at their homes or in such collective dwellings as men's houses. At this time of day, when work has ceased, gossiping and story-telling is the normal activity, so that interviews were not disruptive. Finally, being resident in the village, I was able to gain deeper insights into local social relationships and the experiences of villagers than would an interviewer living elsewhere. Consequently it was possible to adopt a consistent approach to questions, discussion and, later, the interpretation and analysis of field data.

Circulation and migration are, as already noted, not discrete types but rather different states within the mobility process. In this study an index of circularity was developed to indicate differences of intensity through time, as follows:

$$\frac{\text{Number of return moves to the village} \times 2}{\text{Total number of moves} } \times 100$$

This index gives values ranging from 100 for maximum circularity, where every outward move is followed by a return to the village, to zero, where no return move has thus far occurred. It can be calculated for either a single individual or a group sharing a common characteristic. While such a characteristic could be socio-economic (education level, estimated annual income) or demographic (age, sex), time is the one used most frequently here so as to indicate changes in level of circularity.

Age is an unsatisfactory surrogate for time, because migrants who leave the village at any one moment and presumably experience similar pressures, may belong to different age groups. As a result, migrants have been classified according to first year of departure and those who left within a given period constitute a migration cohort. An individual A, who makes two outward moves and returns to reside in the same village after each, has a circularity index of 100.0 (Fig. 9.3). A migration cohort B includes ten men, totalling thirty moves of which thirteen constitute a return, for a circularity index of 86.7. There are six men in another migration cohort C, with six return moves out of a total of twenty-five and a circularity index of 48.0. In these examples, if migration cohort B included men who first departed 1950–54 and C other men from the same village who did the same 1960–64, then these indices suggest that circularity has diminished in recent times.

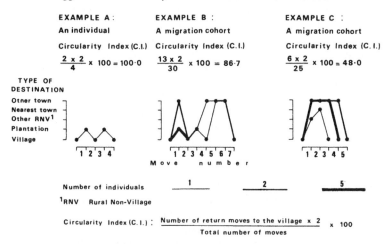

FIGURE 9.3 *Hypothetical examples of population movement*

Such differing intensities can also be illustrated graphically (Fig. 9.3), when A directly departs and returns to the village whereas cohorts B and C each indulge in some intervening movement. Types of migrant destination can be classified as levels within a settlement hierarchy (village, plantation, other rural non-village such as mission or school, nearest town, other town) to standardize the territory covered by individuals who comprise each migration cohort and to reveal if the circularity is more intense when the major destinations have employment opportunities that are limited (plantations) or varied (towns).

Finally, levels of circularity can be compared in the degree of choice which individuals exercise over their mobility. This is closely related to employment, here grouped into either contract or independent. Contract moves include all those made under signed agreement with officially-organised schemes of migrant labour. All others are termed independent, to some extent a misnomer because many employees may be involuntarily transferred as part of their duties: for instance, those who work for government or for private companies like banks. A further example of directed mobility is to attend institutions of secondary and tertiary education. In the course of any lifetime, contract and independent moves need not be mutually exclusive, as is evident from Tables 9.2 and 9.4.

The degree of circularity from Simbu villages

The migration of Simbu villagers is intensely circular. During their adult lives, the index of circularity for men was 86 and for women 51 (Table 9.1). This difference reflects the dominance of male employment in the Highlands Labour Scheme and other kinds of wage work, where transport was provided both to and from the destination and repatriation largely enforced. Women rarely migrated to enter the labour force but rather as dependents of their husbands, and usually only if their spouses held a permanent position like policeman or medical worker, which guaranteed family accommodation. There was less circularity in the migration of such employees, however, because transfer by the employing authority was frequent. Hence Simbu women had a lower index of circularity than men.

Korul village (Fig. 9.1) males recorded the highest overall index (91) and those from Bomkan the lowest (75), a difference not explicable solely in terms of access, economic growth, or social change. Nevertheless, the Bomkan men were the better educated and thus more qualified for permanent employment, which facilitated an economic independence beyond the local community. Conversely, Korul men had worked mostly on plantations under the Highlands Labour Scheme or held casual jobs in towns. The movement of recent school-leavers, the best educated group in every village, was markedly circular and did not follow the expected relationship between highest levels of education and lower indices of circularity. This is because, as recent migrants, they had made only one or two tentative journeys from the village and also because the few jobs available for the partially trained was discouraging school-leavers from more permanent absence.

The migration cohorts of the 1960s reveal a decline in circularity from Simbu villages (Table 9.1). Unlike their predecessors, not all men

TABLE 9.1 Indices of circularity, by sex, for Simbu migration cohorts

| | Before 1946 | Migration cohorts | | | | | | Total | Number in sample |
		1946-49	1950-54	1955-59	1960-64	1965-69	After 1970		
Males									
Burukngaumo	89	100	91	69	72	87	90	83	47
Bomkan	90	–	82	81	55	79	90	75	51
Korul	100	100	89	100	93	78	91	91	64
Emai	97	84	80	94	86	70	88	87	66
Angangoi	100	100	93	93	81	79	72	88	67
Total	96	96	86	89	78	79	85	86	295
Females									
Burukngaumo	22	100	50	42	36	48	67	43	11
Bomkan	75	67	–	75	40	71	89	53	15
Korul	57	–	–	60	67	80	91	62	14
Emai	80	50	67	47	100	36	57	49	16
Angangoi	–	100	60	67	50	–	67	49	8
Total	44	67	30	49	48	53	73	51	64

Index of circularity: $\dfrac{\text{number of return moves to the village} \times 2}{\text{total number of moves}} \times 100$:

Source of field data: Young 1974–75.

in these cohorts had signed labour contracts, because such work was not remunerative and improvements in road communication permitted inexpensive travel beyond the highlands provinces. Another likely factor in the decline of contract migration was the planting throughout Simbu of coffee as a cash crop which, once established, provided men with the money to finance their journeys.

As a result there is a clear relationship between degree of circularity and kind of employment (Table 9.2). Most men who had participated solely as contract workers registered the maximum index (100), signifying direct return to the village, as did another sixty per cent of those who took up jobs independently. However, lower indices tend to be recorded for the latter and for about thirteen per cent were below 50. This is especially characteristic of those who hold permanent positions which involve frequent, involuntary transfers by their employers, although the number captured in the sample is so few that this result should not be over-emphasized.

While the circularity from rural Simbu has recently become less intense, it is still the dominant and most persistent characteristic of the migration process. One reason is the difficulty workers have in obtaining family accommodation in towns or on plantations particularly if, as with most Simbu, they are the lower paid. If dependants must remain in the village, then wage earners clearly will circulate. Second, tribal life still exerts a strong influence on most Simbu and traditionally-defined status remains an important avenue to economic and political power. A migrant who resides in the village for considerable periods can control the family's investments of capital, as in livestock or trade stores, or of money and traditional valuables given to kin and friends. He may thus enhance his status and, with age, become a powerful leader. This may be a far more attractive alternative than continuing as wage labourer, which commands little status. As a corollary, third, unskilled workers in town often feel insecure and by leaving for Simbu they become rejuvenated among their own kind and within their own society. The vague feeling that life in centres beyond the village is dangerous and lonely probably underlies why many migrants were unable to state precisely the reason for their return. Finally, population pressure in Simbu has not yet forced migrants to relinquish land rights and many have relatives tend cash crops as a guarantee of local income when not employed elsewhere. Currently high rates of population growth will greatly influence this balance; lack of cultivable land and low incomes may partially explain the level of circularity recorded for Bomkan, in monetary terms the poorest of the sample villages.

Case studies of migration cohorts from two villages at different periods better indicate the meaning of such varying levels of circularity (Fig. 9.4). Four Korul men, for example, who first departed before 1945 had no formal education and spoke little pidgin. Many were

201

TABLE 9.2 Indices of circularity and employment type for Simbu males

| Circulatory Index | Type of employment | | | | | | | | Total number |
| | Contract | | Contract and independent | | Independent | | Independent, but involving frequent transfer | | |
	No.	%	No.	%	No.	%	No.	%	
0.0	–	–	–	–	4	4.3	–	–	4
0.1–24.9	–	–	–	–	2	2.1	2	28.6	4
25–49.9	–	–	1	1.2	7	7.4	3	42.9	11
50–74.9	2	1.9	12	14.6	13	13.8	1	14.3	28
75–99.9	2	1.9	18	22.0	11	11.7	1	14.3	32
100	103	96.3	51	62.6	57	60.6	–	–	211
Total	107	100.0	82	100.0	94	100.0	7	100.0	290

Source of field data: Young 1974–75.

traditional leaders, involved marginally with cash cropping and the monetary economy. The men in this cohort left Korul in 1943 to be construction labourers on Goroka airstrip in the Eastern Highlands (Fig. 9.1). Following between six months and two years spent there, they were repatriated back to the village. Three of the four subsequently migrated again: one completed a work contract on a rubber plantation in Central province, one became a mission employee in the Eastern Highlands, and all three paid extended social visits to younger members of their families living in towns. All men, with one exception, engaged in both contract and independent work but, irrespective of its different conditions, always returned directly to their families who remained in Korul. Thus the cohort registers the maximum index of circularity.

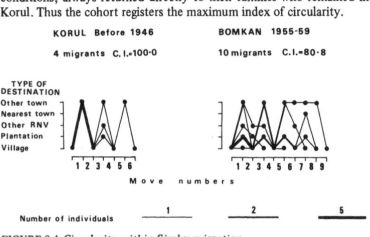

FIGURE 9.4 *Circularity within Simbu migration*

Ten men from Bomkan who first departed between 1955 and 1959 (Fig. 9.4) had little formal education but spoke pidgin and eagerly participated in various business ventures during the late 1950s and 1960s. Seven initially signed contracts to work on plantations in the New Guinea islands and in coastal provinces. On return, four married and did not migrate again, occasionally finding local jobs with the near-by mission or local government council. At different times, the others took up independent employment, either on highland plantations or as casual workers in coastal provinces, including Port Moresby (Fig. 9.1). All acquired some skills in the course of this experience: one became a plantation foreman, another a carpenter, the third a driver; and some-times they moved to other jobs before return to Bomkan. Three of the original ten in the cohort never signed contracts, nor did they always return to the village between one destination and another; two found jobs with the Roman Catholic Mission and the third with the Depart-ment of Health. The circularity index for the Bomkan cohort (80.8) is considerably lower than that for Korul but still sufficiently high to demonstrate the dominance of return to the local community. Whereas movement to plantations was almost always directly circular, transfers between village and town, many at the time quite small, were equally direct for Korul migrants but less so in the case of Bomkan, some of whom lived in other towns before going back to their village.

The degree of circularity from New Ireland villages

During their adult lives, male migrants from New Ireland villages had a much lower index of circularity (49) than their Simbu counterparts (Table 9.3). Not only had very few still alive ever worked as contract labourers and been subsequently repatriated, but also most had attended mission and government schools, acquired skilled jobs, and obtained sufficient economic security to allow more permanent absence from the village. In the course of these jobs, their employers might transfer them from one post to another as well as provide adequate accommodation for their families. Thus different colonial histories for Simbu and New Ireland are reflected in provincial contrasts of socio-economic change and of the significance of circular flows among males. In comparison, New Ireland women, like those in Simbu, normally moved as dependants and their indices of circularity are virtually identical (cf. Tables 9.1, 9.3).

Men from Bom/Ratabu (Fig. 9.2), the most isolated community whose inhabitants received the least formal education, had the highest index (56) whereas the other three villages recorded 46 or 47 (Table 9.3). In this respect, Bom/Ratabu males were more like the Simbu: they had obtained unskilled jobs as dock-workers and labourers in both New Ireland and nearby East New Britain (Fig. 2.1). They also returned often to safeguard their local interests but, compared with the Simbu, most had remained after marriage. Bom/Ratabu was the only New Ireland village with insufficient land for cash cropping, so that local incomes were low and the wages to be earned elsewhere at unskilled jobs thus more attractive. Most men who lacked skills in the other three sample villages preferred to stay and rely on income from cash cropping, supplemented at times by casual work. As in Simbu, the uncertain job market meant that school-leavers maintained high levels of circularity and in each community there were several young men who were cash cropping and had yet to become migrants.

In recent times, compared with Simbu, there is no steady decline in the circularity of successive New Ireland cohorts (Table 9.3). Since some men who first left their village before the end of the second world war had signed work contracts, indices were higher for this early period, but even those indentured in the 1930s to work the gold-fields of mainland New Guinea did not always circulate and shifted from one destination to another with the same or other recommended

TABLE 9.3 Indices of circularity, by sex, for New Ireland migration cohorts

| | Migration cohorts | | | | | | | | Number in sample |
	Before 1946	1946-49	1950-54	1955-59	1960-64	1965-69	After 1970	Total	
Males									
Kaselok/Putput	59	33	47	43	19	14	–	46	41
Fatmilak	49	49	22	50	50	50	67	47	33
Bom/Ratabu	43	46	31	70	71	88	100	56	42
Piliwa/Lapai	50	30	50	58	30	38	100	47	61
Total	52	41	38	55	34	55	88	49	177
Females									
Kaselok/Putput	43	50	80	–	29	83	100	55	27
Fatmilak	67	75	51	52	75	–	100	61	22
Bom/Ratabu	34	54	50	17	37	50	80	44	29
Piliwa/Lapai	61	–	–	24	69	73	67	51	27
Total	47	56	53	31	54	73	81	53	105

Index of circularity: $\dfrac{\text{Number of return moves to the village} \times 2}{\text{Total number of moves}} \times 100.$

Source of field data: Young 1976.

205

employers. Some were also trapped on the mainland when hostilities erupted. From the 'fifties, the degree of circularity for migration cohorts fluctuates according to type of employment: thus the 1955–64 group from Bom/Ratabu contained a high proportion of casual workers with no employment security who soon returned to the local community. Conversely, many New Irelanders who first departed between 1946 and 1954 were taking advantage of the opportunity to attend schools or training institutions and later became teachers, medical orderlies, and store-keepers, with long periods of absence.

Despite this, closer inspection of the New Ireland data reveals no clear relationship between the degree of circularity and kind of employment (Table 9.4). Few men had gone away on labour contracts and most who did subsequently worked at other kinds of jobs. Very low indices of circularity are recorded for similar proportions of men (about 25%) who signed work contracts and found employment themselves. Job transfers, which accounted for 18.3 per cent of all New Ireland moves compared with 9.2 per cent for Simbu, are more significant than appears from Table 9.4 (col. 4), because many classified as independent workers had at some time held occupations which meant involving transfers.

The broad contrast between the circular flows for Simbu and New Ireland males suggests that these diminish in importance with greater economic opportunity without, however, ceasing to dominate the migration process. Both New Ireland and Simbu absentees retain strong links with their local groups, which less successful workers exploit by returning home. The economically more achieving may remain away for many years, but some evidence suggests they also eventually go back and capitalize upon available resources.

The complex nature of migration from New Ireland communities is exemplified by six men of Piliwa/Lapai (Fig. 9.2), who first left between 1946 and 1949 (Fig. 9.5). This cohort was the first to benefit from government sponsorship of formal education and vocational training following the end of the war. After twenty or thirty years' service as teachers, clerks, and skilled workers, some have resettled in their villages and established local businesses. Four of the six men attended school in Kavieng (Fig. 9.2) and three received further training in Rabaul as, respectively, teacher, agricultural officer, and Methodist pastor. Both teacher and pastor served for many years in distant parts of Papua New Guinea before their recent return but neither the agricultural officer nor the other man ever became wage earners.

By contrast, the other two males from Piliwa/Lapai lacked formal education. One worked several times on copra boats belonging to a local plantation owner. The other, born in Morobe province (Fig. 2.1), became a policeman, later a carpenter, and moved to the village after marrying a Piliwa woman he met while working in Rabaul. No one in

TABLE 9.4 Indices of circularity and employment type for New Ireland males

	Type of employment											
	Contract		Contract and independent		Independent		Independent, but involving frequent transfer		Not employed; move for formal education		Total number
	No.	%	No.	%	No.	%	No.	%	No.	%	
Circularity Index											
0.0	—	—	3	16.7	10	7.6	1	9.1	—	—	14
0.1–24.9	—	—	—	—	2	1.5	1	9.1	—	—	3
25–49.9	—	—	2	11.1	17	13.0	6	54.5	—	—	25
50–74.9	—	—	7	38.9	47	35.9	2	18.2	5	35.7	61
75–99.9	—	—	3	16.7	17	13.0	1	9.1	—	—	21
100	3	100.0	3	16.7	38	29.0	—	—	9	64.3	53
Total	3	100.0	18	100.0	131	100.0	11	100.0	14	100.0	177

Source of field data: Young 1976.

this migration cohort was ever separated from their families and, being often transferred by their employers, the index of circularity is low (30.0).

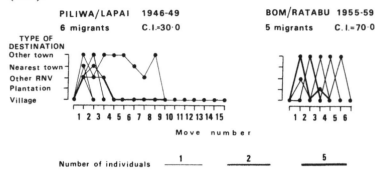

FIGURE 9.5 *Circularity within New Ireland migration*

Five male migrants from Bom/Ratabu, who belong to the 1955-59 cohort (Fig. 9.5), had little formal education and were without marketable skills or training to facilitate outside employment. All went to Rabaul, in nearby East New Britain; three made multiple trips, with periods of absence ranging from six months to four years. They obtained a variety of jobs: domestic service, plantation work, and unskilled labouring on the roads and wharf. Only one, a domestic servant, travelled farther afield by accompanying his employer who was transferred to Port Moresby. Migration ceased at marriage with one exception, who was accompanied by his wife and child. Movements of this cohort were thus strongly circular (index 70.0) and dominated by a single urban destination, Rabaul, which constitutes the primary focus for residents of this part of New Ireland.

Simbu and New Ireland migrants in town

The main disadvantage of a village-based study is that the presence of those interviewed demonstrates their strong links with the rural area. To obtain a zero index of circularity is therefore impossible. Village residents cannot be expected to report accurately on how long migrants have been away. Ideally, absentees should be interviewed at their various destinations: a mammoth task which would have resulted in drastic reduction of the present research and made impossible any assessment of intra-provincial differences. Evidence of the migration experience of some Simbu and New Irelanders living in towns was, however, available from another source: the urban household survey of 1973-74 (Conroy/ Curtain, this volume).

Migration histories collected during the Urban Household Survey were secondary to its major purpose and are thus less comprehensive and less accurate than those available for Simbu and New Ireland villages. They consist of a list of locations where individuals had previously stayed, along with the year that place was reached. Since socio-economic questions were not asked, migration and employment histories cannot be compared. The main discrepancies in these histories are inaccuracies in the assessment of time, omissions of some mobility (especially intervening stays in villages of origin), poor definitions of place, and inconsistencies in which types of movement — notably visits or changes of residence — should be included. A further difficulty is that December 1973, the month most interviews were conducted, was the period of annual vacations, when movement is extreme, and happened also to coincide with self-government, an event reflected in greatly increased mobility between villages and towns because of rumours of social disruption (Young 1977b:145-52).

These deficiencies apart, the migration histories recorded in the urban household survey contain important details about people then resident in towns. They are available for 5,133 individuals, of whom 385 (193 men, 192 women) were born in Simbu and 83 (43 men, 40 women) in New Ireland. This difference in sample size partly reflects the national ratio of about four times as many Simbu-born as New Ireland-born, and partly how the samples were drawn in different towns (Young 1977b:147-8). The New Ireland sample is too small for disaggregated analysis. Level of education was a differentiating characteristic of Simbu and New Irelanders living in town. On average, Simbu were less educated than other provincial groups whereas more than 35 per cent of New Ireland men had received secondary or tertiary training. In the absence of details about nature of employment, it can be assumed that many New Ireland males held permanent jobs and without resigning from them would be unable to return to their province of birth.

Indices of circularity for both Simbu and New Irelanders were low (Table 9.5). Since departure, few men or women had paid more than a brief visit to their local communities and, although moving frequently, remained within the urban sector. Even if all those interviewed went back to reside in their villages without moving elsewhere, thereby doubling the index values, the amount of circularity would still remain far below those revealed by the village studies. This suggests that many Simbu and New Irelanders in town have become quite permanent residents, irrespective of whether they later retire to their home province. Thus migrants either remain village-based and move in essentially circular fashion or establish an urban base and, although usually retaining contact with the local community, are unlikely to resume rural residence while of working age.

Urban-dwelling Simbu and New Irelanders who held permanent government positions had much lower indices than those in casual government or private employment (Table 9.5), being often transferred

TABLE 9.5 Indices of circularity and employment type for Simbu and New Irelanders in town, 1973-74

Type of employment	Simbu males	New Ireland males
Not employed	21.5	36.8
Private	20.5	11.8
Casual government	26.0	0.0
Permanent government		
primary education	2.8	0.0
secondary/tertiary	0.0	0.0

Total indices of circularity			
Simbu males	13.4	New Ireland males	15.3
Simbu females	10.7	New Ireland females	7.3

Sources of data: Urban household survey 1973-74; special tabulations, Young.

from one town to another without intervening returns to their natal village. Although the relationship between degree of circularity and permanent positions can only be shown in general terms because of the coarseness of the original data, the urban survey still provides some confirmation for the rural study, in which job transfer was a main reason for lower indices of circularity among New Ireland males. Individual histories from the urban household survey therefore suggest that migration from Simbu and New Ireland is not dominantly circular for all groups, despite fundamental differences in their demographic and socio-economic composition. This conclusion raises some important issues for both theoretical discussion and urbanization policy in Papua New Guinea.

A transformation in levels of circularity?

Circulation and migration in Papua New Guinea, as in other third world countries where rural-urban links remain strong, are not two distinct processes. They coexist to form a system of population mobility within which individuals describe patterns in response to a complex mix of physical and social factors, variable through time.

Evidence from Simbu and New Ireland indicates the significance of formal education for changes in circularity. In general the unskilled, here represented mostly by Simbu villagers, move to casual employment or to visit friends and relatives and have insufficient resources or job security to take their dependants with them. Their level of circularity is high because for them the village, where they maintain rights to land and through relatives continue to share in a mixed economy of subsistence activities and cash cropping, remains all important. Those with some formal education and marketable skills, which accounts for most New Ireland villagers, are more likely to obtain secure positions and qualify for suitable accommodation for themselves and their families. Although village links remain strong, they have the confidence to migrate between alien locations without intervening periods of local residence and the intensity of their circularity is less. The marked contrast between the movement patterns of New Irelanders from Piliwa/Lapai and from Bom/Ratabu (Fig. 9.5) illustrates this change, as do the differentials in the circular flows of rural and urban dwellers.

Yet formal education, by itself, cannot explain these variations. Informal learning, the acquisition of language skills or the ability to drive a vehicle, increases a villager's capacity and improves the chance of success in the outside world. Additional factors are demographic, in particular the age and gender of potential migrants; socio-economic structure of the community, such as the resource base; and the operation of external controls which may lead to involuntary movement.

In both rural Simbu and New Ireland, young adults who have completed their formal education and just entered the labour force record higher levels of circularity than older persons of similar educational achievement. While this stems partly from the brevity and hence incompleteness of their migration experience, it also results from strong local ties to parents and friends and from tenuous contacts with other regions. In general, the circularity of village women is less than for their male counterparts, reflecting the fact that they migrate with husbands who hold skilled jobs with government. Family accommodation is provided but the possibilities thus diminished of return to the local community.

Favourable ratios of people to land in rural areas also promote circular flows, since return to a village can only occur if social and cultural ties have been well maintained. Although Simbu is densely populated for Papua New Guinea, most villages are able to provide resources for both absent and resident members. Retention by absentees of rights to land for subsistence and cash cropping is possible even in Sinasina and Gembogl, here represented by the villages of Emai and Bomkan (Fig. 9.1), where local densities of more than 100 people per square kilometre have been recorded. Thus Simbu and New Irelanders return from wage employment to oversee local responsibilities and discharge their obligations in tribal society, although this situation may change as rural populations increase rapidly. Longer-established towns such as Port Moresby or Rabaul (Fig. 2.1) already contain considerable numbers of urban born, some of whom do not establish links with the natal villages of their parents and never assert their place in traditional life.

External controls, notably schemes for contract labour and government employment, have affected the migration patterns of Simbu and New Irelanders as of other Papua New Guineans. Contract work, introduced to provide unskilled labour for the plantation sector, greatly stimulated circular flows, as is clearly apparent from the early experience of Simbu men. Although this system probably meshed well with the aspirations and needs of its indigenous participants, there is no evidence from Simbu and New Ireland that the resultant circularity evolved directly from that of pre-contact times (cf. Chapman and Prothero, this volume). Conversely government employment, where workers are frequently transferred between urban-based posts, prevents residence in the village. Thus skilled workers, in this case mainly from New Ireland, move within the urban sector and effectively become permanent absentees from the local communities for the duration of their working lives.

Direct migration, noted for Uganda (Jacobson 1973) and Kenya (Elkan 1976) as well as for New Zealand (Keown 1971) and Australia (McKay and Whitelaw 1977), does not wholly accord with levels of formal education, since it affects both the elite and the less skilled: policemen, health workers, artisans. As Mitchell (1959) has suggested, such persons become increasingly divorced from the village and even when links are maintained through letters and remittances, as with the Simbu in Port Moresby (Whiteman 1973:80-1), there is no guarantee they will ever return as residents. Forms of directed movement effectively mask patterns which result from the voluntary decisions of individuals and make extremely difficult a coherent analysis of transformations in levels of circularity.

While the preceding analysis indicates the possibility of demonstrating a decline in circularity with advancing modernization, its results cannot be statistically extrapolated to other parts of Papua New Guinea. This field study, like others, employs its own definitions and is placed firmly within a particular cultural context. Nevertheless it also reveals some characteristics of the migration process which have received insufficient emphasis and which must be incorporated within more general theories of population movement.

Acknowledgments

I am grateful to the Australian National University, which financed the fieldwork in Simbu and New Ireland villages, and to those who acted as guides and interpreters: Witne Michael, John Kagl Maina, Kwiwa Bagle, Ambai Guade, James Kari, Emmanuel Muro, Taiya Nisa, Bite Ngurpeo, Tike Moris, Tare and Sara Taufi. I also thank Ross Garnaut, one of the organizers of the Urban Household Survey, for permission to use uncoded data contained in the migration histories.

10 Patterns of Fijian return migration in the Wainibuka River valley: Viti Levu, Fiji

Sakiusa Tubuna

My father is from the village of Suvavou, a village about three miles from the centre of the city of Suva on the opposite side of Suva harbour (Fig. 10.1). My mother is from Wailoku, a village six miles from Suva on the Tanavua side of the city. I am the youngest of three children.

After I was conceived, about seven months in my mother's womb, my father died. I was named after him. It is a Fijian custom in marriage that if the husband dies, the relatives of the wife will make presentation of *tabua* (whale's tooth) to the relatives of the deceased. This formally releases their social obligation to the wife so that she can return to her father's home, even though she has children. This was done in my mother's case and we were brought up in my mother's household in Wailoku.

When I grew up, I knew nothing about my own village until after attending primary school at the age of six. My grandmother (father's mother) usually came to visit us, bringing with her gifts of food and money. She at times took me back with her, but I was a stranger in my own *koro* (village).

My mother's relatives paid for my primary and secondary education until I entered the Teachers' Training College in 1961. I was aged nineteen then. At this time I had a deeper feeling for the people of my mother's village and always felt that I was one of their kinsmen. The biggest barrier that separated the two clans was that my father was a Fijian and my mother a Solomon Islander, being the daughter of one of those Solomon Islanders who had been indentured to Fiji during the time of the labour trade (see Frazer, this volume). Since my mother's mother was a Fijian, her family adopted many of the customs and culture of Fiji.

My aunt (mother's sister), who had been taking care of us since our early childhood, died in 1969. By then, I had been a school teacher for six years. At her death, the close relatives of my father came for the

213

mourning and at the same time presented *tabua*, mats and drums of kerosene for my family's return to our original homeland. These were acknowledged by the relatives of my mother.

A day was fixed for our formal return. I had been away from my own people for twenty-seven years, not because of any personal motives or desire but rather because of customary obligations on the part of my mother's relatives. When the day came for our return, it was very sad. I was sad because, when my mother left, there were three of us; now, there was only me. With formal presentations of *tabua*, mats, clothes, and kerosene, I was formally returned to my people. I had left my own village when I was in my mother's womb; now, my sister was not with me because she had married and my brother had died. But I was not alone when I returned, for I had a wife and three sons. That day was one of mourning but also one of great joy. As it is written in the Bible, 'He was lost and was found again; he died and has come back to life'.

My own experience has made me wonder if it is the same with other Fijians. This paper is concerned with first, the nature of return movement among Fijians born in the Wainibuka River valley of Viti Levu, Fiji; and second, with the reasons why village people, having migrated to other areas, subsequently returned to the valley.

Since it was difficult to survey the entire area, five villages were selected; Nasautoka, Naveicovatu, Wailotua, Naseva, and Dakuivuna (Fig. 10.1). From each of these villages, four people were interviewed who have returned after being absent for several years. A questionnaire was the main method of obtaining information, but considerable time also was spent around the bowl of *yaqona* (kava) discussing with villagers the topic of migration. From this information, as well as my experience as a Fijian who has lived in various parts of the country, I had drawn some conclusions about the nature of and reasons for population movement among Fijian villagers.

The Wainibuka River valley

The Wainibuka River valley, one of the richest agricultural areas in Viti Levu, is located in the north-eastern part of the island (Fig. 10.1). The Wainibuka valley borders the river for some 30 km. and the lowland plain consists mainly of rich alluvial and colluvial soil. All the lowland areas have been cleared for farming but most of the highest land is still thickly covered with secondary growth and evergreen tropical rainforest.

Villages are mainly situated on the bank of the Wainibuka River, which was the only means of transport before the construction in the early 1930s of King's Road (Fig. 10.1). A village consists of about 30–40 families, each traditionally living in a *bure* (house with a thatched roof). Nowadays there are very few traditionally-styled *bure* in a Wainibuka village. Houses are constructed of concrete blocks and wood,

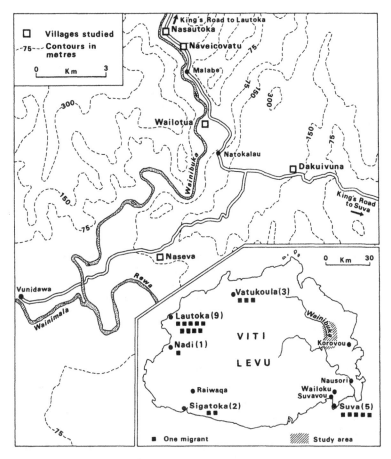

FIGURE 10.1 *Lower Wainibuka River valley, Viti Levu, Fiji, with inset of destinations of twenty migrants*

and roofed with corrugated iron, because they are more resistant to hurricane damage and flooding which occasionally occur in the area.

Life in a Fijian village in the Wainibuka valley is very simple. At one time, all the people were subsistence farmers who attended the root crops in their food gardens. Today, many have leased their communal land and established sizeable cocoa plantations, or run beef cattle, or have planted cash crops such as *dalo* (taro) and cassava for the local market. At times, people in villages are still required to do such communal tasks as weeding the central part of the settlement or tidying the school compound.

The rich, alluvial lowland of the Wainibuka River valley was once famous for the production of bananas for export to New Zealand. In

215

the late 'sixties, because of widespread banana diseases throughout Fiji, followed by devastating hurricanes in the early 'seventies, many farmers turned to other cash crops, particularly cocoa. Cocoa is now the most preferred of all cash crops for, apart from being perennial, it was fetching in early 1979 $Fijian1,800 a ton in the local market ($F1.67 to £1.00; $F1.00 to $A1.00). This is a price which is far higher than any other, locally-grown commodity. Although bananas no longer are exported, many Wainibuka farmers still produce sufficient to meet local demand.

Today, the Wainibuka valley is well served by roads and villagers have no problems in transporting their produce to local markets. The nearest town is Korovou, 16 km. away, but villagers prefer to send their cash crops to either Suva or Rakiraki markets, in the hope of obtaining higher returns. Many villages located along King's Road (Fig. 10.1) have small stalls, at which they sell local produce to car and bus passengers.

Despite the availability of good agricultural land and the accessibility of Wainibuka villages to transport and marketing facilities, many people prefer to leave the local community and find work in the urban places of Viti Levu. The question is not only why these people moved out but also why, later on, they decided to return to the valley and permanently resettle in their villages of birth.

Twenty return migrants from five villages

The twenty people interviewed, all male, are between thirty to seventy-six years of age. Four are very old and retired in the village, but the rest are in their middle ages. There were no women who had returned to Wainibuka after some years away.

Nearly everyone interviewed grows cash crops and three of the twenty raise cattle for beef. Six have quite large areas planted with cocoa, most of them being helped by their adult children. The very high price of cocoa fetched on the world market has led almost everyone with large parcels of land to plant seedlings, in the hope that future prices will remain high.

The level of education reached by migrants made a great difference to the types of jobs they could find away from the village. Eighteen out of twenty did not go beyond village school, which meant that better-paying jobs in urban centres, or with government in rural districts, were not open to them. At the time they went, only seven out of twenty, or far fewer than might be found in a village today, were aged less than twenty-five. Perhaps the fact that none of these had been educated in town explains why most remained in the village. In fact, two of the three migrants aged fifteen to nineteen did not leave independently but accompanied relatives and, after marriage, returned to Wainibuka.

Far more villagers in the older age groups (30-40) moved to obtain paid employment to support children whom they had sent to high school. Although the Wainibuka valley is an area of rich agricultural land, the decline of the banana industry left many farmers short of cash and the only solution was to get a job outside the village to meet their financial commitments.

The local social organization, with its traditional customs and obligations, is also a great disincentive to those who want to be fulltime farmers. As a result, parents often preferred to find wage employment in urban centres while their children attended school. Once this formal education had been completed, the parents returned to their village.

Jobs are not readily available to village migrants in the urban places of Viti Levu. In many cases, relatives already working in such towns as Lautoka (Fig. 10.1) informed villagers of openings; in others migrants stayed with relatives in urban centres for as long as two years before obtaining employment. Some became burdens on the families with whom they lived, but Fijian custom requires that the obligations to kinsmen be met regardless of the hardship involved.

The social burdens and obligations of village living, plus the lack of capital, are two of the major factors contributing to the rural-urban drift of Fijians and were mentioned as the primary reason for leaving by five out of twenty men interviewed. Dissatisfaction with village life had led many not only to move to other areas but also to settle on their own *mataqali* (communal) land, thus becoming independent farmers (*galala*). Another half (10 out of 20) looked for better economic opportunities in urban centres, partly because of the perceived difficulties of finding ready cash in the village, and partly because of the increasing desire of Fijians to educate their children.

Some, when younger (5 out of 20), were attracted by the 'adventure' and 'excitement' of town life. Some later found jobs, but many ended in such undesirable places as gaol. Most therefore returned to the village within five years for they realized that, when in difficulty, there was always food and shelter at home. Half the migrants were in town far longer (5-10 years), mainly so their children could complete a formal education.

Although permanently resettled in Wainibuka villages, at times Fijians leave their community to visit and take local produce to relatives who live in the towns and urban centres. In return they receive groceries, such as sugar, salt, soap, and kerosene. Villagers also stay overnight with relatives when they go to sell their farm produce at town markets.

Since return migrants have lived in both town and village, of greatest interest is what they think to be the main advantages and disadvantages of being resident in either place. Discussions with individuals, as well as around the *yaqona* bowl, produced a list of likely replies. The twenty persons being interviewed were subsequently asked to note their opinions from these separate lists of advantages and disadvantages (Tables 10.1, 10.2).

TABLE 10.1 Opinions about living in town: twenty return migrants, Wainibuka valley, 1978

	Multiple responses
Advantages	
Wage-paying jobs	20
Greater educational opportunities	20
Easy women	8
Freedom from communal duties	6
Freedom from customary obligations	5
Disadvantages	
High cost of living	20
Social problems	20
Congested homes, especially for migrants	20
Lack of social ties with the village	5

Source: Field Survey

TABLE 10.2 Opinions about living in the village: twenty return migrants, Wainibuka valley, 1978

	Multiple responses
Advantages	
Much land for farming	20
Abundant availability of food	20
Social life better	20
Not so many social problems	18
Easy life: not having to take orders	16
No financial worries or difficulties	15
Disadvantages	
Social obligations	16
Too much time taken for communal work	15
No wage-paying jobs	10

Source: Field Survey.

All were attracted by wage-paying jobs in town. Within the villages there had been rumours that it was easy for anyone to find employment but, to the surprise of many, this was not the case and often they searched for months. In the sugar-cane growing areas of western Viti Levu (Fig. 10.1), the situation is quite different, especially during the harvest season. There is always a shortage of labourers during cane cutting and at times villagers will go in large groups to earn money, not for personal benefit but for communal projects such as building a church or school.

The biggest problems faced by these village workers are the poor and congested accommodation together with the higher cost of food and other essentials. Had it not been for such communal projects, many would have been unwilling to remain in the quarters provided. For those individually employed, poor living conditions caused them to look for other jobs or else return to Wainibuka.

Everyone mentioned educational opportunities as a great advantage of the towns and urban centres. While true for many migrants from rural areas, poor housing often presented difficulties for students. In town, many village families depended on relatives who have already rented houses in the low-cost housing suburbs or have become squatters and built shacks on unused land. The congestion in these homes does not provide the educational climate envisaged from the village; social problems often arise and as a result many parents decided to return to their communities. In recent years, local school facilities have been greatly improved by the establishment of junior secondary schools in rural areas, three of which are located in the Wainibuka River valley.

Prostitution has become prevalent in many urban areas of Fiji and made villagers think that these are places where women can be got easily, provided they have money. Many who took goods to sell in the markets spent some if not all their proceeds on liquor and women. A few return migrants said freedom from communal duties and customary obligations was an advantage of living in town. There is, however, no complete freedom for any member of the Fijian community irrespective of where one lives. The burden of custom is sometimes much lighter away from the village but obligations to kin, especially financial ones, must still be met.

Although many people are greatly affected by the communal tasks and social obligations of village life, on balance they still feel much more comfortable in the rural community than in town (Table 10.2). There is abundant land for food crops, local produce can be sold to obtain money, and everyone is related, so there is more communal activity and fewer social problems. On days when there are no collective tasks, villagers can please themselves what they do. At times, instead of going to their food gardens, they sit around the bowl of *yaqona* all day and tell local stories.

As is evident from the opinions of twenty migrants who returned to Wainibuka, living in either village or town has its own advantages and disadvantages. It is possible that their reactions are weighted towards a

219

rural viewpoint, because many did not have well-paid jobs in the urban centres; many experienced hardship while away from their village; some worked mainly as skilled labourers; and many did not achieve their goals after leaving the local community. Fijians are a product of their own social system and have their own way of doing things, but being on a time schedule is not one of them. Consequently many villagers, when faced with the dilemma of town lifestyles, returned to the more familiar environment of Wainibuka.

Four case studies

No one migrates from their own village without some reason. By means of four case studies, it is possible to show the kinds and range of motives that lead people to both leave and return to their birthplace.

Ulaiasi Matagasau of Nasautoka

Ulaiasi Matagasau, aged thirty-nine from Nasautoka village (Fig. 10.1), is married and has four children. He is a farmer and owns a large plantation. He is the youngest in a family of four brothers and two sisters.

Ulaiasi was only ten years old when both his parents died. Fortunately his eldest brother was a school teacher and became responsible for his upbringing. Ulaiasi left primary school when he was fifteen and went to stay with his brother who was then teaching in Raiwaqa, a village about ten miles from the town of Sigatoka on the southeastern coast of Viti Levu (Fig. 10.1). When he was eighteen, his brother found him a job in Sigatoka as an apprentice in a motor-repair garage. Because of the distance between Raiwaqa and Sigatoka, he lived near the town with a cousin who was a civil servant.

Ulaiasi worked as an apprentice for three years, after which he became a mechanic in another repair shop for a further three years. To make it easier to travel to his job, he bought a bicycle. One day, a cousin employed at Nadi International Airport (Fig. 10.1) was visiting and decided to take his bicycle on the understanding that it would be returned the following week. One, two, three weeks passed, but still there was no sign of the bicycle. At last Ulaiasi decided to go and reclaim it.

When he arrived in Nadi, his cousin persuaded him to work at the airport instead of returning to Sigatoka with his bicycle. Ulaiasi did not even send back news of this intention to his brother. He was happy. There were many young men in the area and they usually spent most of their wages on drink, going to nightclubs, and on women. He worked at the airport for nine years, by which time he was receiving $Fijian50.00 every two weeks. He married in 1968 at the age of twenty-nine and his first child was born in 1969.

Yet it was hard for Ulaiasi to provide for his family on $F100.00 a month ($F1.90 to £1.00; $F1.03 to $A1.00). He had to pay rent, electricity, water rates, as well as for his cigarettes and liquor. He stopped drinking liquor in the hope of saving some money but that did not work. The same happened with smoking. At last he decided there was only one solution to the problem: they all must return to their village.

Ulaiasi and his family returned to Nasautoka in 1972. Fortunately, close relatives provided root crops until they could plant their own. The bundles of *dalo* (taro) which he usually bought for three dollars in Nadi were freely given; for Ulaiasi, it was a great relief! The day I saw him, he was in a grand mood. He said never again would he leave the village to find employment in town. He felt freer in Nasautoka: no rent, no water rates . . . When he worked for wages he never had ten dollars in his pocket for more than a day, but in the village it could be kept for a whole week. If there was urgent need for money in Nasautoka, then a bundle of *dalo* could be placed near the main road and sold to those travelling past by car or bus.

Ulaiasi left Nasautoka at the age of fifteen and had not previously experienced village life as an adult. I confronted him with the question as to what he would tell young people about the problem of rural-urban drift. He commented that because of his experience he would never encourage anyone, especially the teenagers, to migrate to the urban centres of Viti Levu.

Viliame Tuitokatoka of Wailotua

The case of Viliame Tuitokatoka is quite different. In 1944, when aged fifteen, he left Wailotua (Fig. 10.1) to look for a job, as the eldest of five sisters and brothers, because his father had been blinded by illness. His mother also caught the disease and was temporarily blinded. This meant great trouble for the family; the only breadwinner had been disabled and the children's school fees could not be paid. Viliame decided to try and find wage employment in one of the major towns of Viti Levu, both to help his family and to pay the school fees of his younger brothers and sisters.

He went to Lautoka (Fig. 10.1), where some Wailotua people already lived, and stayed with his uncle. It was two years before Viliame was able to find employment at the mill of the Colonial Sugar Refining Company. Luckily, his uncle had a well-paid job and looked after him. But Viliame found that his wages as an unskilled labourer were so low that he could send little money home. For fourteen years he wandered around the western side of Viti Levu, from Sigatoka to Vatukoula (Fig. 10.1), working in a particular area for a short period and then moving to another. Although often far away from Wailotua, he kept in touch with his parents by letter.

In 1958, Viliame realized he should return to his village, for his father was getting older. He has remained in Wailotua ever since and worked hard to develop his father's land. Today, he has a large banana plantation and regularly supplies fruit to the Suva market. Cultivating the land has given him greater economic returns than being employed as an unskilled labourer. Although occasionally Viliame goes to town, he vows he would never again leave his village to settle there.

Sailasa Magiti of Naveicovatu

Sailasa Magiti, an old man aged seventy-six, was met while drinking *yaqona* with the villagers at Naveicovatu (Fig. 10.1) and could recollect many experiences as far back as the 1920s. He left his village school in 1918 to attend the Davuilevu Carpentry Training Centre, which was run by the Methodist mission near the town of Nausori, northeast of Suva (Fig. 10.1). Sailasa spent six years there before receiving his certificate from the principal, Mr R. A. Derrick, who is well known throughout Fiji for his contribution to technical education.

In 1927, as a trained carpenter, Sailasa was employed by Whan Construction, a co-partner of Marlow Construction Limited, a major firm in Fiji that still survives. He worked in the capital of Suva for about ten years before being asked by villagers from the Yasawa Islands, west of Viti Levu (Fig. 7.1), to direct the construction of their church. During six months, news of his good workmanship spread and rather than returning to Suva, he stayed another eighteen months helping various villages build new churches and schools. From 1940, he was employed for six years in the gold-mining town of Vatukoula. Because of the scarcity of trained carpenters, he also was called upon to help villages in the Wainibuka area build churches and schools.

In 1956, Sailasa decided to retire to his own village, Naveicovatu, for he had spent most of his lifetime away. Now a very old man, he is proud to tell his grandchildren about these past experiences. Although village people usually gave him much money and gifts in return for building their churches, schools, and modern homes, always he realized that it was more peaceful to be in Naveicovatu, to plant your own food, and follow your own desires.

Sakiusa Vakasilimi of Wailotua

In another part of Wailotua lives Sakiusa Vakasilimi, aged fifty-five and with four children. He always has leased land, on the bank of the Wainibuka River, and supported his family from the sale of farm produce. In 1965, there was a great flood and Sakiusa was one of many farmers, who worked the lowland areas and lost all their crops. When the flood receded, the land was covered with silt and badly water-logged. Nothing could be grown on it.

Money was needed to support the education of his eldest son, who was then at Queen Victoria School near Suva, a large secondary boarding school for Fijians. Sakiusa therefore decided to look for paid employment in Lautoka but promised he would not return to Wailotua until his son had successfully completed secondary education. He went and stayed five years harvesting sugar cane with Indians. In between one harvest and another, he tried to find labouring jobs. It was a hard life. The food was bad, the shelter poor. But Sakiusa was interested only in saving enough money for his children's education. Occasionally, he would send money home to his wife.

Sakiusa finally returned to Wailotua when his son had successfully completed secondary education and been admitted to the Fiji College of Agriculture. Thus he had fulfilled his promise. The eldest son has continued studying at government expense, and his father is happily working the land as before. Had it not been for the flood, Sakiusa would not have left his village. Never again would he repeat his experience in the cane fields. 'I have learned the hard way', he said.

Conclusions

This study of twenty return migrants living in five villages of the Wainibuka River valley has identified several major characteristics of the link between rural and urban environments. Young people often accompany or join older relatives who are employed in urban areas. Many find town life not nearly as attractive as anticipated, for various reasons such as the inability to save money, dissatisfaction with the urban lifestyle, or getting into trouble with the police. They express their disappointment by returning to the village.

Some families who went to town to care for school children and earn money to pay their fees, subsequently returned to Wainibuka when that formal education was completed. For some, the educational opportunities in town were not realized because of poor conditions for study in crowded homes and resultant social problems. The establishment of a network of junior secondary schools in rural areas also meant that good education became available closer to the village.

The higher a person's educational achievement, the greater the chance to find a more permanent job and to stay longer in town. Some villagers have learned skills that mean they are now more successful in the rural context; others have worked in town until they reached retirement and then chose to return to the village. By contrast, those with less education usually obtain lower-paying jobs and are provided with cramped, less hygienic accommodation. Not surprisingly, many return much more quickly to their local communities.

For many men, the strong ties with the village reflect that their wives remained behind when they moved to town to find employment. Others married after leaving Wainibuka and developed new ties with their wife's community. Once back in their own village, new circulation networks have been established through the visits made by the wife's relatives to the husband's household. Since social ties, especially obligations to kinsmen, also operate in the urban area, many villagers return because in rural places these ties are more intense and expressed less in terms of money.

My personal life-history shows that some 'go home' through the re-instatement of traditional birth rights to residence, after the proper ceremonies have been performed according to Fijian custom.

There are many different patterns of movement found throughout Fiji. In the Wainibuka area, and perhaps in other rural areas of Viti Levu with reasonably good access to Suva by road or river, there is a considerable population of return migrants who, at one time or another, have resided for extended periods in urban areas. This in some ways calls for a re-examination of Nayacakalou's (1963:34) belief that 'there is little doubt that the great majority of Fijians living in Suva are destined to be fairly permanently settled there'.

Despite the inherent bias of focussing upon return migrants, this consideration of five Wainibuka villages points to the need for greater study of the nature of and reasons for the circular movement of villagers and for an improved understanding of the advantages and disadvantages of both rural and town life. Such research would prove of considerable value to national planners, who seem obsessed with the problems of rural to urban migration.

11 Circulation and the growth of urban employment amongst the To'ambaita, Solomon Islands

Ian Frazer

Often unnoticed in reports on the persistent and widespread nature of circulation throughout Melanesia is evidence for marked alterations in the pattern of social mobility amongst specific groups of people. Some of these changes involve varied and complex transformations within a continuing circular form, the understanding of which is now a key issue in mobility research (Bedford 1981; Curtain 1980b; Ward 1980). Through rural-urban movement, for instance, Melanesian populations are being redistributed but what remains unclear is the permanence of these changes, their meaning for the people involved, and their implications for rural and urban society. There is increasing evidence about more permanent movement towards urban centres (eg. Garnaut, Wright and Curtain 1977; Morauta 1979) but uncertainty remains about the strength of this trend. Aggregate data show that movers are staying away longer from the village; however such data permit no distinction between extended absence and permanent separation from rural places of origin. Case studies do not usually resolve this difficulty, revealing instead indecisiveness amongst individual movers and ambiguity between behaviour and intent. Simultaneous with extending their involvement in wage employment, many migrants profess the intention to return to the natal community (cf. Strathern, this volume).

Such complexity highlights what Mitchell (1985) terms the epi-phenomenal nature of population movement and the importance of examining the social, economic, and political forces that constitute its context. Throughout Melanesia, labour circulation derives from the need to take advantage of opportunities for earning money in widely separate locations (Chapman 1969:134). For many groups, as with the To'ambaita of Malaita, in the Solomon Islands (Fig. 5.1), this has been possible without having to choose between living permanently in one location or another. Travelling between Malaita and places of employment over several generations, the To'ambaita have established ways to combine opportunities in multiple locations without excluding one or other of them as places in which to work and live. In To'ambaita movement, the choices concern the length of time spent in particular locations and the appropriate time to shift between them.

The main concern of this chapter is with To'ambaita movement between 1960 and 1974–5, during which period it grew more complex in response to wider political and economic influences. Most important was the growth of public-sector employment in the capital of Honiara (Fig. 5.1), concomitant with an overall increase in both the size of the urban workforce and the opportunities for upward mobility available to Solomon Islanders. By taking up employment in town and with government at a time when conditions of work and of urban residence were gradually improving, many To'ambaita became regular workers and relatively stable residents in Honiara during the years when there was also a steady increase in the numbers travelling between the islands of Malaita and Guadalcanal. In combination, these changes constitute a new phase in To'ambaita mobility experience, within which circulation is still the dominant mode but based on a quite different relationship between places of origin and destination. Amongst the To'ambaita, there are now long-term residents in town who appear committed to an alternative urban lifestyle based on regular wage employment.

The To'ambaita

The To'ambaita of Malaita are one of twelve language and dialect groups on that island whose total number, in the 1976 national census, was 6,506 (Solomon Islands, Statistics Division 1980:Table 3). At least twenty-two per cent were enumerated outside their island of birth, most of whom lived in and around the capital of Honiara. On Malaita itself, the home territory of the To'ambaita people is a small headland, in area about 140 square kilometres and composed largely of steep hills that rise to more than 700 metres and extend southwards to join the mountain chain traversing the island's length (Fig. 11.1). Between sea and hills, a narrow coastal shelf encircles the peninsula; it is here that most To'ambaita villages are now located.

Amongst the people of Malaita, the most heavily populated island in the Solomons, there is an important contrast between those who, in the past, were principally coastal dwellers, had a maritime orientation, and were dependent on the sea, and those who lived mostly in the hills and whose subsistence base was horticultural. Traditionally, the To'ambaita were a hill people (*to'a tolo*) and still so regard themselves, even though most have relocated to the coast. In 1933, when Hogbin undertook research among them, half still lived in the hills (Hogbin 1939). Linguistically and culturally, they are closely related to their nearest neighbours: the maritime Lau, of the northeast coast (Ivens

1930), and the horticulturalist Baelelea and Baegu (Ross 1973), who inhabit the hills and coastal slopes to the east and south. Intermarriage is common amongst these dialect groups and they interact regularly through produce markets and other kinds of exchange activity.

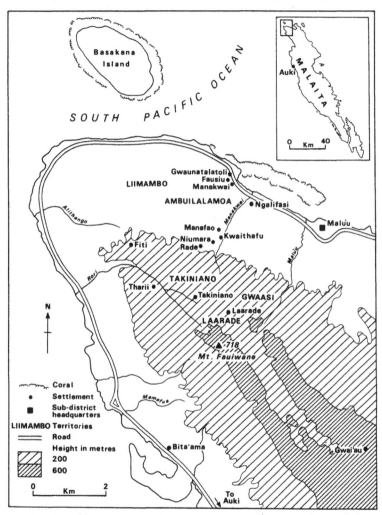

FIGURE 11.1 *Northwest Malaita*

A crucial date in recent To'ambaita history is 1894, when a Queensland labour migrant, Peter Ambuofa, returned to his island and began a christian settlement at a place now known as Malu'u (Young nd 1925: Fig. 11.1). The introduction of christianity by such labour migrants, joined later by Australians from the Queensland Kanaka Mission, set in

227

process in north Malaita a succession of social, political, and economic changes. The first christians, and the missionaries that followed them in the early 1900s, met resistance from a society that, until then, had gained strength from experience of European contact. Conversion thus became a slow, gradual process that is still incomplete and created a deep division within To'ambaita society between christian and pagan – a division that signifies important contrasts in outlook and attitude towards foreign influences (see Keesing 1967).

This division is most apparent in the organization and distribution of To'ambaita settlement. As people joined christian missions, they relocated from their small, dispersed hamlets, in most cases from the hills to the coast. Gradually increasing in size and number, christian villages now occupy and use all available coastal land. They tend to be much larger than the pagan, forming substantial communities of up to thirty households held together by the church and its organization. While kinship and descent are important factors in both christian and pagan villages, the former contain many more distantly and not related persons. In 1971 christians accounted for eighty-nine per cent of the To'ambaita population in residence. The strongest sect remains the South Seas Evangelical Church (SSEC), which traces its history to the early work of Peter Ambuofa and the Queensland Kanaka Mission.

Christianity and coastal residence are synonymous with numerous economic and social initiatives during this century, the most important of which have involved cash cropping and formal education. Coconut planting began with a few individuals in the decade 1920–30. All former labour migrants and early converts to christianity, they modelled their groves on those of the large commercial estates. Today, coastal land is extensively taken up with coconuts, most of which however have been planted quite recently (since 1960) and whose potential in the 1970s was still to be realized. In 1971, a survey of plantings in the large coastal village of Manakwai (Fig. 11.1) revealed an average of 550 trees per household, only a quarter of which were mature and bearing fruit (Frazer 1971:Table 3.5). To'ambaita also have extensive holdings of cocoa, which they planted in enthusiastic response to a vigorous government campaign between 1959 and 1965 to have the crop accepted throughout the island. In Manakwai, they established more than thirty hectares in five years. Although the largest producers of cocoa on Malaita, the crop never met expectations and future prospects are not good. Per capita income from cash crops in 1971, in Manakwai village for example, was $Australian 16 ($A 2.16£1.00):$A5 from cocoa and $A11 from copra (Frazer 1973:Tables 3.5, 3.8).

Formal education amongst the To'ambaita began with church schools run by missionaries and followers of the South Seas Evangelical Mission, which also provided a two-year course in bible training to prepare its followers for leadership and active church work. In 1950, the government established a primary school at Malu'u (Fig. 11.1), thus widening the kind of education available locally. The curriculum was broader and more secular, eventually offering seven years elementary education, advancement to secondary school, and subsequent specialized training. Attendance began very slowly, with children drawn only from christian settlements and in far fewer numbers than the eligible total. Within a generation, however, formal education has become one of the main influences of increasing To'ambaita participation in wage employment.

The To'ambaita economy combines the cultivation of both subsistence and cash crops. It is heavily monetized, except that residents continue to produce a large proportion of subsistence requirements. Both the sizeable amount of cash cropping and a high rate of population growth means that subsistence production is becoming less practicable and suitable land is in high demand, particularly in the most densely populated areas along the coast in either direction from Malu'u. Apart from agriculture, To'ambaita operate many small businesses, as individuals, families, cooperatives, and small companies. These include retail trading, processing and marketing cash crops (copra, cocoa, chillies), raising cattle, and transport ventures. Since the late 1960s a new road has connected all coastal settlements with Auki, the main government centre and port on Malaita, eighty kilometres to the south (cf. Figs. 5.1 and 11.1). Within a few years of completion, numerous vehicles were offering freight and passenger services and transport enterprises became a popular means of local investment.

Population growth is a critical factor for continued socioeconomic development amongst the To'ambaita. Since 1957, a clinic and rural hospital at Malu'u has served much of north Malaita and the people's overall health has improved accordingly. An intensive campaign to eradicate malaria began in 1970. Between then and 1976, the *de facto* population grew from 4,411 to 5,110, at an average annual rate of 2.6 per cent (Groenewegen nd 1972:Table 2A; Solomon Islands, Statistics Division 1980:Table 3). This growth was confined to the densely-populated coastlands and compounded by continuing movement down from the hills, whose villages suffered a net loss of population during the same period (1970-76). Demographically, the To'ambaita population is young and in 1976 forty-eight per cent were aged less than fifteen.

Large-scale disparities within and between different sections of the population, notably in income and access to productive resources, indicate that during this century not everyone has participated in socioeconomic change in the same way or to the same extent. The greatest contrast remains between those living on the coast (christian) and in the hills (pagan). The latter have fewer cash crops and sell some garden produce, but rely mostly on wage employment, taking casual work on the coast or more commonly going to Honiara and other centres. There, because of poorer educational qualifications, they are confined largely to unskilled and semiskilled jobs. Coastal villages are relatively more prosperous, have extensive holdings in cash crops, and are engaged in various forms of business activity, none of which moreover has diminished their involvement in and dependence on wage employment. As the coastal economy has become more heavily monetized, so going away to work has continued, fuelled by local opportunities for formal education that has enabled many migrants to accept higher paying jobs. Today, the disparities and bases of socioeconomic status among the To'ambaita are firmly established; they reflect varying success in agricultural production, rural business, and wage employment.

Labour circulation and urban employment

Movement between Malaita and centres of wage employment extends more than one hundred years to the beginnings of the Pacific labour trade (Corris 1973). Although labour circulation is not the only form of movement in which To'ambaita have been involved, it is distinct from all other, has shown a high degree of continuity, and has made the greatest contribution to socioeconomic change in north Malaita since the beginning of European contact. After the Queensland labour trade was concluded in 1903, recruiting continued within the Solomon Islands, helped by a British administration that had strong interests in the expansion of commercial plantations. Compared with Queensland, employment conditions and treatment of workers were poor but from 1911, when the length of contract was reduced, the wage-labour movement changed little until the Japanese invasion in 1942, thirty years later (cf. Connell, this volume). Labourers were recruited under two-year contracts, picked up and returned to the anchorage nearest their village, and directed to all major foci of commercial development – particularly north Guadalcanal, the Russell islands, and the western Solomons (Fig. 5.1).

Little in the structure and organization of paid employment during this period encouraged long-term commitment to wage labour. Few Solomon Islanders, let alone To'ambaita, were offered positions of responsibility or made any advancement on the job. Between places of employment, *fanua arai kwao* (place of the white man), and *fanua kia* (our land) there was almost total separation, each being associated with different kinds of work, food, human relations, and social environments. Some migrants stayed away for extended periods, some eventually severed their ties with Malaita, but most completed one or two contracts and then returned. Selling one's labour abroad became both a source of income and a form of achievement confined to one period of the male life-cycle: adolescence and early adulthood. An integral part of the experience accepted and sought by both christians and pagans, this circulation of young adults contributed heavily to cash incomes among the To'ambaita but beyond that was considered something quite separate and distinct.

Between 1942 and 1945, when the Japanese invaded the Solomons and an American counter-offensive took place, regular commercial activity ceased. Recruitment of labour continued for the British Solomon Islands Defence Force and the Solomon Islands Labour Corps, while the uniqueness of wartime employment with American personnel, higher wages, and additional benefits from ready access to goods or the sale of curios ensured that the level of To'ambaita participation was high, particularly among young men. Between 1946 and 1949, more than ninety per cent of the To'ambaita became strong followers of Maasina rule (Keesing 1978), which among other actions imposed a boycott on labour recruiting from Malaita until a demand for increased wages was met. This was the first prolonged interruption to To'ambaita involvement in wage work abroad since the 1880s and also the first time they had joined in large-scale and collective action. In fact, the issue of conditions of work beyond Malaita constituted only a small part of the protest being made through Maasina rule and ultimately had little effect on Malaitan dependence on wage employment. Labour conditions nonetheless did improve and this protest came to represent a critical turning point in their historical involvement of moving to *fanua arai kwao*. For the To'ambaita especially, new and greater economic gains were made from this time, which paradoxically meant that they became ever more deeply committed to going away for paid employment.

From 1949, opportunities for wage work steadily increased and the range of opportunities widened. For the first fifteen years, commercial plantations continued to be the main source of employment. Contracts were shorter, one year instead of two, but rather than working less To'ambaita males simply left the village more frequently, often varying their destination from one year to the next. Several became overseers, recruiting and supervising their own work gangs, which at the time constituted as far as any Solomon Islander could advance. Although less oppressive than before the war, plantation work remained poorly paid and islanders had to suffer many iniquities.

The transition in To'ambaita job preferences and employment patterns reached a major turning point in 1964. Many men were then employed by Lever's Pacific Plantations Limited, the largest private company operating in the Solomons, on their estates in the Russell islands (Fig. 5.1). Large numbers of the company workforce had joined the British Solomon Islands Ports and Copra Workers' Union, the first trade union formed by islanders (in 1961) and largely through the initiative of Malaitan workers. In 1964, the union attempted to negotiate with Levers for a wage increase, the company refused, and in

231

September most of the Russell islands workforce went on strike (Frazer 1981b). Although a short-lived protest that failed to extract concessions from the company, labourers from north Malaita decided to leave and many To'ambaita men took up other jobs: on Malaita, in Honiara, and at other places of commercial development. This strike set the seal on a trend that was already well advanced and from that time the numbers in plantation employment steadily decreased.

Honiara became the new capital of the Solomon Islands in 1945, when it formed a part of a huge American military base that had been constructed after the Japanese were defeated on Guadalcanal and was used as a staging area for the rest of the Pacific campaign. Rather than restore the old capital of Tulagi (Fig. 5.1), which had been destroyed, the British administration decided to establish new headquarters from the extensive facilities that already existed. At first, only a small number of To'ambaita were employed by the administration and its officers, as policemen, clerks, labourers, and servants. As more work became available, numbers increased, the movement of labour became more regular, and by 1949 work gangs were being recruited in the To'ambaita area for short-term employment in town, which often was fitted between longer periods on plantations. Stevedoring and port work became popular when two To'ambaita brothers began acting as regular recruiters for the Port's Authority, offering guaranteed employment on three- and six-month contracts. Some men stayed in town longer, taking on other jobs as available, and in time the pattern of movement for wage work became far more variable. The practice of securing a job through direct approach to the employer grew increasingly common and by 1960 had largely replaced formal recruiting.

Starting in the late 1950s, the labour market in Honiara underwent major changes that had long-term implications for To'ambaita movement to town. Until then, urban growth was very slow and Solomon Islanders working in Honiara constituted a small, highly transient, and relatively undifferentiated labour force. Large allocations of British aid from the late 1950s led to increased government expenditure in town (Bellam 1970:72); major development projects were initiated and by the early 1960s a construction boom was underway. The urban workforce grew steadily from 1,411 in 1960 to almost 5,000 in 1969 (British Solomon Islands 1970: Table 11.4) and measures were also taken to increase the level of skill and cumulative experience amongst employed islanders. Although recruitment of overseas people to fill higher-level positions was still common, expanding opportunities for Solomon Islanders gradually appeared.

A career structure for civil-service employment already existed, attracted those with the highest educational qualifications and offered, in turn, the best wages and conditions. Another career structure, established in 1960 for daily-paid workers, aimed to encourage manual skills and on-the-job experience. It incorporated graded wage scales, with promotion based on continuous service, trade qualifications, different levels of expertise and responsibility, and operated through the Public Works department (PWD) — then the largest employer of labour in town. Having worked in Honiara through the 1950s, the To'ambaita were well placed to take advantage of these new opportunities. In addition, simply being in town was attractive, and permitted more scope for the kind of lifestyle that young men pursued through labour circulation. Thus the process of settling into regular employment represented a response to both sets of influences. Large numbers of To'ambaita joined others who had been with the PWD since 1949. The better educated and more experienced qualified for promotion; as a rule they did not advance very far or fast and the resultant increase in wages was not large, but upward mobility was now a feature of paid employment.

As the size of the workforce expanded and people remained longer on the job, there was increased demand for housing suitable for married men with families. Until the late 1960s, the few such houses were allocated to and part of the benefits enjoyed by civil servants and other higher-level employees, while bachelor quarters remained the usual form of accommodation offered to workers. Changes occurred about 1968, when government initiated a programme for low-cost house construction (Hughes 1969). Although only higher-level employees could afford to either buy or rent these dwellings, the scale of construction ensured their availability to a far greater number. Being well established in the upper grades of the workforce, the To'ambaita quickly applied for such houses and began a new form of residential mobility, spreading themselves more widely throughout Honiara.

Whether it was government, commercial companies, or private individuals who employed To'ambaita until the late 1960s, it was never Solomon Islanders and that reflected the limited role of the latter in the urban economy (Bellam 1970). This entrenched pattern began to change and during the next decade To'ambaita became involved, in a small way, in independent business activity: petty trading, produce marketing, taxi operations, building construction, and trade work. Such activities were often combined with a regular job and few entrepreneurs were entirely self-employed. Opportunities for further expansion are favourable, because To'ambaita have the capital, the experience, and place a high value on entrepreneurial activities.

To'ambaita experience since 1960s shows varying involvement in the Honiara labour market and increasing differentiation amongst migrant wage-earners. At the same time as most men were working for short periods at unskilled and low-paid jobs, others with formal education and vocational experience were responding to opportunities for advancement, promotion, and fringe benefits, and becoming increasingly committed to more regular and long-term employment. As has been noted in Papua New Guinea (Levine and Levine 1979:30), age, formal education, and occupational success have become the basis for differentiation amongst migrants in town. One of the more obvious ways in which these differences are being accentuated is in alternative living arrangements, for young To'ambaita in poorly-paid work continued to remain in bachelor quarters while the older and more successful occupied their own houses with families.

Employment and living arrangements in Honiara, 1974

About 500 To'ambaita were estimated in 1974 to be living in Honiara and slightly fewer (434) in the hinterland of rural Guadalcanal (Frazer 1981a:254–7). Even in town they formed a widely dispersed and mobile population, not concentrated in any one location or kind of work. Migrants came from the most isolated as well as the most accessible of villages in north Malaita. Detailed enquiries in Honiara were conducted between March and August 1974 amongst To'ambaita originating from localities well known on the basis of previous field research (Frazer 1973). These included Manakwai village and all christian and pagan hamlets nearby, several christian hill villages around Gwai'au south of Malu'u, and christian settlements around and including Bita'ama on the southern coast of To'ambaita peninsula (Fig. 11.1).

Information about movement, employment, and living in town was gathered from 199 To'ambaita migrants, who were estimated to account for thirty-nine per cent of the total then resident in Honiara. The population structure of this group summarizes certain characteristic features of moving for paid employment (Fig. 11.2). Males in town outnumbered females, with the greatest number of both sexes being in the age range of fifteen to thirty-nine years. The presence of both married women and children is relatively recent. Of men and women aged more than fifteen, only forty-one per cent of the former are married compared with eighty-two per cent of the latter (43 of 105 versus 27 of 33). Thirty of the forty-three married men were residing in town with wives and/or children. Although migrants constitute a younger population than the To'ambaita in general, this demographic profile reveals that their age range is becoming progressively wider and that migrants remain in Honiara after marriage and to raise children.

In May 1974, 82 of the 105 adult males in town were gainfully employed (Tables 11.1 and 11.2). The proportion without work (22 per cent) is not unusual and includes visitors, others newly arrived and searching for employment, and those formerly with jobs who had been dismissed or left of their own accord. Unemployed To'ambaita always stay with those who have both work and accommodation, usually close kin and other people from the same villages on Malaita.

234

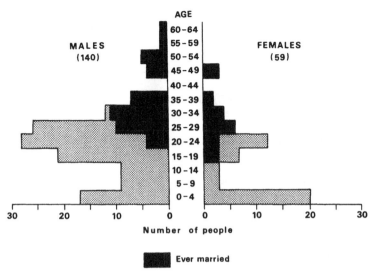

FIGURE 11.2 *Age-sex structure of To'ambaita in Honiara, 1974*

Some are quite mobile and shift their domiciles frequently between various relatives; most are young men with little formal education and few skills who seek whatever casual work they can find. The older men amongst the unemployed are in town for short periods, looking for a job to either offset urban expenses or meet some particular need for money.

Amongst To'ambaita men with jobs, there is a distinct difference between the public and private sectors of employment. Half of them work for government (Table 11.2) and, as already noted, it is here that To'ambaita have achieved greatest success and advanced farthest in the workforce. Better educated and occupying the professional, semi-professional, and skilled levels (Table 11.1), their prospects for further promotion are good because of strong political pressure to increase the participation of Solomon Islanders throughout the urban workforce. Given their success, government employees tend to be more stable than those found in private companies. Of forty-one working for government, eleven are civil servants (27 per cent) and twenty-two daily workers (54 per cent; Table 11.2). Whereas the latter are employed mainly by PWD, the former are spread throughout several departments, involved mainly in administrative and professional activities, and average 6.4 years of service. Amongst workers paid on a daily basis, all but four are skilled and semiskilled, engaged mainly in such trades as plumbing, painting, carpentry and mechanics (Table 11.1). Three of twenty-two have been with PWD for more than ten years and another eleven for between five and ten.

235

TABLE 11.1 Occupations of To'ambaita males in Honiara, May 1974

		Occupations	
	No.	Total	%
Professional			
Government, church (SSEC) administration	2		
Agricultural research, extension	2		
Dentist	1		
Building contractor	1	6	5.7
Semiprofessional			
Clerical worker, bookkeeper	5		
Police	3		
Museum assistant	1		
Salesman/representative	1		
Telecommunications technician	1		
Building supervisor	1		
Church-evangelical	1		
Power station operator	1	14	13.3
Skilled			
Taxi (5), bus (1), other driver (7)	13		
Carpenter/joiner (4), painter (1)	5		
Mechanic (2), painter's assistant	3	21	20.0
Semiskilled			
Store (9), electrical assistant (2)	11		
Classified worker (PWD)	6		
Domestic	2	19	18.1
Unskilled			
Labourer	8		
Store worker	7		
Factory hand	7	22	21.0
Unemployed (including visitors)		23	21.9
Total		105	100.0

Source of data: Field survey.

Most civil servants (nine of eleven) and half the daily-paid workers (eleven of twenty-two) are married and have their families with them in Honiara. One is a house owner, having taken advantage of the special scheme launched in 1968, while the rest live in married quarters and low-cost houses rented from government. Such housing is allocated through a point system based on years of service, level of income, and

TABLE 11.2 Employers of To'ambaita males in Honiara, May 1974

		Employers	
		No.	*%*
Government and statutory authorities			
Daily-paid workers:			
classified (18), labourer (4)		22	
Civil servant		11	
Ports Authority		4	
Electrical Authority		2	
Housing Authority		2	
	Total	41	50.0
Churches			
South Seas Evangelical Church		2	
Watchtower Movement		1	
	Total	3	3.7
Expatriate companies and individuals			
Retail and wholesale trading		17	
Bakery		5	
Tobacco factory		3	
Domestic service		2	
Other		5	
	Total	32	39.0
Solomon Island entrepreneurs			
Taxi operator		3	
Casual work		1	
	Total	4	4.9
Self-employed			
Building contractor		1	
Taxi operator		1	
	Total	2	2.4
Total		82	100.0

Source of data: Field survey.

size of family. Although To'ambaita have had considerable success in securing these houses, they have little choice in their location and thus are now widely dispersed in ethnically diverse neighbourhoods.

In contrast, government workers who are either unmarried or married but without their wives and families are concentrated in bachelor quarters at Kukum suburb (Fig. 11.3), the site of a great number of labour compounds for many departments. The standard dwelling is a barrack-type building, divided into four rooms, each with an outside

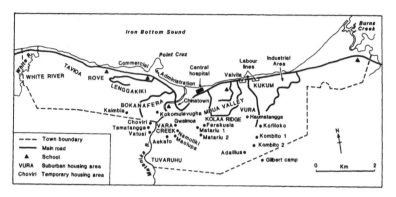

FIGURE 11.3 *Honiara, 1974*

doorway. Originally allocated to rooms with workers from other language groups, the To'ambaita have deployed themselves so that they share their's with kinsmen and other migrants from the same localities on Malaita. This practice is so well established that several such rooms are identified in terms of common villages of origin; each is responsible for its own cooking arrangements and to this extent constitutes a self-contained bachelor household. Although the main occupants are PWD workers, these rooms are used for varying lengths of time by many other 'wantoks' (close kin, plus people from the same locality). As distinct from the regular occupants, the total membership of these households is extremely variable and highly mobile.

In the private sector, To'ambaita concentrate in factory work, retail and wholesale trading, while others hold jobs as domestic servants, bus and taxi drivers, labourers, and tradesmen for commercial companies. Much of this work is unskilled, low paid, and taken on as easily as it is left. Turnover is far greater than in government employment and highest among factory workers and store assistants. At the tobacco plant and a large bakery, the main factories where To'ambaita are employed, older migrants obtained supervisory jobs during the year when each was established (1968 and 1972), which in turn provided the basis for continuing To'ambaita involvement. One, on the staff of the tobacco factory as salesman and supervisor, has the privileges and income equivalent to higher-paid civil servants but is the sole person with such career possibilities. Both plants have labour compounds, where To'ambaita workers can form bachelor households that are closely tied through 'wantoks' to those in the PWD labour lines (Fig. 11.3). Two of the few married quarters that the tobacco factory has to allocate to managerial staff and experienced workers are occupied by To'ambaita families.

Jobs in the retail and wholesale trade (Table 11.2) mostly involve working for Chinese businessmen in Chinatown, a small, concentrated,

238

commercial area on the eastern bank of the Mataniko river (Fig. 11.3). Since individual traders do not employ many persons, To'ambaita are widely spread, often work alone, and usually live on the premises or in the surrounding yard. Some employers offer half rations in lieu of a full wage but the work, generally the lowest paid in Honiara, offers very little scope for promotion, is undertaken mostly by the young and recently arrived, and involves quarters that are small, makeshift, and of very poor standard.

Given the size of their national organizations and the scope of their activities, the main christian faiths are major employers both throughout the Solomons and in Honiara. The South Seas Evangelicals have their administrative headquarters and a religious book store in town. With the precedent of being involved fulltime in church activities for several decades and the continued strength of religion in village and regional affairs, To'ambaita are well placed to enjoy a wider role and some have pursued careers in church work after attending special training institutions termed bible schools. Many more To'ambaita in Honiara are employed by religious organizations than indicated in Table 11.2; they include teachers in church schools, fulltime evangelists and pastors, executive officers, administrative and clerical staff. National directors and secretaries of various churches, positions often held by To'ambaita, generally reside in town and are provided a house for their wives and families.

Although a building contractor and taxi operator were the only self-employed To'ambaita interviewed, the former is one of the most successful Solomon Islanders in Honiara business. He began independently in 1972, after working many years as carpenter and foreman with a private construction company. He employs several tradesmen and labourers, owns a house in town, and in early 1974 bought a large truck to start diversifying his activities with cartage work. The taxi business is operated at a much smaller scale by a man who worked many years for an overseas shipping line and, on returning to Honiara, used his savings to buy a taxi. At first he was the sole driver but later employed another To'ambaita man to share the work. As a sideline, he grows sweet potatoes for the Honiara market. Land on the outskirts of town, of which he has temporary use, has been cleared and planted with the help of other To'ambaita migrants paid on a casual basis.

Responding positively to opportunities as they arose according to previous experience and qualifications, the To'ambaita are now an integral part of the expanding urban economy. Working primarily as wage and salary earners, they fill positions in a range of occupations at all levels of the workforce. Gradual improvements in conditions of employment, opportunities for advancement, and the circumstances of town residence have been the primary incentive to continue with paid jobs. Compared with the past, much greater attention is given to relative advantage and comparative success, which in turn has become a major influence on the decision whether to remain in town. For those with the requisite formal education and on-the-job experience, regular paid employment in Honiara has become the means to assume an alternative way of life to that of the To'ambaita headland of north Malaita.

Segmentation and social ties

To'ambaita experience in Honiara between the years 1960–75 reveals emerging differences in job stability and continuity of employment. Some migrants have settled into regular jobs and become long-term employees; others have approached employment more casually, working for short periods before taking another job or returning to Malaita. To'ambaita experience confirms the point made by Young (this volume), supported by Morauta and Ryan (1982:39), that circulation and permanent migration are not 'two distinct processes'. While many migrants appear to be committing themselves to regular employment and separating more permanently from Malaita, it is not something about which most have made a definite decision. Working in town devolves around a choice of how long they will stay, but that the choice remains is taken for granted.

Increasing differentiation among To'ambaita migrants in Honiara becomes clearer when those interviewed in 1974 are grouped in four categories according to length of time spent in their present job (Table 11.3). Casual workers, represented in the first and second categories, include the unemployed mostly looking for work along with those in a job for less than two years. Most are in unskilled work and inclined to remain for only short periods. In contrast, steady long-term employees (category IV) are men who have held the same job for at least five years.

TABLE 11.3 Length of time To'ambaita in present job: Honiara, May 1974

	Employment category	To'ambaita migrants	
		No.	%
I	Unemployed	23	22
II	Employed in present job less than two years	21	20
III	Employed in present job 2–5 years	34	32
IV	Employed in present job for five or more years	27	26
	Total	105	100

Source of data: Field survey.

All are older, mostly married (21 of 27), mainly government employees (20 of 27), and very few have not moved into the better kinds of accommodation: mainly married quarters or family housing. Some migrants in their present job for two to five years, the intermediate category (III), will remain and become long-term employees whereas others will not. Included in this group are civil servants and classified workers, but more than half (19 of 34) are in semi- and unskilled jobs with few prospects for advancement.

Given continued expansion of the urban labour market and of the job opportunities provided migrants, the overall number of steady long-term workers will increase and To'ambaita will display an ever-increasing commitment to being employed in town. In the mid 1970s, however, up to three in five To'ambaita were unemployed or held semi- and un-skilled jobs. They earned a bachelor wage, lived in quarters designed for single persons, and had limited prospects for advancement. Without greatly increased remuneration for wage labourers, there will be little encouragement for migrants employed at these levels to commit them-selves to longer-term involvement and reduce their high rate of turnover.

Amongst To'ambaita, being stable or mobile in employment are relative conditions that vary closely with age, length of urban experi-ence, and educational status. High turnover and constant shifts within and between workplaces is something specifically associated with the young: *daraa* (bachelors) or *kala wane* (young men). A To'ambaita male newly arrived in Honiara might begin with frequent changes of job and of residence, then gradually become more stable as a particular kind of work is done over a period. Although the longer that migrants remain away from the natal village the more likely are they to become committed to one job, the critical factor is formal education and the prospects that it provides for upward mobility. In Solomons pidgin, a man with educational qualifications is referred to as 'wane sukulu'. On Malaita and in town, 'sukulu' has two levels of meaning: one which designates a person as being from a christian settlement, the other as having formal education. In Honiara, reference to a 'sukulu' migrant stresses the educational dimension, with christian background implied. 'Sukulu' migrants normally have at least five to seven years of formal schooling and in 1974 included twenty-nine per cent of all To'ambaita in town.

There are two exceptions where upward mobility can occur without the training received from primary education. One embraces the positions of overseer or supervisor, including being 'bosboi' of a work group. Being placed in charge of labour, on plantations, in factories, on the waterfront, or in building and construction, is a prospect equally open to both educated and uneducated. The other is becoming driver of a taxi, small truck, or bus. Whilst not easy to secure the training and experience necessary to qualify for a driving licence, so many To'ambaita migrants have achieved this that it is one of their more common skills, irrespective of prior formal education. Driving is considered to be good work, carries much prestige (particularly amongst younger men), and is a skill that is increasingly more marketable − both on Malaita, where the number of vehicles is rising by the day, and in Honiara, where transport businesses have experienced rapid growth.

These exceptions apart, it is 'sukulu' status that underlies the main differences in achievement amongst To'ambaita migrants in Honiara and defines the prospects of higher pay. This process has its parallel in Papua New Guinea (eg. Ward 1980), where the workforce has become segmented between those who are formally educated, upwardly mobile, and move between the towns; and those who remain confined to low paid, unskilled jobs and continue to circulate between rural and urban places. Making formal education more widely available on Malaita will not necessarily remove this divergence amongst migrant workers in Honiara, since the standards for entry into career employment have risen the past twenty years. Nonetheless formal education, as a criterion for divergence, is much influenced by the wider social, economic and political environment and there exists in this fact the conditions for a major transformation in the persistent circulation of To'ambaita labour.

A stronger commitment to urban wage employment and more extended absence from Malaita raises the possibility of a severance or weakening of rural-based social ties, a reduced interest and lack of involvement in village affairs. Thus far there is little evidence for this amongst the To'ambaita. On the contrary, older migrants strongly committed to working in Honiara retain close links with kinsmen and others on Malaita, continue to affirm their attachment to rural communities and in many cases take an active and responsible interest in local activities. Not only do they remain as much a part of town groups formed by close kin and people originating from the same Malaitan localities as do more recent migrants, but also are often leading members of them. Longer-resident To'ambaita represent a stable and responsible influence amongst their younger, more numerous, and itinerant 'wantoks' offering advice about living and working in town, intervening in case of trouble, and organizing the occasional 'pati' or other gatherings for kin.

In fact, by taking up more regular employment and obtaining more secure living quarters, the older and more successful To'ambaita in Honiara have created a situation that helps promote and sponsor the shorter-term movement of others. Some new arrivals are seeking employment, others come only to visit, but all are dependent on more established kin for a place to sleep, for meals, and other kinds of assistance. Consequently a much wider range of people circulate nowadays than previously. Able-bodied males, especially those aged between fifteen and thirty, still predominate but are joined by dependent relatives: parents, younger brothers, sisters, wives, children, even more distant kin. To'ambaita describe contemporary movement to Honiara as *liu* or *liliu* ('walkabout', to wander around). The word *liu*, common to several Malaitan languages, is now part of Solomon Islands pidgin, in which it also refers to unemployed persons.

To'ambaita use this term primarily to describe those who move around seeking work, as well as those who are visitors. *Liu* thus emphasizes the mobility element in behaviour and has acquired pejorative connotations deriving from the fact that mobile persons who frequently change jobs usually depend on others while pursuing that particular lifestyle. As a phenomenon that has emerged from residing and working in town, *liu* movement not only has been made possible by all the To'ambaita away from north Malaita but also reflects and accentuates the socioeconomic differentials that have arisen amongst those in Honiara. This divergence spreads to include villagers living in To'ambaita territory who, with increasing frequency, visit town for short periods and in so doing keep migrants in touch with each other and with kin on Malaita. The complementary processes of increased circulation between the home territory and Honiara, and of To'ambaita long resident in town committing themselves more strongly to regular wage employment, ensures the maintenance and continued strength of social ties between both migrants and those who remain in the village.

When To'ambaita first arrive in Honiara, they draw upon a wide network of preexisting social relationships: the immediate family, people with whom they grew up, descent-group relatives, wider consanguineal relations, affinal kin, school friends, and other peers. Amongst close kinsmen, there is a moral obligation to provide material support and assistance required: meals, a place to sleep, gifts of money. Such kin, as well as persons from the same descent group, are 'my people' (*to'a nau ki* or *imole nau ki*) and will never be ignored or neglected no matter how long To'ambaita remain in town.

Young men maintain closest social connexions with peers, particularly if from the same localities on Malaita, and thus form loosely-integrated groups linked to different places of work and residence. They maintain solidary ties through constant visiting and sharing of urban interests, the strength of which do not depend on organizational structures; there are no formally-instituted leaders, no hierarchy of statuses, no institutionalized meeting places. The durability and importance of such ties derives from moral obligations shared as kinsmen and close friends, extending to regular material transactions between them.

For some time after To'ambaita come to town, their exclusive social world revolves around rural-based groupings. With longer experience, new ties are formed, particularly with persons belonging to the South Seas Evangelicals and the Jehovah's Witnesses. To'ambaita are prominent in both churches, although in neither case is attendance and participation as strong as on Malaita, and their continued association follows directly from the important role of the church in the village. In Honiara, maintaining contact with 'wantoks' through the church is sufficiently well established that people from the same village or locality form distinct entities within the broader congregations. Some migrants have become actively involved in church affairs and thus become members of groups that are ethnically diverse and defined by a common faith.

Migrants maintain rural-based ties with unemployed kin by the constant sharing of their material advantage through gifts of cash, food, clothing, and numerous other goods. The less earned in town, the greater the proportion required for urban living. Thus many younger migrants, usually with lower incomes, confine sharing to each other as a way of pursuing their particular lifestyle. Drinking, gambling, movies, dances, and buying clothes − the kinds of enjoyments they prefer − are a constant drain on their resources. Yet unemployed relatives are not entirely disregarded. In one case observed for two months, a young To'ambaita with monthly earnings of $A26 ($A1.60 to £1.00) shared forty-seven per cent with other kin, some of whom were in Honiara and some in north Malaita.

The ways in which older and more successful migrants deploy their earnings is a recognized test of the importance attached to rural-based ties (cf. Strathern, this volume). Most acknowledge this and go some way to meeting demands made of them (cf. Connell 1981:234). Since to attempt to satisfy everyone would be impracticable, expenditures are meant to sustain good relations with kin who have no job. This means providing hospitality and cash to visitors and new arrivals, as well as remitting money, food, consumer goods, clothing, and building materials to people in north Malaita. The kinds of investment decisions made by long-term employees are another key indicator of rural commitment for options exist both in Honiara (establishing a business, buying houses or land) and on Malaita. Thus far, as illustrated by Basi'oli of Manakwai (Fig. 11.1), long-term workers combine both urban and rural interests.

Basi'oli is aged twenty eight and has been working for PWD for nine years, three on Malaita and six in town. He is a classified worker, son of a church teacher, and the eldest of four children. His father, stepmother, a younger brother with wife and family, and a younger sister all live in Manakwai village. All have visited and stayed with Basi'oli in Honiara; another younger brother living with him is seeking work. Basi'oli's wife, whom he met in town, is from another language group on Malaita (Kwara'ae) and they have three children. Early in 1974 the family lived in a low-cost house rented from government in the suburb of Bokanafera (Fig. 11.3), immediately south of the town's centre. All that time, depending on overtime and special bonuses, Basi'oli earned between $A70 and $A80 each month.

The active ties in Basi'oli's social network include a wide range of people: from Manakwai village, To'ambaita migrants, and from different language groups. They include past and present PWD workmates, people met during years spent in the labour lines, neighbours from Bokanafera suburb, town members of the SSEC congregation, and many other acquaintances. Amongst his friends are periurban migrants from Baelelea, linguistic and cultural neighbours of the To'ambaita on Malaita, through whom he obtains firewood, sweet potato, and other vegetables. One of his close friends, another PWD worker, comes from Tikopia in Temotu province (Fig. 5.1).

Being highly paid and having accommodation in town, Basi'oli is well able to assist other To'ambaita. Close kin stay regularly, others visit for meals, new arrivals are given small amounts of money, and near relatives may receive more substantial assistance when required. Adult villagers in Manakwai, where Basi'oli was born, are amongst the more prosperous To'ambaita in north Malaita, since most receive a regular income from cash crops. Although his family do not have extensive holdings, their income is sufficient to meet all basic requirements. This does not mean that Basi'oli spends less on rural ties or regards them as less important; rather that different kinds of assistance are given.

Having been raised in a christian village and attended primary school for several years, Basi'oli belongs to a peer group with much success at wage employment in Honiara. The men with whom he most closely associates include a higher-level civil servant (a second cousin, of his descent group), a trainee salesman and staff member at the tobacco factory (distant member of the same descent group), a bosun with the Marine Department, and a corporal in the police force (both peer friends from his natal village). All have held urban jobs for at least five years and all earn more than $60 a month. Meeting regularly but informally, this group has launched several investment projects both on Malaita and in town that are jointly financed. Largely initiated by Basi'oli and his second cousin, these projects carry the name *tabi*, meaning 'to branch or fork'.

Initially Basi'oli began a hawkers business, based in his house at Bokanafera, selling tobacco, soap, rice and other food stuffs to neighbours and To'ambaita migrants. About the same time the investment group started collecting money for a cattle project on the southern side of the To'ambaita peninsula (Fig. 11.1), where Basi'oli and his family claim ancestral land. In 1973 two further schemes were initiated: one involving the construction of a family house at Manakwai village using imported building materials, the other a taxi business in town. In November, Basi'oli took his annual leave and returned to build this house with the help of other villagers. During the same month, the business partners bought a car and began a taxi business in Honiara, employing another To'ambaita man to be driver and inviting him to hold shares. This arrangement was still flourishing at the end of 1975.

Taking these successes further, in late 1974 Basi'oli purchased a newly-constructed house at Vura (Fig. 11.3) through the Solomon Islands Housing Authority. Originally intended for the taxi driver, Basi'oli decided to move from Bokanafera since a much larger dwelling had more room for visitors.

Through these investment projects, as much as through the close and visible association with To'ambaita from the same village or locality, Basi'oli is affirming rural ties simultaneous with ensuring more security and greater success in town. His efforts are indicative of a general pattern amongst To'ambaita in Honiara, since at least half those holding the same job for five or more years are involved in business projects on north Malaita, investing in agricultural schemes (coconuts, cocoa, cattle), trade stores, and vehicles used in transport services along the Malaita road. Not all projects are rural-oriented, however, for long-term workers also invest in Honiara, buying houses and initiating business ventures. In this, To'ambaita migrants are exercising options that are not exclusive to urban or rural lifeways but mean continuing involvement in both.

Conclusion

The To'ambaita are now firmly established in Honiara, having exhibited considerable success at seizing opportunities as they became available. This process has been slow and occurred against the background of continuing circulation between Malaita and town. As paid employment grew more attractive and upward mobility became possible, small but increasing numbers became regular and stable workers, often remaining after marriage and especially since 1968, when the family-housing scheme was initiated. Early in the 1960s a form of segmentation became apparent, based largely on differences in formal education and distinguished by a comparative divergence in job mobility and socioeconomic success.

Moving to places of employment arose out of the separation of them from the To'ambaita territory on Malaita. Wage work has always been attractive because it offered more income than was available locally — a differential that still persists despite the growth of cash cropping and other money-earning activities on the home island. When To'ambaita undertook short-term labour circulation, these inequalities of earning capacity were temporary; with more continuous involvement in paid employment they became more sustained; with upward social mobility they are also much greater and definitive. Even so, short-term circulation for low-income and unskilled jobs will continue amongst two kinds of To'ambaita: the young, for whom there is the added attraction of urban life styles; and villagers from the poorer settlements who have few opportunities to earn money. If present trends continue, stratification will likely increase amongst To'ambaita and all Solomon Islander migrants living in Honiara, as differences in socioeconomic status become more sharply defined. Much prestige and respect already attaches to the most successful government employees, skilled workers, and businessmen.

Families and kin share in the successes of To'ambaita migrants, who continue to affirm rural-based ties and to demonstrate attachment to their natal communities. Even with the advantages the To'ambaita now enjoy in town, they are not totally committed to working there and do not regard Honiara as their permanent place of residence. Research in Papua New Guinea, amongst peoples with more extended experience of urban residence (Morauta 1981; Morauta and Ryan 1982), suggests that as these social ties change through the death of parents and siblings and as children born and raised in town become more used to urban rather than rural environments, then the context of decision making changes and it becomes more inevitable to stay in town. At this stage of To'ambaita experience, the option of returning to north Malaita remains important even if not exercised with the same certainty as in earlier decades of wage-labour circulation. Even as this option is expressed, many are making themselves more secure in Honiara (cf. Strathern, this volume). In the mid 1970s, the balance between going back 'home' and staying 'in town' ('long taon') for an extended period turned on success in both employment and business. As more To'ambaita experience such success the process of segmentation, described here at a very formative stage, will not only continue but also become more pronounced.

Acknowledgments

This chapter is based on field research in the Solomon Islands undertaken in 1971-72, 1973-74, and 1975-76, with financial support from the Ministry of Overseas Development, United Kingdom, through the Victoria University of Wellington, New Zealand, and from the Research School of Pacific Studies, Australian National University. I thank Ray Watters, for making the original enquiry possible, and all those To'ambaita whose help was indispensable: especially Arnon Atomea, who first introduced me to north Malaita; and to Jack Buafau, Kwai-mani Festus and Robert Linganafelo, for valuable assistance during fieldwork in Honiara.

Part IV
Rural–urban flows, urbanization, and circulation

12 Bilocality and movement between village and town: Toaripi, Papua New Guinea

Dawn Ryan

Anthropologists discussing circulation and migration are frequently hampered by the fact that detailed field data refer only to a short period and so they must extrapolate in an attempt to discern long-term patterns and trends. By contrast, the material presented in this chapter has been gathered over twelve years (1960–72), with successive censuses making it possible to trace movement and changes in population structure among a particular people, the Toaripi of Malalaua district, Papua New Guinea (Fig. 12.1; cf. Fig. 2.1). Placing these data in broader context, it becomes clear that movements of Toaripi people since 1884, when a mission station was established, occur within an enduring framework of rights and obligations among socially-recognized kin. This network of kin relations is the basis of land rights, and the degree to which obligations are fulfilled or neglected determines whether land claims are likely to be recognized.

The Toaripi number about 5,900 (Seiler 1974) and live in nine villages in Malalaua district. Eight of these are located on the coast, at the mouths of Lakekamu and Tauri rivers; the ninth is inland on the Kapuri river (Fig. 12.1). The terrain is swampy, an ideal environment for the staple sago, but one in which it is difficult to find dry land for village sites. This area never attracted large-scale commercial interests and copra is the only cash crop. Although Toaripi have been in contact with Europeans since 1881, government officials, missionaries, and the occasional trader comprise the few resident expatriates.

People who have moved from village to town since 1945 have not necessarily lost interest in maintaining their rights or claims to land and pursue various strategies to make this possible. In the 1960s most Toaripi in Port Moresby, capital of Papua New Guinea (Fig. 2.1), occupied sites to which they had no legal or customary claim; very few of the men had secure jobs; the people generally were poor and there was no provision for social services (cf. Ryan 1968). Most migrants

251

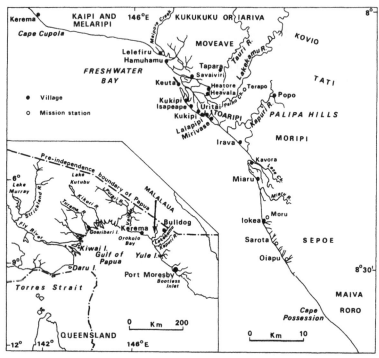

FIGURE 12.1 *Toaripi and neighbouring groups, 1962*

regarded the rural community as a haven and attempted to maintain a place in village-based networks of reciprocity. Those who remained in the village, on the other hand, needed their urban kinsfolk for support and aid, especially on ceremonial occasions. Thus there evolved a bilocal social system, in which people in widely separated locations continued to interact within a framework of kin-based rights and obligations. The degree to which individuals followed a mixed rural-urban strategy in exchanges of goods and services depended as much on the composition of the kinship network as on narrowly-defined economic interests. Unless a person had close kin in the other location, there was no one with whom to maintain ties and the network of exchanges had to be either completely rural or completely urban.

As those born or reared in town grow to adulthood, the continuation of rural-urban links within a bilocal social system becomes uncertain. Since the Toaripi were among the earliest people in Papua New Guinea to move in family groups to the towns, a considerable proportion are likely to become permanent urban residents. On both academic and practical grounds, it is important to monitor changes in population dynamics and observations of Toaripi society over twenty years make

252

it possible to analyze the interaction of traditional structures and new institutions, as well as provide insights about the processes of urbanization and their connexions with population mobility.

Ethnographic material was collected mainly during the six years 1960-65, supplemented by brief periods of fieldwork in 1959 and 1971-72. Between March 1960 and May 1962, twenty months were spent in Uritai village (Fig. 12.1) engaged in research on social change (Ryan 1965) and a further eighteen in Port Moresby (1963-64) studying Toaripi who had moved to the largest town in Papua New Guinea (Fig. 2.1). Aided by field assistants, a social census was conducted of those in town who spoke Toaripi (Toaripi, Moveave, Moripi: Fig. 12.1) to provide a context for more detailed enquiries.

The largest concentration of people from Uritai village was at Vabukori (Fig. 12.2), where adults were questioned about their mobility and work experiences, the history of the settlement and of individual dwellings was recorded, patterns of interaction within the community were observed, and networks were traced of contacts among the various concentrations of Toaripi speakers in town. Documenting processes of social change was helped greatly by the fact that, while in Port Moresby, the Toaripi speakers founded a periurban settlement, so that it was possible to observe how people long absent from their natal villages coped with such problems as the onset of old age (Ryan 1970).

In 1965, brief periods were spent in several district centres – Lae, Madang, Wewak, Rabaul (Fig. 2.1) – enumerating Toaripi speakers and collecting movement and work histories (Ryan 1968). Despite the gradual dispersion of Toaripi from the Malalaua district to Port Moresby and smaller towns, the *de facto* population of Uritai village did not alter dramatically in the period 1960-72, although it fluctuated. In May 1960, it numbered 552; the *de facto* totals in March 1962 and January 1972 stood at 577 and 521 respectively. Material from administration reports and summaries of official census patrols were used to provide context for these successive field censuses, as well as information on broad population trends. Such materials, however, must be treated with caution, since figures reported for *de facto* populations often are inaccurate and the criteria for the definition of both *de jure* and *de facto* populations seldom stated (cf. McArthur 1966).

The Toaripi, 1881-1964

The first recorded European visit to the Toaripi occurred in 1881, when the missionary James Chalmers travelled from Port Moresby along the coast of the Gulf of Papua and called at several villages, including the two adjacent ones of Uritai and Mirihea (Fig. 12.1). Chalmers (1886) cites no figures but the settlements apparently were sizeable, each containing numerous large and imposing men's houses. The Toaripi were not unaware of the presence of Europeans before this visit and were also in contact with other Papuan societies.

Before European settlement at Port Moresby, there had been an established trade (*hiri*) between the Motu (the indigenous people of the area) and groups from the Gulf of Papua, including the Toaripi. The Motu travelled west beyond Toaripi, taking cooking pots and armshells to be traded for sago and canoe hulls. Toaripi travelled east and made the same exchanges. Chalmers had met some Toaripi near the mission station at Port Moresby and his journey to the west apparently reflected the fact that he wished to persuade them to stop disrupting the important *hiri* expeditions by ambushing Motu trading canoes (Chalmers 1887; cf. Barton 1910; Groves 1973). Toaripi obtained bows and arrows from the people of Orokolo Bay (Fig. 12.1: inset) in exchange for cooking pots from the Motu, and string bags and fishing nets from the Tati and Maiva to the east in return for armshells also originating from the *hiri*. In addition they traded with Kovio and Iariva inland, and with the Kaipi, their close neighbours (cf. Chalmers 1898:327; Seligman 1910:204).

Despite the importance of these trading relationships, they seldom led to intermarriage. The Toaripi had a reputation for aggressiveness and most trade between groups was undertaken in a formal manner, as truces in endemic hostility. At the time of Chalmers' visit, the Toaripi were at constant war with the Moveave (Fig. 12.1); they fought, and much feared the Iariva; they occasionally battled the Moripi; and from time to time raided Roro and Motu, as well as ambushing trading canoes. At contact, their social relations were similar to those reported for many Melanesian societies: interaction with other groups was marked by hostility and suspicion, and frequently involved fighting.

Kinship, age-grouping (*heatao*), and affiliation with men's houses were the bases of social life in Uritai and Mirihea, the two original Toaripi villages. These communities were linked by many ties of kinship and affinity, with frequent cooperation between individuals, although as communities they did not necessarily combine for purposes of warfare. In many senses, village members were isolated from the rest of their known universe, despite frequent contact with outsiders in warfare, trading and, occasionally, intermarriage. Toaripi laid claim to all territory between Maipora creek and the Kapuri river (Fig. 12.1) but endemic hostility with neighbouring groups prevented its effective assertion. After Chalmers established a mission station in 1884, pacification followed very quickly; Toaripi were able to range more widely and to establish hamlets in which they lived while gardening and fishing.

Several factors influenced the dispersion of Toaripi from the two original settlements to the nine they inhabit today: patterns of land holding, pacification, increases in population, and erosion at the mouth of the Lakekamu river. Land is divided into named blocks, associated with which are several persons said to have been the first to have laid claim and/or have used it. All who can trace descent from one such person may claim the right to work or live on that piece of land, but acceptance of such a claim reflects several factors: the number of people already exercising rights to the particular plot of ground, alternatives available to the claimant, and above all whether the claimant has behaved like a kinsman.

Toaripi kinship is cognatic: persons of common ancestry may be regarded as kin. If a genealogical tie is to become a socially-recognized relationship, then those so linked ought to participate in a continuous exchange of goods and services. Since within the language group any individual is theoretically kin to virtually everyone else, it is impossible to honour all kinship obligations and the exercise of land rights becomes an important determinant. Most people claim only land to which parents had unequivocal access, but seeking such rights from genealogical kin with whom there has been sporadic exchange of goods and services means that a claim is more likely to be disallowed (cf. Ryan 1970: 61-74).

As pacification enabled Toaripi to settle their territory, land rights tended to become vested in those who inhabited and worked on a particular portion. Gradually, fishing and gardening hamlets ceased to be temporary domiciles and become separate villages. Those at Kukipi (Fig. 12.1), for instance, viewed the surrounding land as belonging to them and the 1921-22 annual report of the Gulf division reported their disputing claims with inhabitants of the parent village. This tendency to form independent settlements as part of a strategy of consolidating land rights was accentuated by population increase. In 1922-23, the annual report for the Gulf division stated that since 1905 the population had doubled in the area between Kerema and Cape Possession. Regular census reports are available from 1920 and these indicate a forty-seven per cent increase in the *de jure* population over the next forty years (Table 12.1). With dry land for house sites generally in short supply, population increases of this magnitude necessitated the spread of settlements throughout Toaripi territory independent of the parent villages of Uritai-Mirihea.

TABLE 12.1 Official de jure *population, Toaripi, 1920-71*

Year	Total	Source
1920	2,387	Patrol Report Kerema 11/30-31
1931	2,750	Patrol Report Kerema 11/30-31
1953	4,019	Village census, Kukipi patrol post, 1953
1962	4,974	Village census, Kukipi patrol post, 1962
1971	5,710	Village census, Malalaua subdistrict, 1971

In 1927 a washout at the mouth of the Lakekamu river destroyed Mirihea and part of Uritai. The fear people had that this might occur was noted as early as October 1919 in a patrol report, and by 1927 permanent houses were being built at such sites at Kukipi, Mirivase and Lalapipi (Fig. 12.1). Partial destruction of the parent villages then removed any possibility of return. Over forty years, up to 1927, the Toaripi population had dispersed considerably. At the beginning, people circulated between the parent villages and fishing or gardening hamlets, as reflected in an official report that in 1914 noted Toaripi at Lelefiru as still maintaining houses at Uritai-Mirihea. With the passage of time, the principles by which land rights are activated and sustained meant that those going regularly to hamlets began to assert overriding claim to neighbouring areas. This tendency was accentuated by population increase and enforced dispersal after the erosion of the Lakekamu river, so that hamlets became separate communities.

Initial land rights to settlement and garden sites derived from membership in cognatic groups in the parent villages, but thereafter shared rights and obligations within the new communities were accentuated at the expense of those with kin who had remained behind. In these circumstances it was unlikely that a multilocal system would evolve. By 1960, these various villages were seen as quite separate, with the vaguely-recognized kin links among individuals residing in them in no way conferring rights to relocate and use land throughout different parts of Toaripi territory. During the period 1962-64, further shifts in the Lakekamu estuary began to erode the site of Isapeape (Fig. 12.1); the response of villagers was to seek a nearby area of dry land which they controlled, rather than attempting to reactivate moribund kin links and disperse amongst other communities.

Wage labour and movement in town

In addition to these changes in population size and distribution, there were external pressures to go away and earn cash. In 1909, gold was discovered at Bulldog (Fig. 12.1: inset) and Uritai-Mirihea was the place through which miners travelled to the field. Despite such opportunities for wage labour, villagers were not especially interested and a government report noted in 1913 that Toaripi were not prepared to sign up to work for longer than one month (Patrol Report Kerema 3/13). Head tax, introduced in 1919-20, forced Toaripi to become indentured labourers and most young men signed one two-year contract. As far as possible, members of the same *heatao* (age group) worked together and, having completed their two-year contract, returned, married, and resumed village life.

The gold at Bulldog was exhausted by about 1918, but three years later exploratory drilling for oil began near Popo (Fig. 12.1). Toaripi were willing to undertake wage labour by the late 1920s but not always able to obtain it, since the oil company and some coconut planters refused to employ them. Commented one administrative official: 'The natives are willing enough to sign on, but their peculiar temperament and disposition make them difficult to handle. They are, it is said, less willing and not more intelligent and capable than the average native labourer who goes to work from other settled parts' (Annual Report Gulf Division 1925-6). During the 1920s and 1930s, Toaripi responded enthusiastically to government schemes for communally-owned groves of coconuts, where each settlement was obliged to set out a 'plantation' as a future source of both copra and cash. Young, generally unmarried men also circulated as wage labourers between their village and expatriate plantations or the town of Port Moresby, and some of their earnings helped meet the needs of local communities.

The Australian army had a camp at Lalapipi (Fig. 12.1) during the second world war and most able-bodied men from Toaripi were recruited as carriers or labourers, while others worked as servants or helpers. From 1946, however, the pattern of wage-labour circulation between villages and towns changed markedly. The reconstruction of Port Moresby, which had been severely damaged, created a tremendous demand for those with trade skills. As a result partly of their association with the armed forces and partly of training schemes sponsored by the mission, numerous Toaripi were able to become subcontractors and undertake tasks ranging from digging ditches to building houses. Reflecting on this experience during the 1960s, these men said that being 'contractors' enabled them to be independent workers and to make good money. Labour was recruited from their own people, by activating socially-significant ties, and squatter settlements arose out of the need to house workers.

The founding of each of nine urban settlements surveyed in Port Moresby in 1963-64 (Fig. 12.2) followed substantially the same pattern. One or more men from the Toaripi language area would ask Motu for permission to build a house on their land. This was a propitious time for such a request, for there was considerable confusion immediately after the war: several Motu villages had been evacuated and their members had only just begun to return; many people were living in abandoned army quarters; there was feverish activity associated with reconstruction of the capital. In thus seeking permission from Motu landowners, Toaripi speakers might call upon long-established trading ties and their consequent links of pseudo-kinship; appeal to recently-shared experiences with the army; or deliberately cultivate residents of a Motu community. If permission was granted, dwellings were constructed of discarded war materials, the men moved in, and later arranged for their families to join them – a sequence that was repeated many times. In the settlement of Konedobu (Fig. 12.2), by contrast, dwellings were erected on crown land. Not all Toaripi males worked either as or for subcontractors, but even those employed by various departments of the administration or private firms tended to join their fellows in the squatter settlements, since no other accommodation was available.

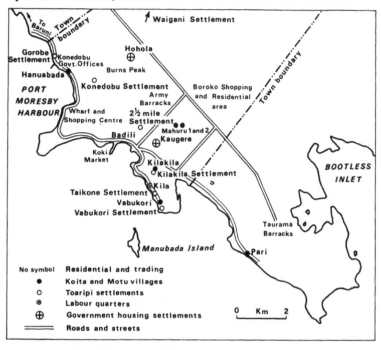

FIGURE 12.2 *Port Moresby, 1964*

Gradually the housing situation in town improved and many employees were able to be accommodated in more permanent living quarters. Yet the squatter settlements remained and their populations continued to increase, to the extent that by the mid-1950s both the administration and Motu landowners expressed their considerable disquiet (cf. Foster 1956). In 1956, for instance, Konedobu settlement had 326 inhabitants (Foster 1956: appendix A); by 1959 there were 343; and the enumerator sheets of the 1961 census reported 497 – a pattern also found in the other squatter settlements. The proportion of Toaripi living away from their natal villages also increased, from thirteen to fifty-eight per cent between 1953 and 1971, although the population totals upon which these calculations are based must be viewed with considerable scepticism (Table 12.2).

The major questions about Toaripi mobility deal with the characteristics of both village and urban populations. To what extent does this movement from the village reflect a shift of Toaripi from their traditional

258

TABLE 12.2 De jure *population and absentees, nine Toaripi villages, 1953-71*

Year	Total population	Absent	
		No.	%
1953	4,019	529	13.2
1962	4,974	1,746	35.1
1971	5,710	3,294	57.7

Source of data: Village sheets, patrol censuses, Malalaua district.

homeland to the towns? To what extent are those who leave rural communities lost to those who stay behind? How much circulation takes place between town and village, and village and town? The field census of all Toaripi-speakers in Port Moresby, including those from Uritai village, provides a context for this discussion. In 1963–64 the *de facto* population was demographically young (Fig. 12.3): 1,358 (55.7%) were less than fifteen (Table 12.3), and of these, 993 (73.1%) had never resided in the village of their parents. Out of 594 males at least fifteen years of age, 141 (23.7%) had never married, but the youngest cohorts (15–19, 20–24) accounted for eighty-five and forty-six of these. On the other hand, 424 (71.4%) were married and with their families in town, whereas only thirteen with wives and children were unaccompanied. These figures, together with the fact that the ratio between the sexes was quite even for each age group (Fig. 12.3), are a quantitative expression of the fact that since 1946 Toaripi have moved to and settled in town as families.

TABLE 12.3 De facto *Toaripi population, Port Moresby, 1963–64*

Age (years)	All Toaripi speakers				Nine Toaripi villages		Uritai village	
	Male	Female	TOTAL		TOTAL		TOTAL	
			No.	%	No.	%	No.	%
0–14	713	645	1,358	55.68	735	56.11	181	57.28
15–49	539	444	983	40.30	533	40.69	128	40.50
50+	34	23	57	2.30	32	2.44	4	1.27
Unknown adults	21	20	41	1.68	10	0.76	3	0.95
Total	1,307	1,132	2,439	99.96	1,310	100.00	316	100.00

Source of data: Port Moresby census, Ryan 1963–4.

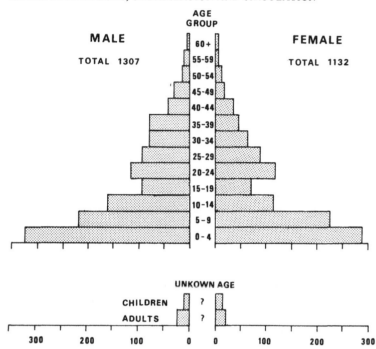

FIGURE 12.3 *Age-sex structure of Toaripi speakers, Port Moresby, 1963-64*

Vabukori settlement

Vabukori squatter settlement, where lived the greatest concentration of people from Uritai, was divided into three sections, each associated with a particular village of origin (Fig. 12.4). Section A, including mostly those from Mirihea villages (Isapeape, Hamuhamu, Lelefiru: Fig. 12.1), was sited on land owned by a particular group of Motu. Although both sections B and C were on land belonging to a different group of Motu, settlers in the former came mainly from Uritai villages (Lalapipi, Mirivase, Uritai, Kukipi) whereas those in Section C were from Mirihea communities.

The original settler in section B had arrived before the war, following which he was joined by kinsmen and affines who soon erected their own dwellings. By 1963, this settler had been dead for some years, so that residents were not necessarily linked to him or to each other. In general, they made no attempt to be friendly with the landowners. In 1963, the original settler in section A still lived there. All householders were related to him or his wife and all cultivated good relations with the Motu. One of the two founders of Section C remained in 1963; he and those closely related to him attempted to be on good terms with the owners. Other families, however, had only tenuous links with this original settler and generally ignored the Motu.

Vabukori was hardly a community. Widely-dispersed ties among cognates and the differing history of each section meant that there was no overarching structure to daily relationships, and the squatter settlement was a locality rather than a distinct social entity. Clusters of people, interacting frequently through ties of kinship, blood or marriage, originated from the same village, as did those interacting with Toaripi beyond the settlement. Visiting often took Vabukori householders to the other side of Port Moresby while neighbours were virtually ignored − a vivid illustration that neighbourhood was not a principle of social organization among Toaripi in town. Long-term residents, those who had left Uritai by 1950, did not return often and then only for a vacation or on special ceremonial occasions. Six out of nineteen households in section B were in this category, yet they frequently entertained village visitors and kept in touch with local affairs.

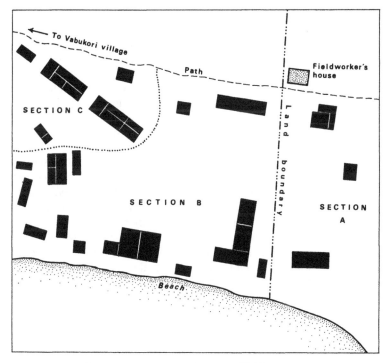

FIGURE 12.4 *Vabukori squatter settlement, 1964*

Bilocality and Toaripi mobility

In Uritai village many important decisions, like the siting of a house or timing of a feast, would be postponed until persons long absent were consulted. Such people were not considered lost to the local community despite their years away and this continued interaction indicated that Uritai villagers should be viewed as members of a bilocal social system, irrespective of whether they remained in their natal place or in Port Moresby (Ryan 1970).

Such an approach is rather different from that usually adopted by anthropologists who, when addressing population movement, assume a particular difficulty in analyzing social structures whose members are widely separated. Beginning with activities and their contexts, social interaction may occur irrespective of the location of actors, so that distance or physical separateness becomes irrelevant. Actual behaviour of people exhibits regularities and general predictability, which in turn reflects its occurrence within a normative and structural framework (cf. Firth 1955:3). Thus Toaripi act within a context defined by cognatic kinship, age-grouping (*heatao*), and affiliation with men's houses, expressed through a continuous exchange of goods and services. In so doing, they constantly make choices among possible actions and adapt to different circumstances (cf. Firth 1954, 1955), no matter whether they are all located in one village, scattered among several, or dispersed between village and town. Toaripi consistently pursue strategies to enable participation in town life without abandoning their position in the natal community (cf. Barth 1966). This is done by fulfilling obligations of reciprocity towards those who are socially important: kinsmen, affines, age-mates. Such obligations and expectations operate irrespective of the location of participating individuals, so that the social system remains unitary although existing in widely separated places.

Movement between Uritai village and Port Moresby exhibits several patterns. A villager goes to visit, stays with relatives a few weeks or months, and then returns with news and town goods or money. Others, most likely adult males and sometimes accompanied by their families, leave for wage work but with no intention of being away for a long time. Some men go to earn money for a specific purpose while others simply desire the experience of living in Port Moresby; these remain for two or three years and then return to the local community. A few even alternate village with town residence throughout most of their lifetimes. Among Uritai adults, the experience of visiting or short-term residence in Port Moresby was common by 1960.

Lengthy absence from the village interrupted by infrequent visits or actual return is a pattern of mobility followed by far fewer families. One, the husband of which was very ill, reappeared after many years in Lae (Fig. 2.1). He died soon after, as did his mother, who had continued

262

to live in Uritai. His widow, originally from Moripi (Fig. 12.1), decided to remain with the children; they had a house but very little else. At mealtimes, their food was consistently of poorer quality than that of any other family studied and often they ate once rather than twice a day. Their food came from a breadfruit tree near the house, from small gardens within the village, and from fishing, while the woman obtained other items by working regularly at local feasts. Although the family had rights to sago and coconut trees, they had no canoe to collect the products, except occasionally when lent one in return for their labour. Such difficult circumstances reflected the fact that the family had no immediate relatives in Uritai and maintained no social contact during their long absence. Eventually they left for the widow's natal community in Moripi. Another family, by contrast, returned to Uritai in 1960 after a similarly lengthy absence and was still in residence eleven years later. In this case, the wife's parents were alive and the returning family lived with them until, subsequently, the husband built a house on land to which he had undisputed rights.

Wives and children may leave Port Moresby to live for a few months in Uritai, especially if the wife is pregnant or an important feast is planned and she can participate in its preparation. One or more children may be sent to grandparents and remain with them for up to five years. In each case, interest is maintained in the village and families resident in Port Moresby can claim to have fulfilled obligations to their kin.

Finally, there are Toaripi who live in various parts of Papua New Guinea, generally employed by government, and who return regularly for holidays. In the 1960s, public servants could receive fares to their natal village once every two years and thus were able to make the journey no matter how great the cost or distance. Toaripi living in Port Moresby often charter a motor vessel to go back for Christmas, during which time many reciprocal payments and feasts are arranged. Although present only a few days, visitors often fulfil important obligations to village members.

Of Toaripi surveyed in Port Moresby in the 1960s, virtually all were scrupulous about maintaining some links with the village and exchanging either or both goods and services, primarily because this was the means by which claims to land were able to be asserted and permission secured from villagers already using the plots in question. Retention of land rights in Uritai was important because the urban environment provided no support for old age or disability. Since, however, most Toaripi in town were very poor and had little spare cash to send to the local community (cf. Ryan 1968), the rural network frequently had to be sustained in other ways: occasional outlays for ceremonial occasions; helping with large projects while on holiday; having family members live and work for periods in the village. Residents of Uritai also had need of their kin in town. Throughout the Toaripi language area, population continued to grow but local opportunities for wage labour were as restricted as ever, though reliance on money had increased. Villagers needed to relocate some of their membership, especially from overcrowded households; to have kin who could contribute occasionally from external sources when large capital inputs were necessary; and also to have a place to stay when visiting town or searching for employment.

The evolution of a periurban settlement beyond the town boundary raises doubts whether this bilocal social system will endure. A Uritai man, who had first left the village before the war, began fishing at Waigani swamp (Fig. 12.2) and eventually built a house there. His daughter married a Motu from Baruni village and he claimed the right of residence because his grandchild was a landowner. He cultivated food gardens, sold fish in Koki market in town (Fig. 12.2) and, as time passed, contacted several other elderly men who were long-term urban residents and suggested they join him. In mid 1963 others, originally from Mirihea and Moripi (Fig. 12.1), began to spend time at Waigani swamp. All these were linked to the Uritai man by distant kinship or through being of the same age group and all had been in Port Moresby for so long that they had few, if any relatives in their natal communities. Having either reached the effective ends of their working lives or been unemployed for long periods, they saw the opportunity to make subsistence gardens and to obtain money from selling fish at the market.

By the end of 1963 there were eleven Uritai families at Waigani swamp, as the older residents had been joined by younger men who had been unemployed for periods of up to two years. Although their food gardens had not begun to yield and dwellings were only partly built, the site had the appearance of a permanent settlement. Two years later there were sixty families: twenty-five originally from Uritai, twenty from Moripi, and fifteen from Mirihea, but usually they maintained houses in various localities within Port Moresby and spent only part of their time fishing and gardening at Waigani. During the rainy season, when the road to the swamp became impassable, families elected either to remain or return to town. The settlement by 1970 had expanded even more and looked long established. Many Toaripi now lived there continuously and only occasionally visited their former homes within the town boundary, a pattern that persisted throughout 1971–72, when several dwellings in Vabukori were occupied by families replacing Waigani residents.

The history of this periurban settlement demonstrated that Toaripi no longer participating in wage labour need not return to the village, but what effect this strategy will have on the persistence of a bilocal social system remains an open question. Another factor likely to influence the persistence of town-village links is the rising proportion of those born in Port Moresby since, as previously noted, 73.1 per cent aged less than fifteen in 1963 had never resided in their parent's community.

From the standpoint of Uritai, these changes are documented by comparing the *de facto* populations for 1962 and 1972 (Table 12.4). Although the total population declined from 577 to 521 during this decade, the number of children remained constant whereas persons aged at least fifteen declined from 254 to 187. As a result, the percentage of children dependent on adults present in Uritai rose from fifty six to sixty four.

TABLE 12.4 De facto *population of Uritai village, 1962 and 1972*

	Males	*Females*	*TOTAL*
Number of residents 1962	280	297	577
Number of residents 1972	259	262	521
Number of adults (15+) 1962	123	131	254
Number of adults (15+) 1972	89	98	187
Per cent adults 1962	43.9	44.1	44.0
Per cent adults 1972	34.4	37.4	35.9
Number of children 1962	157	166	323
Number of children 1972	170	164	334
Per cent children 1962	56.1	55.9	56.0
Per cent children 1972	65.6	62.6	64.1

Source of data: Village censuses, Ryan.

Only one family absent in 1960–2 had returned by 1972; conversely, five families formerly resident had departed for town and another sixty-three people who, in 1960–2, were single and living in Uritai had married and gone to Port Moresby. The fact that Toaripi in town still originate from the village increases the likelihood that exchanges of goods and services will continue. The critical question, however, is whether young people moving from Uritai and similar communities will influence their peers born in town to believe that the ancestral village remains important and that links with it ought to be maintained.

Toaripi in district centres

Censuses and interviews undertaken in 1965 among Toaripi speakers living in district centres — Lae (322), Rabaul (213), Madang (201), Wewak (96; Fig. 2.1) — provide some comparison with the Port Moresby data. Generally, they had travelled first to the capital and their subsequent movement reflected employment opportunities or transfers as members of the public service. Of the eighty-two from Uritai resident in these towns, mainly Lae and Wewak, only government workers had revisited the village, since they alone could afford the fares, whilst links maintained with kin tended to be with those in Port Moresby rather than others who remained in the local community.

The numbers living in the various district centres were so small that it was impossible to define one's urban network in terms of village ties. Instead, Toaripi speakers interacted with each other but showed preference for the immediate social group: that is, Toaripi rather than Moripi or Moveave (Fig. 12.1). Apart from work, there was no association with people of different language or culture, while in Wewak and Madang many Toaripi-speaking men had joined together and become independent contractors. Those from Uritai professed a continued interest in the village but seldom wrote letters, visited, or invited kin to stay with them; in short, there was no network of interaction comparable with that existing between Port Moresby and the community of origin. Perhaps it is significant that two out of three families who returned to Uritai in 1960-62 and who failed to reenter the social system had spent many years in Lae.

Economically, Toaripi in district centres were far better off than their counterparts in Port Moresby. Unemployment was virtually nonexistent, husbands were employed either by government or as tradesmen in the booming construction industry, and living costs were lower. Consequently there was little concern about having to go back to the village for subsistence and children, having been born in these towns or spending the greater part of their lives in them, had no concept of what the rural communities were like. As they become adults, it is difficult to imagine their retaining even tenuous links with the natal places of their parents.

Conclusion

Based on the history of periurban settlement at Waigani swamp and of Toaripi living in district centres, it appears that village-town and town-village links will weaken as the economic rationale for them disappears and as a generation born and raised in town become adults. Yet the situation in Port Moresby itself is highly complex. Many Toaripi speakers there will remain extremely poor and it is uncertain whether they will become urban dispossessed or continue to look to the village for security. Nor is the future necessarily in their hands for Papua New Guinea is now independent, with a government greatly concerned about urban problems. Should official policy define Toaripi living in town as temporary sojourners, then certainly their village communities will continue to be very important.

In more than seventy years to 1972, Toaripi have engaged in several kinds of mobility, each and all of which may be viewed as strategies that enable people to adapt to changing circumstances within a relatively stable framework of expectations (cf. Firth 1959:340-2). Pacification made possible their dispersal from the two original settlements and in turn enabled them to cope with population increase, itself largely a result of introduced health services, and with the partial destruction of the original site. In moving away from the parent villages, Toaripi were able to exercise land rights inherited from their ancestors and they clustered into discrete communities because of the norm that those genealogically linked are expected to validate their kinship through social interaction. What was once circulation between temporary hamlet and residential village became migration from parent settlement to new community.

Between 1920 and 1940, young Toaripi men went away as contract labourers to earn cash for their head tax – a strategy of circulating for money that was supplemented locally by the sale of copra. Wherever possible, these workers signed up with their age group (*heatao*), thus preserving an important social institution in a new environment (cf. Ryan 1970:53–6). Since 1945, the pattern of mobility has become increasingly complex. Rather than constantly circulating between village and town, especially Port Moresby, in response to greater economic opportunities and a growing desire for cash, many Toaripi men had their families join them. In the capital they utilized age-old links with Motu landowners and settled established communities based on village-defined ties of kinship and affinity. Over time, these urban settlements and the distant villages evolved as foci of bilocal social systems. Within this context several forms of mobility occur: short- and long-term circulation from a village base; circulation from an urban base, by which absentees maintain local interests without actually returning; and permanent relocation from village to town. Such movement behaviours embody strategies that permit both rural and urban Toaripi to cope with increases in population density, desire for cash, and the insecurity and poverty felt by most in Port Moresby, but according to village-based linkages.

Whether these bilocal social systems persist will depend on how far they meet the needs of increasingly long-term residents, especially as children born or raised in town reach adulthood (cf. Morauta and Ryan 1982:49–53). If the establishment of the Waigani squatter settlement is indicative, then a completely urban-based strategy may develop. On the other hand, the Toaripi-speaking population in Port Moresby is sizable and includes persons of greatly differing prosperity. Those who are poorly paid and sporadically employed may well continue to regard the village as offering some form of security, irrespective of time away. Conversely, the more affluent have no need for such insurance and may well ignore local ties after a few years' absence. As was the case in the 1950s and 1960s, bilocality is unlikely to occur or persist unless there is complementarity of interests between village and town (cf. Nair this volume).

The social composition of communities also has great bearing upon the maintenance of natal links. Among Toaripi, recognition of kin is cognatic and the spread of an individual's network depends on both the desire and the ability to participate in the exchange of goods and services. By definition, then, urban residents without relatives in the village have no one with whom they might retain meaningful ties. Yet brief revisits to both Uritai and Port Moresby in 1980 and 1981 indicated that bilocality remained quite important, despite the increasing socioeconomic differentiation of Toaripi in both village and town.

Both circulation and migration feature in Toaripi mobility, as a result of which a bilocal social system focussed on both local communities and Port Moresby evolved and has endured. Little evidence exists of similar bi- or multilocal social systems that connect Toaripi villages with towns other than Port Moresby, link the capital with district centres, or district centres with each other. This reflects the fact that comparatively few migrants reside in other towns, that little circulation occurs between district centres and either Port Moresby or rural communities, and that no strategic interests are served by the existence of bilocal social systems. Among Toaripi, patterns of mobility reflect the interaction of constraints and opportunities provided by principles of social organization operating within the context of a rapidly-changing social, economic, and political world. As such, they exemplify processes occurring throughout Papua New Guinea, and probably throughout Melanesia as a whole.

Acknowledgments

The fieldwork upon which this paper is based was funded by a Commonwealth Post-Graduate Research Scholarship, the Research Committee of the University of Sydney, the Australian National University, and Macquarie University, and I thank those bodies for their support. Gratitude is expressed to the many field officers who gave generously of their time and allowed access to district documents.

13 Rural village and periurban settlement: circulation among the Kuni of Papua

Morea Vele

> The forms that migration has taken over time and the changing physical and cultural environment within which it has taken place must be seen as complementing aspects of a unitary system: change in one produces compensatory change in the other (Baxter 1973:xi)

Population movement in Papua New Guinea is not new, particularly since its first inhabitants are believed to have been groups who relocated about 50,000 years ago from the Asian mainland. Scholars who came much later to study these so-called 'primitive societies' found that, in most places, people were able to relate the origins and routeways of their pioneer ancestors in largely mythological revelations. In pre- and early-contact days, mobility was a basic feature of local societies, necessitated by the lifestyle of being hunters, gatherers, or shifting cultivators, as well as by having often to escape the head-hunting expeditions of neighbours (cf. Watson this volume). Much of this movement took place within a limited and clearly defined area; it was therefore largely oscillatory and circulated 'around the centre of gravity of the clan territory, although in some areas there was a long-term trend in a particular direction as more dynamic groups occupied new territories' (Skeldon 1978a:1).

The nature of and reasons for circular mobility throughout Papua New Guinea have changed significantly since European contact. In particular, the pacification of warring tribes has permitted circulation to expand in time and over space, while such introduced activities as formal education, wage employment, urban settlement, and long-distance communications have all contributed greatly to changing its form (Skeldon 1978a:20; Oram 1976:108). Evidence of these influences on contemporary circulation can be seen in the selectivity of movement streams, in the dominance of younger men with formal education, and in the importance of economic considerations for those who move.

This paper, a case study of the Kuni of Central province (Fig. 2.1), attempts to show that contemporary circulation is not an extension of historical movements but rather reflects the significant influence of external forces. It focuses on the mobility connexions between Anasa, a periurban settlement located beyond the city boundary of Port Moresby; the village of Deva Deva from which these migrants originally came; and the city core itself. A survey of Anasa settlement was undertaken in February–March 1978 to establish the amount of circulation, here defined as any move that both began and ended in the same location and involved an absence of at least twenty-four hours, but not more than one month. These data were related to that previously collected from this periurban group during November–December 1976 (Vele 1977) and to a detailed anthropological study of the Kuni homeland made during the mid 1960s (Van Rijswijck 1967).

The Kuni of Papua

The Kuni are mountain people located on the foothills of the Owen Stanley ranges about 130 kilometres northwest of the capital of Port Moresby. They live in an area of 733 square kilometres and according to the 1973 village directory number 2,560. They are about midway

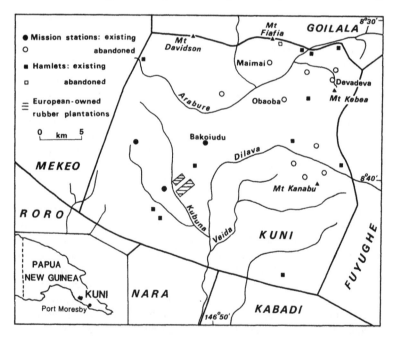

FIGURE 13.1 *Kuni territory, 1963*

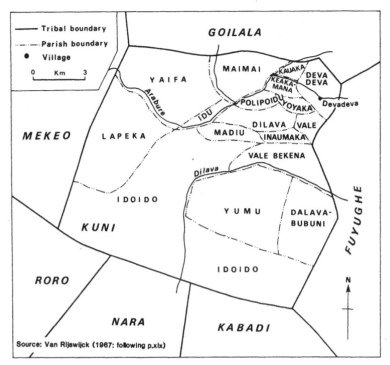

FIGURE 13.2 *Distribution of Kuni parishes*

between the mountain-dwelling Goilala and Fuyughe in the north and northeast (population 28,000); the Mekeo and Roro of the west and southwest, who number 6,284 and 3,823 respectively; and the Nara and Kabadi to the south and southwest (populations 629 and 5,665; Fig. 13.1).

The mythology of the Kuni relates that their first ancestors originated from Fuyughe territory and migrated southwards in two main groups, Dilava Meaba and Dilava Doidoi, settling at Deva Deva and Obaoba. Contemporary Kuni are said to be descended from the Dilava Doidoi who, after a quarrel with their Dilava Meaba forebears, dispersed to live throughout Kuni territory. The parish, the most sizable local group among the Kuni, is primarily a territorial unit within which there exists conspicuous socioeconomic cooperation and political identity (Fig. 13.2). Parish membership is highly flexible and determined mainly by continued residence as a result of either marriage or birth. This flexibility of sociopolitical arrangements made the Kuni very mobile, so that residential instability was a prominent feature of both pre- and early-contact society. Such high territorial mobility in turn reflected the flux in patterns of parish membership, the socioeconomic reciprocity between parishes linked through history or blood, and the fear of attacks from neighbouring tribesmen.

Precontact Kuni, like most tribal societies in Papua New Guinea, engaged in warfare across tribal and parish boundaries, although some clashes occurred within parishes. These conflicts reflected disputes over women, pigs, chieftainship, and land, but did not prevent them from acting as intermediaries in trade with their neighbours. Salt, shell ornaments, and fish were bartered by coastal tribes through Kuni to the mountain peoples of the north, who in turn exchanged plumes of the bird of paradise (*yobu*), dogs' teeth, stone axes, and flints through the Kuni to the coastal groups (Fig. 13.3). These trading patterns involved the Kuni and their neighbours in much circulation, particularly after contact and pacification, thus reinforcing their mobility patterns within and between adjacent parishes because of residential flexibility.

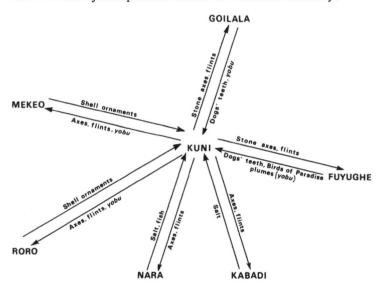

FIGURE 13.3 *Precontact exchange with Kuni*

Around the turn of the twentieth century, the Kuni were contacted by Catholic priests, notably a Father Jullien who in 1896 accompanied some Nara (Fig. 13.2) to a feast they were hosting. Two years later, a mission station was established at Obaoba (Fig. 13.1) and Catholic priests were 'virtually the sole representative of western civilization [and authority] in Kuni, and the main agent of pacification . . . and change in the area from the time that the first resident missionary was appointed at Obaoba in January 1900' (Van Rijswijck 1967:78). Among the many socioeconomic changes that occurred subsequently were the decline in tribal warfare, in the practice of magic and sorcery, and of polygamy; the introduction of such cash crops as rubber and coffee; and the establishment of the Kuni Catholic club. The mission also provided medical services and began schools. Most of these attempts to improve the socioeconomic condition of the Kuni failed, largely because of distance, the rugged terrain of their territory, lack of transport and marketing facilities, and the constant movement of people within and between parishes.

Few Kuni left their homeland before the second world war, but this trend soon changed drastically. From the 1940s, most who went were youths or able-bodied men, many of whom stayed away. Thus, in October 1964, a sample survey of sixty six Kuni living in Port Moresby found that seventy-one per cent had never returned (Van Rijswijck 1967:465). Administrative patrols to Kuni in the late 1940s were aware of this drain on the rural labour force but seemed unable to lessen it. In 1953, according to Van Rijswijck (1967:139),

the rate of male absenteeism in the 16–45 year age group was recorded as 36 per cent of males, or 178 out of a total of 495 able-bodied men. Sixteen hamlets were noted to have fewer adult males than females, and the proportion of pregnant women to those of child-bearing age was recorded to be as low as 28:499.

Both the Catholic mission and the colonial administration were alarmed at this high rate of outmigration and its concomitant socioeconomic repercussions, particularly on village food production and on the everyday activities of local communities. The introduction by the administration in 1958 of an annual head tax of two Australian dollars on every adult male only reinforced this exodus of the young and able-bodied in search of wage employment (cf. Ryan this volume). Indeed, 'by 1961, the situation had reached such a point that old Kuni men appealed explicitly to visiting [administration] patrols that young men should forcibly be brought back to the homelands' (Van Rijswijck 1967:140).

Complicating this situation of high mobility was the resettlement in July 1961 of Kuni at Bakoiudu (Fig. 13.1), under the auspices of the mission but with administrative support. The area of resettlement, situated in the middle of Kuni territory, covered about 16,200 hectares at an elevation of 397 metres. This scheme aimed particularly to relocate those, including the Deva Deva, from northern and less accessible parts (Fig. 13.2) as a way of reviving attempts to improve the people's socioeconomic condition. Located at the centre of Kuni territory and within easy reach of the coast, Bakoiudu emerged as the most likely site. Following negotiations and agreement with the two principal landowners, the mission station relocated from Obaoba to Bakoiudu along with ten volunteer settlers from the parishes of Vale and Inaumaka (cf. Figs. 13.1 and 13.2).

The reaction of the Kuni to news of this resettlement was mixed, with men from Maimai and Deva Deva objecting most strongly because they would be farthest removed from their traditional homeland. As Van Rijswijck (1967:153–4) noted, this scheme confronted people with a dilemma:

> To most, the posited advantages of resettlement were negligible compared to the security and attraction which the homelands held for them. On the other hand, the homeland Kuni took their religious obligations seriously and to many, despite their love for their land, the removal of the Mission from the homelands was irreconcilable with their remaining there. Many informants state that had it not been for the withdrawal of the Mission from the homelands, they would never have resettled at all.

This transfer of the station was further emphasized by the relocation of mission schools and medical services, and subsequently by the concentration at Bakoiudu of other socioeconomic activities jointly sponsored by both the Catholic and the colonial administrations.

Circulation between Bakoiudu and the homelands was most frequent during the first years of resettlement, because the earliest residents depended on their village lands for sustenance while making new gardens and returned continuously to acquire both food and rootstock for planting. Although such trips decreased as the relocated households became more self-sufficient, periodic visiting continued as before because the homelands represented an alternative way of life. At any point in time, this constant mobility made difficult the enumeration of the *de facto* population of Bakoiudu. In December 1961 there were an estimated 200 settlers and in 1962 about 400, while an administration census in 1964 recorded some 850.

Relocation continued from the northeastern parishes of Kuni, including Deva Deva, from which sixty to seventy settled in 1965 on a block of land purchased from the administration two years previously. In 1964, Deva Deva in Port Moresby sent $Australian63 ($A1.70 to £1.00) to buy food for those intending to relocate, who would clear their block of land, while in 1967 a further $A30 was transmitted to the new settlers. Most of the periurban group now living at Anasa (Fig. 13.4) agree that virtually all their kin relocated to Bakoiudu for primarily economic reasons (to grow rubber as a cash crop or work for money) and secondarily so that children could go to school there. Some families who remained behind in Deva Deva village have their children board with relatives so that they can also attend. These and other links mean that there is constant circulation between Bakoiudu and Deva Deva. People in the homeland villages visit both children and relatives in the project area while those in Bakoiudu make periodic returns to maintain land rights, because they suspect that their's in the resettlement are not entirely secure. Should they ever be asked to leave Bakoiudu, they would have their Deva Deva land to which to return.

From the advent of Europeans in Kuni to 1973, when the Hiritano highway reached Bakoiudu, people from Deva Deva and other northern parishes had to travel on foot to the coast via the mission station and proceed to Port Moresby mainly by boat. After the second world war, most migrants consequently walked four days and then travelled by sea at a cost of three Australian dollars. Once a highway had connected Bakoiudu with the periurban settlement of Anasa and with Port Moresby itself, the time spent in travel was greatly reduced. Since 1973, people from Deva Deva village can reach the mission station within two days and then take a light truck (passenger motor vehicle) at a fare of five *kina* (K0.83 to $A1.00; K1.41 to £1.00) to complete the journey to town.

Anasa periurban settlement, 1968–78

Anasa, located on the outskirts of Port Moresby and eleven kilometres from Boroko post office (Fig. 13.4), was founded in 1968 by Olaba Guluma, a man from Deva Deva village. Guluma was one of many Kuni who migrated to town during the late 1940s and 1950s, principally to seek wage employment. He also wanted to look at the large-scale changes that Europeans were said to have initiated and which had occurred, on a very much smaller scale, at the Catholic mission station in Obaoba (Fig. 13.1) from the turn of the century until the 1950s. Just before the second world war, Guluma and several other Deva Deva men were employed at a rubber plantation in Nara for a monthly wage of ten shillings, but returned to their village with the outbreak of hostilities. Guluma first heard of the changes in Port Moresby from workmates at the rubber plantation, later confirmed by six fellow villagers who had been carriers for the allied forces (cf. Ryan this volume). Twelve months' experience of wage work had not satisfied his curiosity of European ways and this, reinforced by stories of employment opportunities in town plus their relative absence throughout Kuni, all contributed to his final decision to try his luck.

The opportunity to leave for Port Moresby came with the visit to Deva Deva of one of Guluma's younger cousins, Aniava Sivali, who had gone away soon after the war ended and had found employment as a domestic servant. Sivali indicated that Guluma could stay in his quarters in Konedobu, near the town centre (Fig. 13.4), while searching for work. Accompanied by two close relatives, they walked four days to the district headquarters of Kairuku, on the coast, where Guluma worked on the local wharf to earn the boat fare of one pound two shillings. Once in Port Moresby, he secured a job as cook in a government mess at Konedobu and resided in the cooks' quarters. These, being close to those of Sivali, guaranteed that both men spent most of their spare time together. After serving two years as cook, Guluma became a laundryman for a European, who provided him with his own domestic quarters nearby.

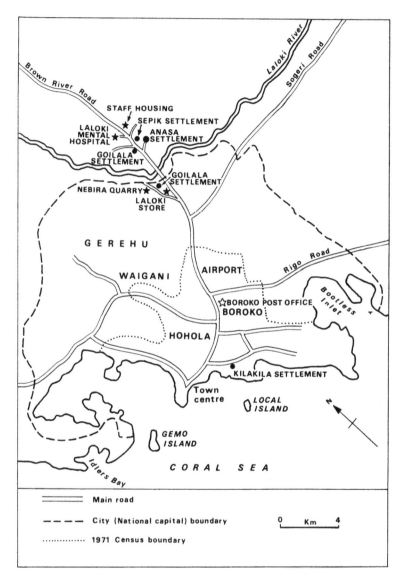

FIGURE 13.4 *Port Moresby, 1978*

All the other men from Deva Deva village resident in Port Moresby in 1978 began arriving in the late 1950s and 1960s, with the last originating from the relocated community of Bakoiudu. Many came as teenagers, while those with wives and children arrived alone, but all went directly to Konedobu and lived with either Guluma or Sivali. As employment and their own accommodation was found, the newcomers left; some to different places in Konedobu and others to Jackson's Airport, Boroko, and Kila Kila settlement (Fig. 13.4; cf. 12.2). When, in 1968, Guluma left his job as laundryman and thus lost his domestic quarters, he approached the colonial administration for a block of land within the urban area on which to settle. Since none were apparently available, he was referred to a man from Hanuabada (a long-established periurban village: Fig. 12.2), who had some blocks of garden land at Laloki, about fifteen kilometres away (Fig. 13.4). On approaching this Hanuabadan, Guluma found the man to be a distant relative through marriage and was given permission to live on that land but asked to keep watch over the food gardens. Guluma was also allowed to have other Deva Deva reside there, who could establish their own gardens if they wished.

The periurban settlement of Anasa dates from this time, beginning with three temporary dwellings and two permanent residents: Guluma himself and an old man from Deva Deva village. All the other men employed in Port Moresby had their own living quarters but spent weekends at Anasa, initially by themselves and in later years with their families. Anasa quickly became the place to which both visitors from Deva Deva and Bakoiudu as well as those seeking wage employment were sent on arrival in Port Moresby. In turn, this greatly reduced the overcrowding of Deva Deva that until 1968 had been occurring in domestic quarters in town and about which many employers, especially in Konedobu, were often dissatisfied. From 1970, several factors led to a steady increase in the number of permanent residents: wage-workers returned to Deva Deva/Bakoiudu to bring back brides, wives, and perhaps their children; families in town relocated to Anasa when men changed jobs and lost automatic access to living quarters; and yet other families established a household while the husband looked for other employment and often grew food crops to earn money. By 1978, all household heads who worked in Port Moresby and remained there during weekdays had established permanent dwellings at Anasa, which were maintained by members of their immediate or extended families and to which they returned on weekends or public holidays.

Anasa, 1978

The periurban settlement, sited about one and a half kilometres north of Laloki River bridge (Fig. 13.4), is reached from the main highway by an unsealed road that can only be used by vehicles in the dry season. It is rectangular in shape, surrounded by a fence for security, and has fifteen completed dwellings that face each other on all sides of the rectangle (Fig. 13.5). As is typical of traditional villages, the settlement's centre is left open for social and political gatherings or recreational pursuits. Most dwellings are sited on posts, underneath which are timber platforms that serve not only as a popular refuge during sunny days but also as places where there occurs much informal socioeconomic and political discussion about the settlement affairs.

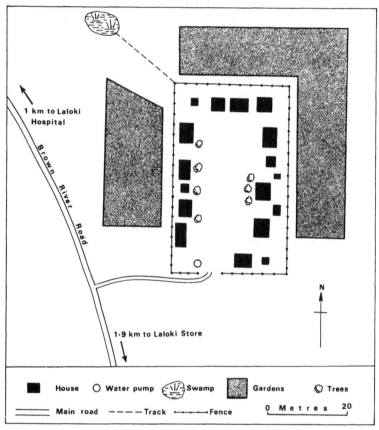

FIGURE 13.5 *Anasa periurban settlement, 1978*

Within the locality are three other periurban settlements, two inhabited by Kunimaipa Goilalas (Fig. 13.2) and the third by people from East Sepik province (Fig. 14.1). Little interaction occurs between these, except for the Goilala settlement situated directly across from Anasa whose residents come for Sunday church services. Also in the vicinity is a large trade store operated by the Kairuku Development Corporation, from which most residents as well as passing travellers obtain trade goods; a quarry owned by the Steamship Trading Company, at which all men from one of the Goilala settlements work; and a mental hospital (Laloki) more than a kilometre to the northeast of Anasa. Minor clinical illnesses among residents are treated at Laloki hospital, while more serious afflictions must be dealt with at the general hospital in Port Moresby.

Unlike their periurban neighbours, the inhabitants of Anasa have paid no cash to the Hanuabadan landowner for the use of his garden area since the settlement was first established. The absence of any payment probably reflects a long-standing Motu practice: that normally there is no such expectation when a person's property is being used by a brother-in-law, which in fact is the kinship link between Guluma (the founder) and the Hanuabadan. Especially in the early 1970s, the people sent garden crops whenever there was a good harvest through Guluma to the landowner, who is now old and does not expect this as much. In particular, the landowner recognizes that most settlers are unemployed and need all their garden produce both for subsistence and to sell in the Port Moresby markets to obtain cash. The Deva Deva of Anasa feel this verbal agreement with the Hanuabadan to reside on and use his land is as good as any written one, especially since it is further cemented by the connections between distant kin. As long as the old man remains alive, their rights to the present site are secure provided, however, that the Koiaris (the traditional owners of the entire Laloki area) do not demand their land back from the Papua New Guinea government. In the 1950s, the colonial administration acquired this land from the Koiaris, subsequently exchanging it with the Hanuabada for the Konedobu area, on which the former administrative centre of Port Moresby now stands. The uncertainty about Anasa's future, whether from the actions of the landowners' descendants or the Koiaris, has led settlers to maintain their land rights in Deva Deva. They do this by actively participating in local affairs, accommodating visitors from both the village and Bakoiudu, and by sending back gifts and money to relatives when these visitors leave.

In general, the Laloki area is most suitable for growing crops for both subsistence and cash, which is the main economic activity for all periurban settlers except the Sepik (Fig. 13.4), who are primarily wood carvers. Unfortunately, Anasa itself is sited about twenty metres from a swamp (Fig. 13.5), which during heavy rains overflows and floods both the residential area and adjacent gardens. When this occurs most gardens are destroyed and people must rely on those with wage employment and on the subsistence crops brought by visitors from Deva Deva village. To alleviate this problem, in mid 1977 the purchase of land three kilometres to the north was successfully negotiated with a Koiari owner and, following an initial payment of K100 (K0.83 to $1.00; K1.41 to £1.00), the people began to establish gardens there.

In November 1976, the *de jure* population was sixty eight (41 males, 27 females) who lived in twelve households. Thirteen months later, in February 1978, there were fourteen households and the total population had risen to ninety three (54 males, 39 females). This increase was due largely to the arrival of new migrants from both Bakoiudu resettlement area and Deva Deva village. Although males outnumbered females (54:39), the population in February 1978 was composed mainly of able-bodied men and women in the 15–44 age bracket (Table 13.1). Most children less than five were born in town (Port Moresby or Anasa) and have not yet seen Deva Deva village or Bakoiudu. The thirteen males in the 5–14 cohort form the most sizable group in the settlement (24.1 per cent), most of whom are not in school and have received less than six years primary education.

TABLE 13.1 Age-sex structure of de jure *population, Anasa settlement, 1978*

Age (years)	Males		Females		TOTAL	
	No.	%	No.	%	No.	%
0–4	9	16.7	6	15.4	15	16.1
5–14	13	24.1	7	18.0	20	21.5
15–24	12	22.2	5	12.8	17	18.3
25–34	11	20.4	7	18.0	18	19.4
35–44	7	13.0	7	18.0	14	15.1
45–54	1	1.9	5	12.8	6	6.4
55+	1	1.9	2	5.1	3	3.2
Total	54	100.2	39	100.1	93	100.0

Source of field data: Vele, February 1978.

The educational level reached by those aged fifteen to forty-four is very low when compared with the same group for other urban populations of Papua New Guinea. For Anasa, 67.3 per cent (42 of 49) have received no formal education, whereas in 1973–74 the urban household survey recorded 53.3 per cent of 15–44 year-olds in Port Moresby as having less than five year's schooling (Garnaut, Wright, and Curtain 1977:36). For the country as a whole, the proportions ranged from 32.3 per cent for Arawa/Kieta/Panguna to 72.5 per cent for Goroka (Fig. 2.1; cf. Conroy/Curtain and Young this volume), thus placing Anasa at the less-schooled end of the range. At the other extreme, only one person in the settlement had received tertiary training, whereas in 1973–74 8.6 per cent of Port Moresby's population had at least ten years' formal education.

All those males in Anasa aged between five and fourteen who were born in Deva Deva village had never been to school and they accounted for almost half (46.2 per cent) in this age category. Although lack of local schools in the northeastern parishes of Kuni is the main reason, it also appears that villagers are not coming to Anasa for the sake of their children's education but rather to seek wage employment for themselves. This is reflected in the fact that during 1978 some parents sent children back to Bakoiudu for schooling. If Anasa's population structure in 1976 is compared to that in 1978, the trend is for older adults without any schooling to return to Deva Deva and for younger, formally-educated men to move into the periurban settlement.

Seventeen out of thirty-two adult males (15-plus years) are in wage employment, all of whom work in Port Moresby. Between 1976 and 1978, several changes have occurred in the distribution of those earning money (Vele 1978: Table 2). The ratio of employed to unemployed has reversed (8:16 versus 17:15); there has been a distinct increase in those with training and skills deriving from six or more year's formal education (1 versus 7); and among adult men without formal schooling, the number who have secured jobs has risen to the same extent as those without has declined (7 versus 9; 12 versus 10). Only two of the seventeen men currently employed in Port Moresby commute each day, while the others remain in town during work days and return to Anasa for the weekend.

Circulation of the Deva Deva Kuni

Since the completion in 1973 of the Hiritano highway as far as Bakoiudu, the amount of travelling by the Deva Deva Kuni between the mission station, Anasa, and Port Moresby has risen sharply. In contrast, visits to Deva Deva itself are less frequent since it is still a walk of two days from the end of the highway to most communities. Even so, most adults at Anasa have visited both Bakoiudu and Deva Deva village more than once since 1973. Journeys by unemployed Anasa residents occur mainly at times of good harvests, when some money has been earned from selling surplus garden produce at markets in Port Moresby. Other persons make independent trips in either direction when burial or marriage payments take place, whereas wage earners in town tend to travel to Bakoiudu and Deva Deva during vacation leave, because they can more easily afford the cost of ten *kina* to make the return journey by passenger motor vehicle. In short, the village of Deva Deva has evolved into a multilocal society, whose members are also to be found at the Bakoiudu resettlement area, the periurban community of Anasa, and Port Moresby itself. Although this process began during the second world war, it has been reinforced by the construction of the Hiritano highway, by which the maintenance of links among kin are not only facilitated but also more easily upheld.

281

To establish the intensity of Deva Deva circulation between the natal community and its various satellites, a mobility register was kept at Anasa for the twenty-nine days between 19 February and 19 March 1978. In this was recorded every move that occurred between Deva Deva/Bakoiudu and Anasa, as well as between Anasa and Port Moresby city, involving a minimum absence of twenty-four hours from the place of origin prior to return (cf. Chapman 1975). During the month of the survey, there were fifteen such instances of circulation from Kuni to Anasa and a further forty-eight between Anasa and the city itself (Table 13.2). It is important to note that people's mobility was much influenced by the death at the periurban settlement of one household head during the first week of field research. Rather than being considered an atypical situation, however, this death served to reveal the close interaction between Anasa settlers and their rural kin, as well as how important events like marriage, death, and ceremonial ('sing sing') result in circulation amongst the Deva Deva Kuni.

Of the fifteen Deva Deva who circulated between Kuni and Anasa (Table 13.2: top panel), two carried news of the death of the immediate family at Bakoiudu. In response eleven close kin, two families with children and three unmarried adults, left to discuss with the periurban settlers when and where the body should be buried. These visitors brought a bag of betel nut for these meetings and stayed for eight days. They left together, just as they had come, once burial had taken place at Nine-Mile Cemetery just beyond the city boundary of Port Moresby (Fig. 13.4).

The other two people (both males) arrived from Deva Deva village in mid March, too late for the funeral because of the time it takes news to reach northeastern Kuni from Bakoiudu. They indicated they would remain at Anasa for the funeral feast and related ceremonies. The decision to bury the household head at Nine-Mile Cemetery rather than at his natal place is of interest, for it suggests feelings of permanency that settlers have towards Anasa. This seems ironic given their uncertain title to the site, especially once the Hanuabadan landowner dies, the deliberateness with which they have maintained land rights in Kuni, and the general uncertainty throughout Papua New Guinea about traditional ownership of land.

Thirty five of the forty eight circular moves recorded between Anasa and Port Moresby (Table 13.2:bottom panel) were made by the fifteen wage-earners and their families, who remain in town during work days and return to their periurban households at weekends and public holidays – a practice that began with Anasa's founding in 1968. Of the other thirteen people who made round trips to Port Moresby, six either went to view the corpse at the hospital or to seek advice about place of burial from the wife and other close relatives of the deceased. One mother took three children, including a sick daughter, to hospital, while three other women went to the city to buy store goods. All these absences were brief, except for the mother and her three children who remained at the central hospital for two weeks.

TABLE 13.2 Circulation[a] of Deva Deva Kuni, by age, sex, and education, 19 February–19 March 1978

Age	Sex	Formal education — None	Less than six years	Some post-primary (plus six years)	Total number circulating	
Persons circulating between Deva Deva/Bakoiudu and Anasa						
0–4	Male	1			1	3
	Female	2			2	
5–14	Male	1		–	1	2
	Female	–		1	1	
15–24	Male	1			1	1
	Female	–			–	
25–34	Male	2	1		3	4
	Female	1	–		1	
35–44	Male	2		1	3	4
	Female	1		–	1	
45–54	Male	1			1	1
	Female	–			–	
TOTAL		12	1	2	15	
Persons circulating between Anasa and Port Moresby city						
0–4	Male	–			–	2
	Female	2			2	
5–14	Male	3	2	2	7	10
	Female	–	1	2	3	
15–24	Male	2	1	8	11	12
	Female	–	1	–	1	
25–34	Male	8	1		9	12
	Female	3	–		3	
35–44	Male	6			6	11
	Female	5			5	
45–54	Male	1			1	1
	Female	–			–	
TOTAL		30	6	12	48	

[a]Circulation refers to a move made from and back to the same point of origin, that involves an absence of at least twenty-four hours but not more than twenty-nine days.

Source of data: Mobility survey of Anasa, Vele, 1978.

For both the Kuni-Anasa and Anasa-Port Moresby circuits of movement, people range from young to middle aged and there are twice as
many men as women. Level of formal education is thus the only differentiating characteristic between the rural Deva Deva and those who
reside in the periurban settlement, since three quarters of the young
adult males (15-24 year-olds) who circulate between Anasa and Port
Moresby have been to secondary school (Table 13.2:col. 5).

Some broader conclusions

Demographically, Anasa has the characteristics of an urban settlement:
there is a surplus of males over females, practically half the residents
are aged fifteen to forty-four, and adults generally have moved to seek
wage employment in town. In terms of the low level of formal education
that prevails, it is a rural settlement, especially when compared to other
urban populations in Papua New Guinea. Other factors that make
Anasa a more rural- than urban-oriented place are its location away from
the industrial and commercial centre of the city, the origin of its inhabitants from one particular village (Deva Deva), the common recognition of the first settler (Olaba Guluma) as its leader and, overall, its
communal atmosphere.

In general, however, the periurban settlement of Anasa illustrates a
positive mix of both urban and rural lifestyles. On the one hand, residents have been much affected by urban wage structures, price changes,
and opportunities for both employment and formal education. On the
other hand, their continued contact with fellow villagers in Deva Deva
and Bakoiudu, which has been further facilitated by improved transportation, enables them to retain certain customary values (reciprocal
cooperation, recognition of a leader) that appear compatible with new
and modern ones. This complementary mix of rural and urban values is
reflected in the huge 'sing sing' (ceremonial feast) staged by settlers in
1976 and by their contribution of K150 in February 1978 towards the
cost of a coffin and burial rites for the household head who died.

This case study also shows the complex nature of contemporary
circulation in Papua New Guinea. Conventional thinking depicts movement behaviour in terms of a simplistic picture of people transferring
from the subsistence mode of production into the urban or capitalist
mode. Broadly speaking, in contrast, the Deva Deva Kuni have transferred from a sector that is wholly subsistence (Deva Deva) to one that
is a mix of cash cropping and wage employment – initially in a rural
setting (Bakoiudu) and later in the periurban environment (Anasa) –
and finally to a production sector that is wholly urban and monetized
(Port Moresby city). The circulation of people links the various settings
of these different modes of production and, in turn, reflects the importance for the Deva Deva Kuni of the introduction of long-distance

transport, urbanization, wage employment, and formal education. Indeed, the construction of the Hiritano highway to Bakoiudu, access to education and employment in Bakoiudu and Port Moresby, and the very lack of these opportunities in the rural village itself have been mainly responsible for both the evolution of a multilocal social system among the Deva Deva Kuni and the people's continued circulation to ensure the maintenance of that system.

Acknowledgments

My thanks to Drs. John Conroy, Richard Curtain, and Louise Morauta for helpful comments on an earlier version.

285

14 Circular mobility in Papua New Guinea: the urban household and rural village surveys of 1973–75

John D. Conroy and Richard Curtain

The time and space measures held by an observer to define a move will tend quite clearly to reflect both the perceptions of the society under observation and the purposes for which research is being conducted. Thus Chapman (1976:139–40), in a study of south Guadalcanal, is concerned with 'meaningful time units that reflect the behaviour of definite groupings of people (tribesmen) who have moved, in continually cyclic fashion, over specified areas and in particular directions'. The 'meaningful time unit' in this case is as little as twenty four hours and the spatial displacement that constitutes mobility correspondingly small. Conversely, Bedford (this volume) reports of the UNESCO/UNFPA project in eastern Fiji:

> primary concern was with movements which necessitated a substantial restructuring of the round of social and economic activities of those involved. Absences from home for a few hours or days were thus not considered relevant whereas a period of at least four weeks generally did result in a locational shift in the primary activities of an individual mover.

Chapman's approach reflects his belief that the study of population movement at the macrolevel must be preceded by 'sensitive studies of individuals' (1976:140) and, more particularly, his perception that circulation lies at the heart of Guadalcanal mobility behaviour. Consequently he sought to expand Mitchell's (1959) concept of tribal mobility as circulation 'to embrace all movements that both begin and terminate in the local community, in that tribesmen return to their homes no matter what the reasons for their departure nor how long they have been absent' (Chapman 1976:131). It is obvious that such an approach to mobility focuses very squarely on village communities and is much concerned with short-term and short-distance movement within rural areas. It is doubtful whether such a frame of reference is equally relevant

286

in situations where social and economic change has proceeded farther than in the Solomon Islands. Evidence from Papua New Guinea suggests that movement from certain districts is beginning to assume, at least for some, a more permanent and migratory character (cf. Young, this volume).

For census purposes, the indigenous population of Papua New Guinea is categorized into three sectors based upon place of enumeration: rural village, rural nonvillage (separate concentrations of less than 500 people as found around schools, mission stations, cash-crop plantations, and patrol posts), and urban (centres with populations of at least 500). At the time of the 1971 census, the vast majority still remained in rural villages but rapid growth rates for other sectors were effecting a significant redistribution of people (Table 14.1). Urban areas had expanded in

TABLE 14.1 *Indigenous population by sector of residence, 1971, and average annual growth rates, 1966-71, Papua New Guinea (%)*

Sector of residence	1971 census			Male/ female ratio	Annual growth rate 1966-71		
	Male	Female	Total		Male	Female	Total
Rural village[a]	76.7	86.0	81.2	96	0.4	1.1	0.8
Rural nonvillage	12.0	6.3	9.2	204	7.7	12.8	9.2
Urban	11.3	7.6	9.5	161	15.8	20.4	17.5

[a]See text for definition of sectors of residence.
Sources of data: Tabulations from 1966 and 1971 censuses, Bureau of Statistics, Papua New Guinea.

size to be as important as the rural nonvillage sector, much of which was proto-urban in nature. Rapid rates of urban growth are not inconsistent with an essentially circular pattern of movement; during the 1960s and 1970s, in fact, this was true of most town dwellers despite increasing lengths of absence from the village. Given these trends in Papua New Guinea, the process of mobility could be studied within time and space dimensions that focussed attention on rural-urban movement, over longer periods than those chosen by Chapman for the Solomon Islands, and without any presumption of circularity. The case for such an approach is stronger when the objectives are to assist in framing government policies about townward movement and urban unemployment (Conroy and Curtain 1982).

National censuses incorporating a total enumeration of the urban population were held in 1966 and 1971 and an urban survey of seven selected towns conducted during 1977. The capital of Port Moresby, with 95,000 inhabitants, was found to be twice as large as Lae (Fig. 2.1) and over the period 1966–77 urban growth rates were high, averaging between eight and eleven per cent a year (Table 14.2). After the 1971 census, demographers made significant upward adjustments to the urban population totals, believing that considerable underenumeration had occurred. If annual rates are calculated from these adjusted data, then urban growth is much reduced for the 1970s (Table 14.2: cols. 2 and 3), but if unadjusted data are used, then the growth is less dramatic during the 1960s and correspondingly less precipitous in its decline over the next decade (cols. 4 and 5).

TABLE 14.2 Size and growth of indigenous population in selected towns of Papua New Guinea, 1966–77

| | Total population 1977 | Average annual rates of growth | | | | |
| | | Adjusted 1971 data | | Unadjusted 1971 data | | |
		1966–71	1971–77	1966–71	1971–77	1966–77
Port Moresby	95,249	13.2	8.2	10.5	10.5	10.5
Lae	40,992	18.6	5.0	16.0	6.9	10.4
Rabaul	11,520	6.4	4.2	3.7	6.2	4.7
Madang	18,667	11.0	4.2	10.0	5.4	8.7
Goroka	8,593	12.1	4.4	9.6	6.9	8.2
Popondetta	5,739	–	5.2	–	7.1	–
Kavieng	4,134	8.8	7.5	6.5	9.4	8.1

Source: Adapted from Skeldon (1978b:Table 3).

Such demographic uncertainty formed the context for two major social surveys with which this paper is concerned and which were undertaken jointly by the University of Papua New Guinea and the Australian National University. The first, the urban household survey, took place in 1973–74 and involved a random survey of almost 4,000 households in fifteen towns, including all the major ones. This was followed in 1974–75 by a rural village survey, which contacted a sample of more than 10,000 adults in fifty villages throughout the mainland of Papua New Guinea. Comprehensive accounts of both the urban household and rural surveys have been published (Garnaut, Wright and Curtain 1977; Conroy and Skeldon 1977), but this paper also draws upon the original field data.

For the rural survey, 'migration' was defined as an absence from the rural village sector for one or more months, excluding hospital treatment. This definition incorporates rural-urban movements, as well as those between the rural village and nonvillage sectors. Given a concern with policy aspects of urbanization, the research design excluded mobility between village communities. Within this framework, a 'migration' is said to have been completed when a person returns to reside in the home community; during the time away, visits may have been made, as by skilled workers on leave from their jobs, before this actually occurs. Thus what has been termed 'circular migration' in this and other studies in Papua New Guinea is not the same as 'circulation', in Chapman's usage, which encompasses all return movements including relatively brief visits.

The rationale for and method of sampling in the rural survey have been described elsewhere (Clunies Ross, Curtain, and Conroy 1975; Clunies Ross 1984). Since villages were not selected randomly, it is at this microlevel that the results are most reliable. Field data also were grouped by clusters of villages within a district to draw attention to differences within and between clusters and to suggest propositions about the particular socioeconomic processes involved. In aggregate, the survey results provide a standard of comparison for those of the village and the cluster, as well as permit the testing of hypotheses about general relationships between phenomena. Here, the emphasis is upon the cluster or district, ten of which were surveyed (Fig. 14.1).

Adoption of such a procedure reflects a growing tolerance amongst theorists of sampling methodology toward the use of purposive schemes in survey research (see Royall 1970; Brewer and Mellor 1973). Often, in third world countries, the imposition of strict requirements for random

FIGURE 14.1 *Ten districts of the rural village survey: Papua New Guinea, 1974–75*

289

sampling has inhibited the collection of much-needed data under circumstances where randomization is a practical impossibility. This limitation applies particularly to field conditions in Papua New Guinea, where rural areas selected for study can range from densely-populated mountain slopes to the lightly settled swamplands of coastal river systems.

The urban household survey, which preceded the village one by a year, included all persons present in cities and towns. Those born elsewhere were classified as migrant and information obtained about their length of stay, employment status and, for those without regular wages, other sources of income. Visitors likely to remain less than three months were distinguished in terms of whether or not they were seeking work. The rural enquiries, in addition to recording information about moves made beyond the village sector, also identified shorter-term visits made for less than a month to other sectors (predominantly provincial or major urban centres). Some details are available about brief revisits made by absentees to the home community. Despite the deliberate exclusion of most short-term and short-distance mobility, 'circular migration' to and from the rural village was found to be overwhelmingly important, although there were some signs of an emerging urban population.

A migration continuum in Papua New Guinea

The district, within which the rural village survey was conducted, is an administrative unit; in 1974–75 Papua New Guinea was divided into seventy-nine such units with an average population of about 30,000. It might have a single language (as in Wabag:Fig. 14.1), a dominant language (Hagen, Ambunti), a prevalent group of mutually intelligible dialects (Malalaua, Gumine), a small number of related languages (Goilala), or a great diversity of languages (Finschhafen, Maprik). Community or similarity of language comes to be reflected in other aspects of culture. Generally, a district also had some common experience with the history of administration and mission contact. All ten districts surveyed are in the mainland of Papua New Guinea: two (Goilala, Malalaua) located in the southern coastal region of what before independence was the Territory of Papua; three (Finschhafen, Ambunti, Maprik) in the northern coastal region of the former Territory of New Guinea; and five (Okapa, Gumine, Hagen, Wabag, Mendi) in the highlands region. Although Goilala and Finschhafen belong to coastal regions, most areas studied are in the uplands.

Three criteria are important to assess patterns of movement for one or more months beyond the rural village sector: current or return migrants as a proportion of the total village populations; the proportion of such migrants who have returned to the natal community; and the

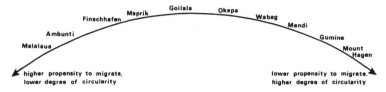

FIGURE 14.2 *A migration continuum of ten rural districts, Papua New Guinea, 1974-75*

masculinity ratio among those absent from the village at the time of survey (Table 14.3; cf. Clunies Ross 1977b). Based on these criteria, there is a continuum of districts that may be conceived as ranging from lower (Malalaua, Ambunti) to greater (Mount Hagen) circularity of movement, and from a higher to lower propensity to migrate (Fig. 14.2). The precise ordering of districts on this continuum was decided by reference to patterns of 'migration' over time, as revealed by age cohorts. Although some rearrangement of districts might be possible, there is a clear-cut range from 'coastal' to 'highlands' districts, with Goilala as a bridge between the two.

TABLE 14.3 *Migrant status of persons fifteen or more years, for fifty villages in ten administrative districts: Papua New Guinea, December 1974-January 1975*

District	Current and return migrants as % of total population				Return migrants as % of total migrants			
	Male		Female		Male		Female	
	%	No.	%	No.	%	No.	%	No.
Malalaua								
Age: 15-24	83.7	227	63.4	175	14.2	190	27.9	111
25-34	91.9	123	72.3	119	29.2	113	34.9	86
35-44	67.5	80	32.8	67	66.7	54	63.6	22
45-54	36.7	49	29.1	62	61.1	18	72.2	18
55+	22.5	49	16.0	50	90.9	11	87.5	8
Total	73.1	528	51.8	473	30.3	386	38.8	245

Male:female ratio of absentees 1.79.

District								
Ambunti								
Age: 15-24	89.3	159	57.7	137	14.8	142	25.3	79
25-34	93.8	112	61.6	86	28.6	105	32.1	53
35-44	83.6	128	45.1	133	42.1	107	35.0	60
45-54	61.3	80	26.6	79	63.3	49	52.4	21
55+	36.7	60	25.0	44	77.3	22	81.8	11
Total	78.8	539	46.8	479	33.9	425	34.8	224

Male:female ratio of absentees 1.93.

291

District	Current and return migrants as % of total population				Return migrants as % of total migrants			
	Male		Female		Male		Female	
	%	No.	%	No.	%	No.	%	No.

Finschhafen

	%	No.	%	No.	%	No.	%	No.
Age: 15–24	68.9	206	84.3	251	24.7	142	25.6	86
25–34	87.9	174	44.0	157	36.0	153	31.9	69
35–44	85.1	94	37.3	102	53.8	80	42.1	38
45–54	83.5	85	20.2	94	54.9	71	47.4	19
55+	63.8	94	14.1	64	86.7	60	66.7	9
Total	77.5	653	33.1	668	44.3	506	33.9	221

Male:female ratio of absentees 1.93.

Maprik

	%	No.	%	No.	%	No.	%	No.
Age: 15–24	73.4	286	41.6	214	23.3	210	51.7	89
25–34	84.7	202	50.3	165	43.7	171	43.4	83
35–44	83.7	172	29.3	181	65.3	144	64.2	53
45–54	72.7	165	11.6	121	84.2	120	64.3	14
55+	37.3	67	14.0	50	92.0	25	71.4	7
Total	75.2	892	33.7	731	51.6	670	52.8	246

Male:female ratio of absentees 2.83.

Goilala

	%	No.	%	No.	%	No.	%	No.
Age: 15–24	62.9	132	27.0	115	31.3	83	29.0	31
25–34	69.6	138	29.8	104	36.5	96	48.4	31
35–44	67.0	91	25.4	59	70.5	61	46.7	15
45–54	41.7	48	12.8	39	60.0	20	20.0	5
55+	13.6	22	10.0	10	*	*	*	*
Total	61.0	431	25.4	327	44.1	263	38.6	83

Male:female ratio of absentees 2.88.

Okapa

	%	No.	%	No.	%	No.	%	No.
Age: 15–24	77.4	133	14.3	92	43.7	103	53.9	13
25–34	69.5	118	11.6	155	67.1	82	33.3	18
35–44	68.1	72	4.9	61	89.8	49	*	*
45–54	36.7	60	–	28	90.9	22	–	–
55+	19.2	26	–	17	80.0	5	–	–
Total	63.8	409	9.6	353	64.4	261	44.1	34

Male:female ratio of absentees 4.90.

Wabag

	%	No.	%	No.	%	No.	%	No.
Age: 15–24	63.4	175	26.2	145	43.2	111	39.5	38
25–34	71.4	119	18.6	129	64.7	85	58.3	24
35–44	35.9	106	9.7	93	79.0	38	44.4	9
45–54	6.1	115	1.1	89	57.1	7	*	*
55+	–	27	2.8	36	–	–	*	*
Total	44.5	542	14.8	492	56.8	241	45.2	73

Male:female ratio of absentees 2.60.

District	Current and return migrants as % of total population				Return migrants as % of total migrants			
	Male		Female		Male		Female	
	%	No.	%	No.	%	No.	%	No.
Mendi								
Age: 15–24	63.5	167	15.3	137	35.9	106	42.9	21
25–34	59.8	92	2.1	141	67.3	55	66.7	3
35–44	32.6	95	–	108	77.4	31	–	–
45–54	12.5	80	–	61	80.0	10	–	–
55+	6.3	32	–	12	100.0	2	–	–
Total	43.8	466	5.2	459	53.4	204	45.8	24

Male:female ratio of absentees 7.31.

District								
Gumine								
Age: 15–24	57.1	84	15.0	107	50.0	48	6.3	16
25–34	61.5	122	3.0	99	65.3	75	*	*
35–44	44.6	74	3.8	52	75.8	33	*	*
45–54	23.9	46	–	35	100.0	11	–	–
55+	3.1	32	–	6	*	*	–	–
Total	46.9	358	7.0	299	65.5	168	9.5*	21

Male:female ratio of absentees 3.10.

District								
Mount Hagen								
Age: 15–24	27.1	214	13.7	161	53.5	58	77.3	22
25–34	21.9	155	7.2	166	82.4	34	75.0	12
35–44	32.0	100	3.2	93	96.9	32	*	*
45–54	12.1	59	1.6	64	100.0	7	*	*
55+	–	26	–	33	–	–	–	–
Total	23.7	554	7.4	517	74.1	131	76.3	38

Male:female ratio of absentees 3.78*.

*With numbers, indicates that four or fewer persons are involved; with ratios and percentages, that the calculation is based on very few cases.

Source of data: Rural village survey 1974–75. Cf. Clunies Ross (1977b: Table 1).

Certain of the 'coastal' districts have been subject to external influences — missionary, commercial, trading, administrative — since the late nineteenth century (Malalaua, Finschhafen), while in others (Maprik, Ambunti) such influences began later but were evident by the 1920s and 1930s. The highlands districts (Okapa to Mount Hagen), at the right end of the continuum, for the most part were brought under Australian administrative control from the late 1940s to early 1950s. Goilala district represents an intermediate case of a highlands region relatively accessible to the coast (Fig. 14.1), as well as to the main town of Port Moresby, and with a history of European contact long before the second world war.

It is possible to discern a time sequence in patterns of outmigration by reference to the proportions of current and previous migrants in each age cohort (Table 14.3:cols. 2 and 4). Assuming that age at first migration is broadly similar from district to district, they may be listed chronologically in terms of when significant absences began to occur from the rural village sector. The ranking that results from this exercise is: Finschhafen; Ambunti (women outmigrating much earlier than from Maprik); Maprik; Malalaua (women departing sooner than from Okapa); Goilala (women leaving earlier than from Okapa, but men perhaps slightly later); and Okapa. In these terms, there is no great difference between Gumine, Mendi, Mount Hagen, and Wabag. This sequence is much influenced by the far greater extent of labour recruitment before the second world war in the Australian mandated territory of New Guinea, compared with the Australian colony of Papua. Thus, the Papuan districts of Malalaua and Goilala emerged relatively late as sources of outmigrants despite the fact that their contact histories are quite as long as for Finschhafen and Ambunti in New Guinea (cf. Ryan, this volume).

To what extent does this chronological ordering correspond with the continuum from lesser to greater circularity within migration? In part, the persistence of circulation reflects how recently significant numbers from each district entered the national migration stream. Thus late entrants, notably from highlands districts, reveal the strongest degree of circularity, whereas in Malalaua and to a lesser extent Ambunti it is much less evident and absences from the natal place are increasingly long-term or even permanent. Perhaps significantly, both Malalaua and Ambunti are swampy lowland districts in which sago is the staple and where governmental provision of services and capital infrastructure has been relatively slight. By contrast, significant outmigration appears to have begun from Finschhafen yet it still displays a quite high degree of circularity.

Villagers who had not migrated beyond the rural sector for at least a month nonetheless could have visited their provincial headquarters for shorter periods (Table 14.4). If such district-level data are ordered as for the migration continuum (Fig. 14.2), then marked discrepancies emerge between districts with similar patterns of outmigration – as for example between Mendi and Mount Hagen or Goilala and Okapa. Since historical, social, and locational factors may well prove significant for specific districts, explanation of these divergences must rest upon detailed case studies. It is puzzling, nonetheless, why Mount Hagen men should be conspicuously less mobile than those from Gumine district, where transport facilities generally are less adequate than in the former. Nor is there a marked difference in the likelihood of engaging in tribal fighting, which might be thought to tie people to their home territories. Unfortunately, the few district-level enquires

TABLE 14.4 *Status of villagers aged fifteen or more years who had not moved beyond the rural village sector for at least one month (%)*

	Not moved for one or more months		Not moved for less than one month		Visited provincial town for less than one month		Proportion been to town within previous year		Don't know; no reply	
	Male	Female	Male	Female	Male	Female	Male	Female	Male	Female
Malalaua	26.9	48.2	7.2	23.7	18.4	19.7	48.5	41.9	1.3	4.8
Ambunti	21.2	53.2	13.7	39.9	7.4	11.9	75.0	82.5	0.0	1.4
Finschhafen	22.5	66.9	7.2	28.0	12.9	34.6	56.0	43.7	2.4	4.3
Maprik	24.8	66.3	9.6	25.7	13.9	36.0	61.3	60.5	1.3	4.6
Goilala	39.0	74.6	32.3	63.9	3.3	4.6	42.9	80.0	3.5	6.1
Okapa	36.2	90.4	11.0	35.4	23.7	50.1	50.5	53.1	1.5	4.9
Wabag	55.5	85.2	17.1	39.2	38.4	46.0	75.5	95.2	0.0	0.0
Mendi	56.2	94.8	1.7	3.9	52.8	89.3	84.2	92.2	1.7	1.6
Gumine	53.1	93.0	10.9	35.8	39.7	54.5	81.0	80.4	2.5	2.7
Mt. Hagen	76.3	92.6	36.5	38.1	37.8	52.6	90.4	83.3	2.0	1.9

Source of data: Rural village survey 1974–75.

were not directed towards answering such questions (cf. Avosa 1977; Kuange 1977; Siaoa 1977).

Another aspect of mobility is the frequency with which villagers return after being absent for at least one month in the rural nonvillage and urban sectors. Data extracted from the rural survey by Skeldon (pers. comm.) for males aged fifteen or more and classified by regional groupings show that about forty percent of Finschhafen men have migrated at least three times (Table 14.5). This result corroborates the view of Finschhafen as a district whose population has a high propensity to migrate but in what is clearly a circular pattern. The Sepik (Ambunti/ Maprik) and highlands regions (Fig. 14.1) have relatively large proportions of once-only migrants, while the Papuan region (Malalaua/Goilala) reflects its access to the main town of Port Moresby in having a far higher incidence of men who have completed five or more such movements.

While the mean number of migrations completed by adult villagers is available for only Maprik and Ambunti, two districts within the East Sepik province, this statistic may be compared with the average number of jobs held during their absence (Table 14.6). In these areas of relatively high outmigration and with tendencies to long-term absence, completed moves averaged rather less than two per adult and jobs obtained, for men, somewhat the same. Since seeking employment is a major but not the only reason why males are away for one or more months from the village, these data reinforce the generalization advanced by Clunies Ross (1977a:43): namely that, in the past, the probability of finding employment was close to certain within whatever period men were prepared to be absent from the local community. Among men, in fact, economic reasons for going away dominate, especially when the basically economic rationale of obtaining formal or vocational education is considered (Fig. 14.3). Although most women accompany male spouses or relatives, perhaps one-fifth act independently to seek some kind of job or an education.

The motives discerned in any situation of course depend very largely on the definition of 'migration' adopted for the purposes of research. The focus of the rural village survey on rural-urban migration and town growth meant attention to movements longer in time and space than those constituting the bulk of observations made in the Solomon Islands by Chapman when examining the small-scale world of tribal circulation. Thus the conclusions of a study 'designed to test the notion that considerations other than economic are far more significant than has often been recognized in migration research' (Chapman 1975:129) are determined largely by its assumptions and bound definitionally to show the dominance of 'social' reasons over 'economic' ones. Perhaps if the data for south Guadalcanal were disaggregated further by time-period of and reason for movement, it would become clear that the importance of

TABLE 14.5 *Incidence of outmigration for adult males, by region of origin, 1974–75 (%)*

Region of origin	No. of males 15+ years[a]	No of return migrations					Don't know; no reply	Total
		1	2	3	4	5+		
Ambunti/Maprik	1,010	52.7	24.1	11.6	5.8	5.4	0.4	100
Highlands	912	49.9	26.2	12.0	5.9	4.4	1.6	100
Malalaua/Goilala	558	39.1	23.1	12.4	4.7	9.1	11.6	100
Finschhafen	471	25.3	23.8	13.0	10.8	16.1	11.0	100
Totals	2,951	44.9	24.5	12.1	6.4	7.5	4.6	100

[a]Excluding males in formal education during the year prior to the survey.
Source of data: Skeldon (pers. comm. 17 September 1975).

TABLE 14.6 Number of migrations made and jobs held by adults[a], Maprik and Ambunti districts, East Sepik

| | Maprik | | Ambunti | |
	Male	Female	Male	Female
Completed migrations	1.96 (595)	1.32 (228)	1.87 (409)	1.97 (214)
Standard deviation	1.24	0.70	1.53	1.73
Jobs held	1.60 (573)	0.61 (228)	1.54 (409)	0.19 (214)
Standard deviation	1.41	0.26	1.42	0.47

[a]Aged fifteen or more years.
Source of data: Rural village survey 1974–75.

economic motivations increase in direct relation to length of absence from the home village.

Throughout Papua New Guinea in the 1960s and 1970s, the pattern and character of outmigration from rural places increasingly involved longer term and even permanent absence, a process that has proceeded farther in districts to the left of the migration continuum (Fig. 14.2). A consequent reduction in the youth and masculinity of the urban population becomes evident when their age and sex structures for 1966, 1971, and 1973/74 are compared with the national proportions for 1971 (Table 14.7). This shift reflects an increase in the reunion of migrants and their families in town, associated initially with substantial rises in the level of paid employment and later in the 1970s of real urban wages. At the same time, such socioeconomic changes as the

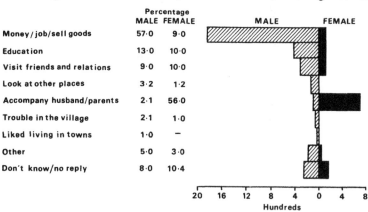

	Percentage MALE FEMALE	
Money/job/sell goods	57·0	9·0
Education	13·0	10·0
Visit friends and relations	9·0	10·0
Look at other places	3·2	1·2
Accompany husband/parents	2·1	56·0
Trouble in the village	2·1	1·0
Liked living in towns	1·0	—
Other	5·0	3·0
Don't know/no reply	8·0	10·4

TOTAL INTERVIEWED : MALE 3,315   FEMALE 1,242

Source of data : Rural village survey 1974-75

FIGURE 14.3 *Reasons for most recent move from ten rural districts, 1974–75*

TABLE 14.7 Age–sex profiles of indigenous population in seven major towns, Papua New Guinea, 1966, 1971, and 1973/4 (%)

Age group	National census June/July 1966		Seven major towns National census July 1971		Urban household survey Dec 1973/Jan 1974		Indigenous population National census July 1971	
	Male	Female	Male	Female	Male	Female	Male	Female
0–4	8.13	7.21	9.17	8.25	9.41	8.94	9.53	9.01
5–9	5.42	4.59	7.11	6.04	7.42	6.19	8.64	7.56
10–14	4.59	2.73	5.58	3.77	7.28	3.87	5.79	4.99
15–19	10.71	4.07	9.60	4.80	6.84	4.22	4.82	3.88
20–24	12.20	4.30	9.34	5.22	8.53	6.26	3.92	4.17
25–34	17.20	4.94	11.32	6.13	11.23	7.29	6.96	7.21
35–44	6.26	2.19	4.46	2.50	5.66	2.64	5.29	5.38
45–54	2.18	0.95	2.24	1.35	1.60	1.04	4.10	3.72
55–64	0.56	0.32	0.68	0.63	0.52	0.41	1.87	1.56
65 plus	0.18	0.10	0.35	0.23	0.34	0.40	0.90	0.64
Male:female ratio	2.07		1.49		1.43		1.08	

Source of data: Adapted from Garnaut, Wright, and Curtain (1977:23).

299

expansion of formal education enhanced the likelihood of independent movement of females.

Along with a trend towards permanent absence during this period there continued an intense circulation of rural visitors to the towns and, to some extent during national vacations, a corresponding reverse flow to the village of long-term absentees living in urban places. Concerning the former, Skeldon (1977b:25) estimated that 'at any one time between 7 and 13 per cent of the total population of the larger urban centres in Papua New Guinea is made up of a short-term floating population', composed predominantly of males aged between fifteen and twenty nine. In 1976, he conducted a census in the highlands town of Goroka (Fig. 2.1) and decided that the annual turnover of these sojourners about equalled its total population (Skeldon 1976)!

Since the rural survey was conducted during the period of annual vacations, it coincided with the time when many urban wage-earners take advantage of statutory provisions for biennial leave, in many cases with fares paid by their employers. When those visiting the home village for less than two months are considered, it is notable that two districts at the far left of the migration continuum (Malalaua, Ambunti) show the most pronounced tendency not only to return for holidays but also towards large-scale and long-term absence (Table 14.8: cols. 2-5, cf. Table 14.3). This is evidence of considerable circulation, in the sense used by Chapman (1976), but without necessarily the intention to live in the village. For many migrants from Malalaua and Ambunti, in fact,

TABLE 14.8 Frequency of migrant visits to home village in districts surveyed

District	Current migrants visited Dec. 1974–Jan. 1975				Current or former migrants visited during most recent absence			
	Males		Females		Males		Females	
	%	No. inter- viewed	%	No inter- viewed	%	No. inter- viewed	%	No. inter- viewed
Malalaua	40.5	269	46.0	150	49.7	330	49.1	220
Ambunti	26.3	281	35.6	146	32.0	416	32.4	216
Finschhafen	12.1	282	19.2	146	56.4	408	54.1	170
Maprik	16.2	328	19.0	116	22.4	606	28.6	241
Goilala	1.4	147	0.0	51	4.1	244	0.0	74
Okapa	15.1	93	36.8	19	31.1	251	55.9	34
Wabag	34.6	104	25.0	40	31.3	233	43.7	71
Mendi	11.6	95	0.0	13	26.1	199	22.7	22
Gumine	3.4	58	5.3	19	6.3	159	6.2	16
Mt. Hagen	29.4	34	22.2	9	33.3	120	17.9	28

Source of data: Rural village survey 1974–75.

the locus of existence has shifted from village to town (cf. Morauta and Ryan 1982; Ryan, Young, this volume). Overall, the proportions are higher for current or former migrants who visited the village during their most recent stay in another sector (Table 14.8:cols. 6-9); nonetheless Christmas and New Year provide the principal opportunity for brief returns in Malalaua, Ambunti, and Maprik, at one end of the continuum, and Wabag and Hagen at the other. Finschhafen is the conspicuous exception, with the proportion of all return visits made by migrants past and present far exceeding those who returned during December 1974 and January 1975.

Aside from brief visits, most absentees maintain considerable contact with village kin, even for districts at the left end of the continuum for which the average time away is longer. Among males from the highlands region (Fig. 14.1), the ratio of such links for the Mendi is markedly lower and likely reflects their continuing signature of labour contracts, compared with the prevalence elsewhere of independent departures for wage work. Perhaps for Mendi men, away for fixed terms and bound to return, maintenance of home contacts was seen as less necessary. In general, the principal means of sustaining such connexions was to exchange letters, remit money, send goods, and accommodate rural visitors (Table 14.9).

TABLE 14.9 Forms of contact by current migrants to villages of origin (%)

District	No. interviewed	Letter[a]	Send money	Send goods	House visitors	Other means
Malalaua	419	68.7	12.4	13.8	18.9	2.6
Ambunti	427	67.7	43.8	15.7	24.1	3.7
Finschhafen	428	59.4	27.8	10.0	26.6	0.9
Maprik	444	52.7	30.2	7.2	20.2	5.4
Goilala	198	5.1	10.6	10.6	42.9	23.2
Okapa	112	48.2	30.4	11.6	37.5	14.3
Wabag	144	50.0	36.1	11.1	51.4	4.2
Mendi	108	31.5	35.2	7.4	18.5	5.5
Gumine	77	54.5	27.3	6.5	57.1	7.8
Mt. Hagen	43	72.1	46.5	30.2	51.2	9.3

[a]Since multiple contacts were possible, percentages sum to more than 100.
Source of data: Rural village survey 1974-75.

Apart from the Goilala and Mendi, among whom literacy levels are low, at least half those currently away keep contact with village kin by letter. Transfers of money, while generally less important, are most common among those from Ambunti or Hagen and negligible only among the Goilala and Malalaua. This latter result probably is misleading, since large numbers from Malalaua return home from the capital city for long vacations (Table 14.8), at which time considerable amounts of cash are injected into the village economy (Avosa 1977; Siaoa 1977; Morauta and Hasu 1979). By contrast, visits by Goilala migrants are minimal. Absentees from such poorer districts as Goilala, Gumine, and Wabag more likely earn lower incomes so that providing hospitality for kinsfolk in town was most important. This is not so for Mendi, alone among highlands districts, because contract labourers cannot offer hospitality to anyone.

Commitment to urban life

A rising tendency for long-term and even permanent absence detected in the rural village sector, especially in Malalaua and Ambunti districts of the Gulf and East Sepik provinces, complements results from the urban household survey (Garnaut, Wright, and Curtain 1977). In Rabaul (Fig. 2.1), thirty-four per cent of male migrants had been there for at least ten years, in Port Moresby thirty-one per cent, and in Madang twenty eight per cent. Among both men and women, an exceptionally high proportion of long-term residents in Port Moresby came from Gulf province; sixty-seven per cent had resided for at least five years and forty-three per cent at least ten (Garnaut, Wright, and Curtain 1977:61). The same pattern was true for Lae, second largest town in Papua New Guinea (Table 14.2), to the extent that people from Gulf tended to have lived there for more years than those originating from any other province, including Morobe in which Lae is situated (Fig. 2.1). Similarly, very high proportions of those from the East Sepik had been in Rabaul (mainly men) and Madang (both men and women) for long periods.

While many people from districts at the far left of the migration continuum revisit the village when domiciled in town (Table 14.8), the urban household survey revealed many absentees who did not. Gulf and Goilala migrants in Port Moresby included far higher than average proportions of such persons: forty-seven per cent of the former had not visited their natal place for three years, thirty-two per cent for five, and eighteen per cent for ten years. This pattern was even more marked among Gulf migrants in Lae − forty-nine per cent had made no return visits in three years, forty per cent none in five, and a quarter none in ten years − leading to the observation that this group 'seemed to have settled down . . . with very slight connections with their area of origin' (Garnaut, Wright, and Curtain 1977:67). Even higher proportions recorded among the East Sepik in Rabaul suggest that they rarely visit their home area.

Similarly, a core of migrants had been in town for long periods without having visited or exchanged gifts with the natal village, the most visible of whom were in Rabaul (22.6 per cent), Lae (15.7 per cent), and Port Moresby (15.3 per cent). From the survey data, it was not possible to establish whether such lack of contact was leading to the loss of access rights to rural land and the emergence of an urban proletariat, but Morauta (1979) argues convincingly that this process is occurring and particularly among Malalaua in Port Moresby. To be sure, this prospect is most likely among children born in town, of whom about a quarter in Port Moresby, Lae, and Rabaul aged ten to fifteen had never visited their parent's community. Summarizing urban-based data on actual links with the village and stated intentions of returning to live there, Garnaut, Wright, and Curtain (1977:79–80) concluded:

It was in the large, old coastal towns — Rabaul and Port Moresby in particular, but also Lae, Wewak and Madang — that there were the strongest signs of firm commitment to urban life. These signs were strongest amongst people from a few provinces — notably Gulf and East Sepik. However, the evidence on dispossession was ambiguous even for these groups, with relatively high proportions claiming to send and receive gifts to and from the home village.

Circular movers: homing pigeons and rational actors?

Circulation over the life cycle of the individual remains the dominant form of population movement throughout Papua New Guinea and will remain so in the forseeable future for most people from most places. Two broad categories may be distinguished among long-term or permanent migrants in town, discovered most commonly among absentees from Malalaua and Ambunti: those who maintained close contact between urban and rural areas through both quite high rates of circulation and a variety of other means (letters, gifts, hospitality); and those who preserved some connexion without circulating between town and village. An emergent third category may comprise children born in town who had never visited the natal place of their parents (cf. Morauta and Ryan 1982).

As Mitchell (1985) argues, any discussion of the circulation of labourers and their dependents needs to include an adequate description of the 'setting' within which that movement is taking place. The fact that earlier cohorts of migrants in Papua New Guinea most often returned to the local community by middle age is no secure basis for predicting similar behaviour among the younger generation, for the latter are subject to quite different circumstances and incentives than accounted for their parents' actions. Various socioeconomic changes during the 1960s and 1970s have altered the 'setting', or context of population mobility, and in directions likely to foster long-term or life long absence. Notable amongst these have been the establishment of an urban wage structure for unskilled and semiskilled workers based on family needs rather than, as previously, those of the individual male; and great expansion in formal education which, reinforced by political evolution, made available to Papua New Guineans the full spectrum of employment opportunities and income levels. Complementary influences were the progressive transfer to middle- and upper-echelon citizens of subsidized housing built originally to attract skilled expatriates from metropolitan countries (Stretton 1979: Chapter 4) and an increasingly urban bias in the distribution of government expenditures (Jackson 1975).

Thus far, these and other changes influential upon the degree of long-term outmigration from village communities have occurred in the towns rather than rural areas and acted to increase levels of commitment to urban life. It is not clear from available evidence, however, whether such higher levels will result in a greater tendency to sever one's ties with the local community or alternatively to extend periods of absence to include one's entire statutory working life — in itself a concept derived from western industrial norms of workforce participation, where the expected career of employment is stretched close to the limit of mean life expectancy. Revitalizing village communities is a difficult process under optimum conditions but, despite much recent rhetoric, little has been achieved in Papua New Guinea effectively to tilt the balance of advantage, as perceived by potential movers, towards the rural scene.

Yet there is no evidence that poorer living conditions in rural areas are making towns more attractive, since the processes of class formation, population growth, and loss of land rights have not proceeded sufficiently far to promote an exodus of landless. Throughout the country, the likely impact of population pressure would be for movers to reduce the frequency and duration of absence so as to protect their access to and use of village land (Harris 1977) and so avoid becoming part of an urban proletariat. Thus migrants from Simbu (formerly Chimbu), a province of high rural densities upon whose experience Harris based the above conclusion, suffered the highest incidence of unemployment among long-term and long-distance absentees in fifteen towns studied. Comment Garnaut, Wright, and Curtain (1977:185): 'From the perspective of excess supply of urban labour as a symptom of rural-urban imbalance, the most disturbing evidence from the UHS [urban household survey] was the very low wage levels and employment rates that Chimbu migrants appeared to be prepared to endure, even when they maintained close links with home villages'.

Despite powerful incentives favouring long-term urban residence, particularly among the most highly educated, the strength of migrant ties to their 'as ples' (natal village) should not be underestimated. This reflects not simply emotional attachment but also the need for an option to return to the rural economy upon retirement or whenever dictated by the uncertainties of a precarious urban existence. Evidence of this dual dependence — in Australian demotic, the two-bob-each-way strategy — is revealed by how far town dwellers invest in rural housing, which the urban household survey suggested was negatively correlated with level of formal education. Even so, such investment was made by fully fifty-four per cent of all male migrants aged twenty-five to forty-four who were in wage employment and had completed primary school (Curtain 1980a:288). Similarly, in 1977, a random survey of public servants in Port Moresby documented that about half had built houses in their home communities and a further forty per cent intended to do so (Stretton 1979:42). Quite often, such a house serves the additional purpose of sheltering parents in their older years.

This rural orientation among urban dwellers may have been influenced by current proposals to end subsidies for publicly-provided housing and shift that responsibility to the individual within a free market (Papua New Guinea, Committee on Review on Housing 1978). Substantial capital gains would likely be won by higher-status public servants, who would be able to purchase the government houses they presently occupy. How urban or rural are the basic orientations of this elite will be reflected in whether these capital gains are used to finance their eventual reestablishment in the natal community or alternatively to cement further their position in urban society. Apart from questions of social equity and income distribution, the longer-term plans of most urban dwellers still are dominated by a basically rural orientation, so that policies designed to create profitable investment opportunities in provincial areas would succeed in diverting both capital and resources to predominantly village populations.

Acknowledgments

We are grateful to Murray Chapman and Elspeth Young for detailed comment on an earlier version, but admit to stubborness on some matters; and to Kundapen Talyaga, for preparing several tables.

15 Fijians and Indo-Fijians in Suva: rural-urban movements and linkages

Shashikant Nair

Field research in Melanesia has emphasized the impermanent character of population movement and in particular its circularity. Various researchers from several academic disciplines have used different indices of movement and even adopted different definitions for terms like 'circulation', 'circular migration', and 'migration'. Regardless of the definitions and indices employed, there seems agreement that a move by a Melanesian does not result necessarily in the severance of all bonds with the place regarded as home and that individuals who leave their natal communities frequently return to live there.

A major goal of field research conducted in Fiji, on which this paper is based, was to identify whether the dominant form of movement from rural areas to the capital of Suva (Fig. 7.1) is a permanent and one-way flow (migration) or impermanent and more circular in nature (circulation). The fact that the country's population includes a similar proportion of Melanesian Fijians and Indo-Fijians of immigrant stock also permitted consideration of the role of ethnicity in prevailing patterns of mobility. The main propositions advanced were that: most people who have moved to Suva intended ultimately to return to their rural areas of origin; over the past generation, movers have been spending greater amounts of time in the urban area and have become increasingly committed to it in social and economic ways; and Indo-Fijian movement to Suva was more permanent than that of Fijians.

Data were collected in Suva over five months (November 1977–March 1978), mainly by means of a questionnaire survey and migration histories. The primary purpose of the formal questionnaire was to obtain information on the characteristics of movers, and residence places of other family members, the amount of visiting and other ties between places of origin and destination, the nature of investments in the rural areas and in Suva, and intentions about future residence. Areas chosen to be surveyed in Suva included both long-established and recent

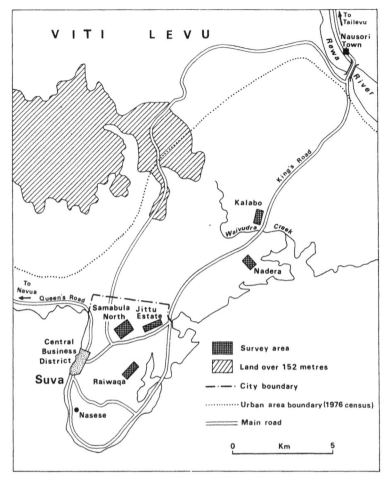

FIGURE 15.1 *Suva urban area, 1977*

migrants, as well as a range of socio-economic situations: of planned housing (Nadera and Raiwaqa), of squatter settlements (Kalabo and Jittu Estate), and one area of higher-income households (Samabula North; Fig. 15.1). Only heads of household were surveyed, and being migrant was defined as having been born outside the urban area but having lived there for at least six months. A total of 400 household heads were interviewed, 199 of whom were Fijian and 201 Indo-Fijian. Most of these were from lower-income groups, to which the majority of all urban movers belong.

Raiwaqa and Nadera (Fig. 15.1) house both Fijians and Indo-Fijians. They include units rented to low-income tenants as well as 'purchase

307

plan' dwellings offered for sale to families with medium-low incomes by the housing authority, a statutory government organisation. Kalabo, equidistant between central Suva and Nausori town, is a Fijian settlement located on native land belonging to the people of the nearby village, from which its name is taken. The Jittu Estate houses predominantly Indo-Fijian squatters, who occupy an area of freehold land for which they pay minimal rent. Samabula North, by comparison, is a privately-developed suburb that consists mainly of single-family, concrete-block dwellings with roofs of corrugated iron. Income levels are relatively high and both professional and commercial occupations represented, along with small business owners (generally Indo-Fijian) and middle-level civil servants (dominantly Fijian).

The medium size of Fijian households surveyed, at 6.5 persons, was greater than for the Indo-Fijian (4.9). Generally, the former were larger because, in addition to the nuclear family, more than half (57%) contained close relatives, most of whom were young, unmarried adults and children who either worked or attended school in Suva. Of the 201 Indo-Fijian households, only one tenth consisted of extended families.

Two thirds of the Fijian heads of household had been living in Suva for ten or more years, eighteen per cent for twenty-five or more, and only two per cent for less than twelve months, whereas comparable percentages for the Indo-Fijians were forty-eight, twelve, and three (for details see Nair 1980: Table 5). This length of residence for Fijians is consistent with results obtained from a social survey undertaken in Suva in 1959, in which seventeen per cent out of 528 Fijian householders were found to have lived there for more than twenty-five years (Verrier, reported in Nayacakalou 1963:34). Most Fijian heads of household had come either from the distant island provinces of Lau (31.2%), Kadavu (17.6%), and Lomaiviti (9.5%), or from the province of Cakaudrove in Vanua Levu and the island of Taveuni (13.6%; Fig. 15.2). The majority of Indo-Fijians were from the coastal provinces of Viti Levu (Ba 25.4%, Ra 16.4%, Rewa 14.9%), and from Macuata province of Vanua Levu (12.1%).

Residential intentions

Although circulation often is distinguished from migration on the grounds that it is intendedly impermanent (Zelinsky 1971:225–6; Gould and Prothero 1975b:42), academic opinion differs greatly about the utility of statements of residential intentions (cf. Goldstein, this volume). Research in both Black Africa and Melanesia suggests that links maintained with rural areas of origin and various indices of urban commitment, as of length of stay and proportion of working life spent in town, are the most appropriate criteria of assessment (Caldwell 1969; Adepoju 1974; Odongo and Lea 1977; West 1958; Strathern 1972).

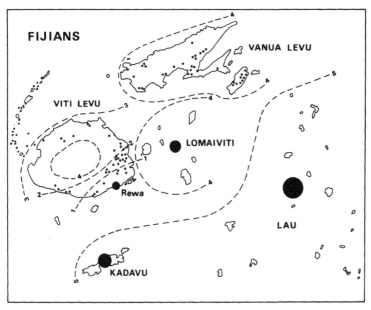

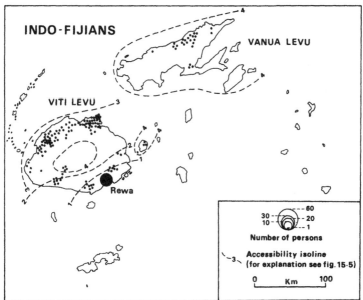

FIGURE 15.2 *Birthplace of household heads and accessibility to Suva,
November 1977–March 1978*

309

Consequently this paper first considers the residential intentions of migrants to Suva, followed by the links that household heads maintain between places of origin and destination, their length of residence in town, and the perceived advantages and disadvantages of living there. With this approach, movements between the capital city and various rural places of origin are incorporated within rural-urban linkages.

Generally, Indo-Fijians had a clearer idea than Fijians of their residential intentions (Table 15.1). Overall, twenty-eight per cent of household heads felt unable to say whether they would remain in Suva and where would be their final place of residence. If this considerable degree of ambivalence is ignored, there is still a statistically significant difference between the intentions of the two ethnic groups. Whereas almost forty per cent of Fijian householders intend ultimately to live in their villages, only six per cent of the Indo-Fijians expect to return to their former settlements. Most Indo-Fijians (66%), and a considerably smaller but still substantial proportion of Fijians (29%), said that they intend to remain in Suva for the rest of their lives.

TABLE 15.1 Intended length of stay in Suva of household heads, November 1977–March 1978

	Fijian		Indo-Fijian		Total	
	No.	%	No.	%	No.	%
Forever	57	28.6	132	65.7	189	47.3
Up to 1 year	3	1.5	5	2.5	8	2.0
Up to 5 years	9	4.5	9	4.5	18	4.5
Up to 10 years	6	3.0	2	1.0	8	2.0
More than 10 years	6	3.0	0	0	6	1.5
Until retirement	29	14.6	2	1.0	31	7.8
Until children are educated	25	12.6	1	1.5	26	6.5
Until enough money	2	1.0	0	0	2	0.5
Unsure	62	31.2	50	24.9	112	28.0
Total	199		201		400	

Fijian/Indo-Fijian difference: Chi square = 98.9 with 8 d.f.
Significance = 0.000.

The replies from Fijian household heads indicate no overwhelming desire to return permanently to their villages of origin; almost thirty per cent intend to remain forever in Suva while a slightly higher proportion were quite undecided (Table 15.1). Of those who expected to return, very few had immediate plans. Most with firm opinions explained that they would leave Suva when they retired or when their children had completed formal education. The fact that such events loom larger than estimated years before return suggests that position in the life-cycle is more influential than elapsed time in the decision to move and that people do not declare residential intentions without an underlying rationale.

For Fijians, residential intentions are affected in varying degree by area of origin, present length of stay in town, household income, and ownership of property in Suva, but far less by age than position in the life cycle. A high proportion from the small island provinces of Lau and Lomaiviti (Fig. 15.2) either intended to stay permanently in Suva or remained unsure. Of all who desired to remain forever, sixty per cent originated from Lau even though Lauans comprised only thirty-two per cent of all householders interviewed. Bedford (this volume) found the same pattern at the Lauan settlement of Qauia where completion of children's education, the primary reason given for movement to Suva in that survey, rarely was followed by the return of parents to their villages.

Of those Fijians who spent less than half their lives in town, fifty-one per cent intended to return permanently to the village, compared with thirty-three per cent of those who had lived there more than half their lives. Of the sixty-five resident for twenty or more years, only a quarter thought they would ultimately go back to their village; of the 168 who had been in town for less than twenty years, only sixty-three (38%) felt this way. Briefly, the greater the proportion of their lifetime that Fijians have spent in Suva, the higher is the likelihood that residence will become permanent.

Income and ownership of high-value property is similarly influential. Fijians who are in the high income groups and earn at least $F60 a week ($F0.59 to £1.00; $F1.00 to $A1.00) were the least interested in returning to their village communities (8 out of 48). Conversely, the highest proportion of potential returnees were among those who earned less than this amount. By itself, it is not the ownership of land or a house in either Suva or the rural areas that affects residential intentions, but rather the individual ownership of such high-value property as concrete homes and freehold land to which there is secure title. Thus the proportion intending to return to their place of origin but who also own homes there (46.4%) and in Suva (46.25%) is virtually identical, whereas only three of those who have concrete homes sited on freehold land in Suva expect to do so.

Area of origin, length of stay in Suva, income, and ownership of property consequently have the most influence upon residential intentions. Although there are some Fijians who were born in the distant provinces of Lau and Lomaiviti, have stayed in Suva for many years, earn high incomes, and still intend to return ultimately to their villages, there are many more who do not. There is no clear relationship between expressed intentions and economic groupings (by income and property ownership), except perhaps for those elites with very high incomes and substantial houses sited on freehold land who declare with certainty that they intend to reside forever in the capital. Only this group may be said to display an unequivocal preference. By contrast the most ambivalent, those who said they were 'not sure', were persons who had been in Suva for comparatively short periods or were from the relatively poor and isolated provinces of Lau, Lomaiviti, and Kadavu (Fig. 15.2). In short, the longer heads of Fijian households live in Suva, the more their preferences will tend toward permanent residence.

Few Indo-Fijians (12 out of 201) expect to return to the rural settlements from which they came (Nair 1980: Table 7). If those who prefer permanent residence in town are compared with others who are unsure or prefer locations elsewhere (132 versus 57), then length of stay, income, and ownership of property are the most important factors. Among Indo-Fijians, the most certain about where they would live in future were those who had been in Suva for ten or more years, had weekly incomes of at least $F60, and owned concrete homes and freehold land. Conversely the least sure had smaller incomes and had resided there for less than a decade. Year of birth has slightly more bearing upon Indo-Fijian than Fijian intentions, since progressively higher percentages of those in the older age groups hoped to remain in Suva forever. Perhaps the younger generation regard their longer life expectancy as providing the opportunity, as well as the time, to search for alternative places of residence – a luxury which older people feel cannot be indulged.

The clear contrast between the residential intentions of Fijians and Indo-Fijians is not adequately explained by parallel differences of income, length of time spent in Suva, or area of origin. Indo-Fijian heads of household come from rural settlements located in the more fertile parts of Viti Levu and Vanua Levu (Fig. 15.2) whereas more Fijians originate from the outer islands. Nevertheless, more Fijians remain longer in Suva and a greater proportion earn higher incomes. Do such contrasts in residential preferences reflect basic differences in the cultural and political backgrounds of these two ethnic groups?

Rural linkages

In addition to statements of residential intentions, the degree to which links are maintained with places of origin may be viewed as an indication of future actions. Visiting is the most obvious means by which urban residents acknowledge ties to their families, their villages, and their provinces of origin, but they may also remit cash, send foods and other items, participate in traditional ceremonies, pay provincial taxes, host visitors from their villages, and contribute to rural projects. Patterns of visiting to rural areas from Suva also help establish whether those who say they will return act in ways to facilitate this intention. Thus household heads were asked several questions about the incidence, length, and reasons for visiting their areas of origin, a visit having to be of at least six hours to be counted. Although almost all Fijians and Indo-Fijians resident in Suva do visit rural communities of origin, there are some significant differences on the basis of ethnicity: whereas Indo-Fijian householders make more frequent visits their absences are of shorter duration (Nair 1980: Table 11).

Of 199 Fijian heads of household, seventy-seven per cent (154) had visited their village at least once since arrival in Suva and thirty-two per cent (63) within a year of being interviewed (Fig. 15.3). Of the forty-five Fijians who had never returned since departing, thirty-one had resided in the capital for more than five years. These figures are comparable with

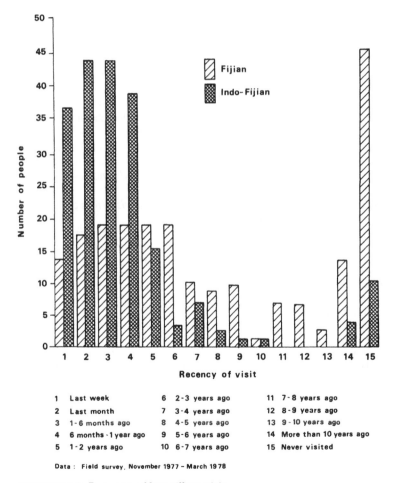

FIGURE 15.3 *Recency of last village visit*

those reported for Qauia, a periurban settlement of Lauans, from which about fifty-four per cent had not made any visit back to Lau (Fig. 15.2) during the six years preceding that survey (Bedford, this volume: Table 16.3). An even greater number of Indo-Fijians visit their settlements of origin: 191 out of 201 had made at least one since residing in Suva and many more regularly made annual visits than did Fijians (73% versus 28%).

The median duration of return visits by Indo-Fijians (1.3 weeks) is shorter than that of Fijian householders (2.1 weeks) but the former make them more frequently. During their last rural visit, sixteen per cent of the Indo-Fijians and forty-seven per cent of the Fijians stayed for more than a week (Fig. 15.4), while of thirty-three who were absent from Suva for at least a month only four were Indo-Fijians.

The dominant reason for most visits to villages and rural settlements relate to kin: spend holidays with relatives, attend weddings or funerals, pay respects to the sick (Table 15.2). Some Fijians (ten out of 154 during the most recent visit) returned for customary reasons, the most important of which is *mataniqone* when a new-born child is introduced formally to the father's village for the first time. Others — twelve during the last visit — who came from the areas close to Suva helped with such village projects as building a church, or tended their food gardens. The long Christmas vacation is the most popular time for Fijians, and to a lesser extent Indo-Fijians, to visit kin. Many Indo-Fijians also return for religious festivals like the Hindu *Diwali* or the Muslim *Eid*, and for vacations.

TABLE 15.2 Primary reason for most recent rural visit from Suva, November 1977–March 1978

	Fijian		Indo-Fijian		Total	
	No.	%	No.	%	No.	%
Spend holiday with kin	85	55.2[a]	145	76.3	230	66.9
Attend wedding or funeral and see sick relative	22	14.3	33	17.4	55	16.0
Traditional ceremony	10	6.5	3	1.6	13	3.8
Food gardening	10	6.5	0	0	10	2.9
Contribute to *soli*	7	4.5	0	0	7	2.0
Unemployed in Suva	3	1.9	1	0.5	4	1.2
Farm work, such as harvesting cane, looking after cattle	1	1.3	2	1.0	4	1.2
Other	15	9.7	6	3.2	21	6.1
Total	154		190		344	

[a]Based on number who visited.

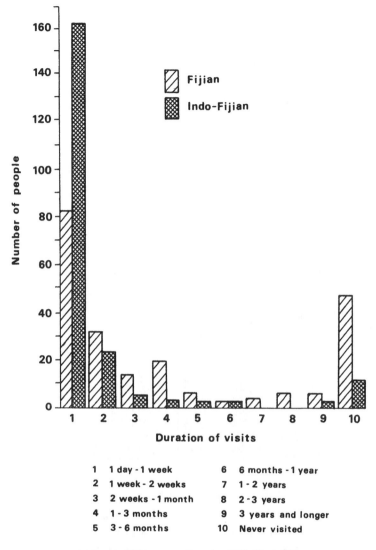

FIGURE 15.4 *Duration of last village visit*

Variations in the number, recency, and duration of return visits are best understood in terms of the distance and accessibility from Suva of the different settlements of origin. Accessibility of household heads to these places was assessed by the combination of distance with regularity of transport links (Fig. 15.5, cf. Fig. 15.2). Rural areas within

315

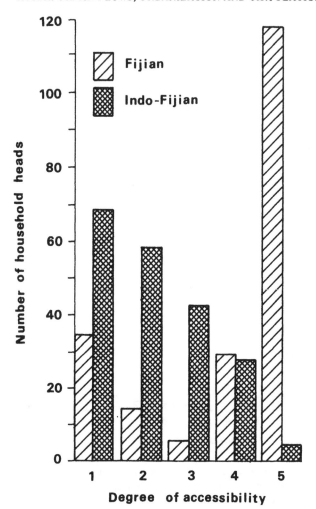

FIGURE 15.5 *Distribution of household heads according to accessibility to Suva of area of origin*

two hours' bus ride from the city were considered the most accessible, and the islands of Lau, Kadavu, and Lomaiviti (excluding Ovalau, to which there is regular transport) the least accessible (Fig. 15.2). Although some of these islands now have air strips, high fares ensure that the most common means of travel is still by cargo boats which, however, only run when there is sufficient business.

Looking at both Suva and different settlements of origin reveals that many Fijians who came from less accessible areas made fewer return visits, whereas for most Indo-Fijians the converse is the case. Of the forty-five Fijian householders who had not returned to their village since arrival in the city, thirty eight (83%) came from the most distant locations compared with only three from settlements accessible by bus. Of 143 Fijians who did not make an annual visit, seventy per cent (100) originated from communities difficult of access. Conversely, eighteen out of twenty eight Indo-Fijians (64%) who averaged more than ten returns every year travelled to places very close to Suva.

Accessibility also affects the duration of return visits, as evident from the Fijians, a higher proportion of whom were born in more isolated locations (Fig. 15.2) and stayed away longer from Suva. Difficulties experienced, such as finding return transport, can mean that those who originate from the outer islands are far less willing to leave town for a brief or single visit, especially if they have a regular job, since lengthy absences from work can result in termination. From the replies about patterns of visiting it appears, most important, that length of urban residence will not lead to any reduction in the regularity of visits made by either Fijians or Indo-Fijians and that such practice remains strong, even after many years in Suva.

As found important in previous research in Fiji (Nayacakalou 1975; Spate 1959; Ward 1965), many links other than visiting exist between places of origin and destination. There is, however, a basic difference between Fijian and Indo-Fijian heads of household, since the former are involved in eight kinds of interaction but the latter in only three (Table 15.3). The first level of interaction for Fijians is the extended family, with whose members there is the greatest degree of attachment and the most intense reciprocity. Most household heads (161 out of 199) said they helped family members who remained in the village and that this assistance was reciprocated (147 out of 199). Cash remittances are the most important form of contribution made by Suva residents, whereas gifts from rural kin are of such traditional products as woven mats and baskets, scented coconut oil, and palm leaf brooms.

The second level of interaction for Fijian householders is the village, but with the difference that contributions received there are far greater than the reverse flow to urban residents. Whereas 161 Fijians noted that they helped their rural community, only seventy-one said this was reciprocated. Although cash is the most common form of assistance

317

usually it is not sent individually but collected from fund-raising activities in town and remitted later. Reciprocal help from the village to Suva residents commonly occurs during traditional ceremonies, especially before (*somate*) and after (*burua*) burial of the dead. Such assistance includes guidance on ceremonial procedures, provision of labour, and supplies of food and materials used for those attending the funeral feast in town.

TABLE 15.3 Forms of contribution to places of origin by Suva residents

Nature of contribution[a]	Fijian (199)		Indo-Fijian (201)		Total (400)	
	No.	%	No.	%	No.	%
Host people in Suva	159	79.9	75	37.3	234	58.5
Send food and materials	153	76.9	1	0.5	154	38.5
Help during weddings and funerals	153	76.9	54	26.9	207	51.8
Remit cash	152	76.4	71	35.3	223	55.8
Take part in *soli* (fund raising for specific project)	138	69.3	na	na	138	34.5
Take part in provincial festivals	135	67.8	0	0	135	33.8
Pay provincial tax	97	48.7	na	na	97	24.3
Pay land rate	28	14.1	na	na	28	7.0

[a]Includes all contributions. Thus a person who did not make one form of contribution (such as hosting people in Suva) may have made another (like taking part in *soli*).
naNot applicable.

The province is the largest administrative unit within the Fijian administration and the third level at which Fijians interact with rural areas. The Fijian administration is responsible for native lands and other affairs, with jurisdiction over fourteen provinces. Almost all operating funds are obtained from either provincial taxes or other kinds of levy; the head tax (48.7%) and land rates (14.1%) were the most common payments made by Suva residents. The former is an annual levy on Fijians and the land rate a substitute that some provincial councils have adopted, whereby each adult male registered as a land owner pays according the amount held by his *mataqali* (sub-tribal land holding unit). Since provincial funds, however obtained, are used chiefly to finance the Fijian administration and for such projects as local area schools, the flow of contributions is mainly from town to village and thus in marked contrast with other kinds of reciprocal exchange. Even so, fifty-seven out of 199 (28.6%) Fijian householders in Suva made no financial payments to their provinces of origin – a far higher ratio than those providing no assistance to either extended family or village community.

Compared with Fijians, the Indo-Fijians maintain fewer and less intensive links with rural areas (Table 15.3), which reflects the fact that they have no residential unit of reference comparable to the Fijian village and that the provincial administration deals exclusively with those who are ethnically Melanesian. Indo-Fijian links to their areas of origin are consequently at the level of the extended family, but only 34.3 per cent (69 out of 201) said they assisted rural kin and a mere 6.5 per cent (13 out of 201) received any reciprocal help. Except for one Indo-Fijian who sent materials bought in Suva, most such links to rural families was through cash remitted. Of the sixty nine who helped kin, fifty four assisted with wedding, funeral, and religious ceremonies, while thirty-seven per cent regularly hosted visitors to town.

These basic contrasts between Fijians and Indo-Fijians in Suva are not explained by their group characteristics but reflect differences of tradition and lifestyles, as well as of administrative context. For both ethnic groups, the presence of parents or children in rural areas generally results in more intense linkages and greater mutual assistance. In addition, Fijians in higher-income groups are expected and do contribute slightly more to their extended family, natal community, and province of origin. Of Fijians earning $F50 or more each week, the links maintained by eighty-two per cent were very intense, in contrast with the lower but still substantial proportion (68%) of those whose weekly incomes ranged from $F30 to $F50. The economically most successful Fijians find that longer residence in Suva, increasing experience, and higher incomes are accompanied by correspondingly greater demands on them from relatives and rural communities. For these Fijians, the greater their length of stay in the city the more intense are likely to be their links with villages and provinces of origin.

Urban commitment

The concept of urban commitment used in this study refers to the degree to which people are involved in urban living. Through such indices as proportion of lifetime spent in Suva, location of nuclear family unit and of property owned, and perceived advantages and disadvantages of life in both Suva and rural areas, an assessment is made of the extent to which movers are bound to lengthy residence in Suva. Information on, for example, time lived in the capital and location of property outlines the actions of town residents compared with their future intentions discussed previously.

In Suva, most Indo-Fijians wish to remain in town forever, and many Fijians for most of their lives although they might return ultimately to reside in rural areas. If number of years resident in the capital is expressed as a ratio of total and working lifetime, then about thirty-six per cent of the Fijians and a quarter of the Indo-Fijians have spent more than half their lives in Suva (Table 15.4). If it is assumed that gainful employment begins at age eighteen, then 47.2 per cent of the Fijians and 30.3 per cent of the Indo-Fijians have worked nowhere else. Three quarters of Fijian heads of household and more than two thirds of the Indo-Fijian have resided more than half their working lives in Suva, a result that is similar to those reported for the Qauia survey of Lauans (Bedford, this volume: Table 16.3). All these proportions suggest a fairly high degree of commitment to city residence since, as Mitchell argues in his research with Zambians, 'if a man has spent more time in urban than in rural areas since he turned 15, then he is more committed to urban life' than someone who has spent more time in rural areas, and 'if a person has spent a comparatively long time in one town (in this case more than 5 years) then there is evidence that he has settled in that town' (Mitchell 1969b:487).

To some extent, the total length of time as well as the proportion of working life spent by Fijians in Suva reflects place of origin rather than simple accessibility or distance. Heads of household from the most distant provinces of Lau and Lomaiviti had spent slightly longer periods in town, whereas those from other relatively inaccessible places like Vanua Levu and Taveuni had not, compared with yet others who originated from communities closer to Suva. Amongst Indo-Fijians, there is no relation between accessibility of area of origin and length of time spent in Suva (Figs. 15.2 and 15.5).

As Mitchell (1969b:487) argues, the presence in town of a man's wife is another useful index of commitment to urban residence, since it indicates that his stay is likely to be lengthy. Most household heads in Suva had moved as part of a nuclear family; less than one tenth of either Fijians or Indo-Fijians interviewed had any such family members still resident in the community of origin. About two in five nuclear-family households became established through either chain migration or marriage: in the former, a few members settled in town and are subsequently joined by others; in the latter, young adults move from the rural areas, marry, and later have children. Such high percentages of mover households with nuclear families in Suva further suggests that most Fijians and Indo-Fijians are committed to lengthy residence there.

Odongo and Lea (1977) have demonstrated in Uganda that actual or intended ownership of property in town and/or rural areas can indicate commitment to one locality or another. Such may be demonstrated by those who have or intended to purchase a house or land in Suva, thus being oriented to a lengthy period if not permanent residence in an urban environment. Because many rural-born residents in the capital neither own nor intend to acquire property and because others, in fact

TABLE 15.4 Time spent in Suva by household heads (%)

Per cent	Total lifetime						Working life					
	Fijian		Indo-Fijian		Total		Fijian		Indo-Fijian		Total	
	No.	%	No.	%	No.	%	No.	%	No.	%	No.	%
0–9	15	7.5	23	11.4	38	9.5	9	4.5	6	3.0	15	3.7
10–19	27	13.6	38	18.9	65	16.3	5	2.5	11	5.5	16	4.0
20–29	34	17.1	37	18.4	71	17.8	13	6.5	14	7.0	27	6.7
30–39	17	8.5	39	19.4	56	14.0	6	3.0	19	9.5	25	6.3
40–49	35	17.6	13	6.4	48	12.0	14	7.0	12	6.0	26	6.5
50–59	32	16.1	20	10.0	52	13.0	11	5.5	22	10.9	33	8.3
60–69	22	11.1	12	6.0	34	8.5	8	4.0	14	7.0	22	5.5
70–79	7	3.5	6	3.0	13	3.3	10	5.0	17	8.5	27	6.7
80–89	9	4.5	11	5.5	20	5.0	17	8.5	18	9.0	35	8.7
90–99	1	0.5	2	0.9	3	0.8	12	6.0	7	3.5	19	4.7
100	0	0	0	0	0	0	94	47.2	61	30.3	155	38.7
More than half	71	35.7	51	25.4	122	30.6	152	76.2	139	69.2	291	72.6
Median	41.9%		32.0%				95.4%		71.5%			

have property in both rural and urban places, it is difficult to reach any firm conclusion about the relationship between the location of property owned and the commitment of migrants to Suva.

With ninety-nine per cent of Fijians owning land in rural areas, it is understandable that a high proportion (71%) also have houses there. The ownership of title to *mataqali* land demonstrates a clear commitment to the countryside, rooted in the traditional culture; but it does not explain why a relatively high (58) per cent of rural house owners prefer Suva. Looked at in conjuction with expressed intentions to purchase property in the city, this figure could indicate an emergent commitment to Suva on the part of Fijian heads of household. About equal numbers of Indo-Fijians have land in the capital and in rural areas, but almost twice as many (74%) own houses in the former as in the latter (39%). Indo-Fijians expressing an intent to purchase and a preference for residence in Suva form a clear majority whether or not they already own property, or will purchase land or houses. For Indo-Fijians, commitment to. Suva thus appears stronger than for Fijians, but the relationship between property ownership and commitment to place is not nearly as clear cut as has been demonstrated for Uganda.

Household heads in town were asked what they viewed as the advantages and disadvantages of life in both the capital city and rural areas (Nair 1980: Table 18). These questions were drawn from African studies which have shown that degree of commitment to city residence can be examined through the attitudes people have towards urban and rural living (Mayer 1961; Southall and Gutkind 1957). For both ethnic groups, good public services such as schools, hospitals, shops, roads, parks, and playgrounds are considered most important and Suva is viewed as the place where these needs are best met. Availability of employment and adequate livelihood is of almost equal importance and, again, the capital most easily satisfies these requirements, although this means a loss of 'free time' and a far higher cost of living than in rural communities. For Fijians in particular, maintenance of culture, customs, and traditions also is considered important and far more easily achieved within the village, although local obligations can be a burden.

For most Fijians and Indo-Fijians resident in Suva, life in an urban setting satisfies many more needs than that in rural areas. Attitudes suggest that most are committed to lengthy residence there, despite the cost of subsistence, the disadvantages of crime and pollution, and the difficulty of obtaining jobs. For Fijians, village living continues to address important cultural and social needs, but these can be met by occasional short visits and the maintenance of reciprocal linkages.

Fijian and Indo-Fijian differences

Before concluding about patterns of movement to Suva, it is necessary to suggest reasons for the contrast in residential intentions and rural-town links among Fijian and Indo-Fijian heads of household. Since Fijians are indigenous Melanesians and Indo-Fijians of immigrant Asian stock, their vastly different cultures and traditions clearly are critical to understanding the strong bonds that the former appear to retain with their places of origin.

The village, or *koro*, comprised of several subtribes (*mataqali*), has special significance for Fijians. While production of food within the *koro* generally is organized on a family basis, exchange between families is common and often defined communally (Spate 1959:77). At times of traditional ceremonies and such major activities as building a church or road, all villagers act collectively under the direction of the village head. A certain form of communalism thus exists throughout Fijian society and reciprocity is emphasized; both are integral to what commonly is known as the Fijian way of life. Fijians are taught both formally and informally that their traditions and lifestyles must be retained at all costs. One demonstrable way for those in Suva to reaffirm that urban residence implies no disavowal of this special identity, is to maintain contact with the natal village, to participate in various traditional ceremonies, and to reciprocate in the exchange of goods and labour. To refuse to contribute, especially when there is a request for help from the village, would be to demonstrate that one is becoming less Fijian and could lead to ostracism.

The system of land holding similarly underlies the strong links that Fijians maintain with their rural communities. Of the country's total land area, about eighty-four per cent is native land owned communally by Fijians. Access to and use of native land depends upon membership in a *mataqali*. The fact that the Native Land Ordinances of 1880 and 1912 do not permit Fijians to sell their land creates a permanent bond to those rural areas in which *mataqali* land is located. Even though some *mataqali* may have too many members, too small a land area, or much land of poor quality, Fijians have both a very strong emotional attachment to and vested interest in their land, which makes it impossible for them to revoke all links with the natal village. Those born in rural communities but resident in Suva may have very few material possessions yet live confident in their knowledge that the *mataqali* land remains secure, a birthright that cannot be revoked by law.

All this is reinforced by bonds of kinship. While entire families may move to the capital together or through a sequence of chain movements, always there are some kin who remain in the village. Such kin, rather than being ignored or abandoned, receive cash, exchange goods, and are provided housing during visits to town. This fact led Nayacakalou (1975:99) to observe that kinship ties were 'the foremost' of all those he noted urban Fijians to maintain with rural areas.

The Indo-Fijians are descendants of people who immigrated to Fiji

323

from different parts of India, mostly between 1879 and 1920, when about 60,000 entered the country, nine tenths as labourers indentured to work sugar plantations and the rest as free migrants. During this period, most resided in barracks or 'lines' near the sugar mills but over the next few decades began to establish their own farms, mainly of sugar cane, wherever they could lease or purchase land. Thus Indo-Fijian settlements came to be dispersed throughout the rural areas of Viti and Vanua Levu (Fig. 15.2), without the rigid code of caste behaviour and village organization that characterized their home country. Apart from ties of friendship and later of kinship, there was little social homogeneity within and between these scattered settlements. Nowadays, the bonds that unite Indo-Fijians of a particular place are those of common interest, such as existing schools and cooperation in cane-cutting gangs during harvest, or result from marriage, the cumulative expansion of kinship, and the inheritance of land and property over generations.

For Indo-Fijians living in a rural settlement, these bonds have none of the emotional base or administrative reinforcement that the *koro* has for the Fijian. Nor are rural groupings necessarily identified with a cultural heritage or viewed as the anchor of one's whole existence. Above all, there is no traditionally sanctified system of exchange amongst kin and no conditioned expectation to contribute to settlement affairs. Indo-Fijians can and often do make sales of land preparatory to relocation to some other place, while some have no option but to move when the lease expires on the native land they cultivate. Once Indo-Fijians sell or lose access to land, they feel no attachment to that area other than those of kinship and friendship with people left behind.

In summary, there are many strong social and traditional reasons why Fijians retain close ties with their rural places of origin, whereas for Indo-Fijians only kinship and friendship are important. Urban-resident Fijians also display a stronger desire to return to live in their natal places and, at heart, regard themselves as villagers. The Indo-Fijians, on the other hand, have no strong desire ultimately to live in local communities. Paradoxically, such basic contrasts are not readily detectable from proportion of total or working life spent in Suva, primary location of one's immediate family, or nature of employment in town. Of all the questions on degree of commitment to urban residence, only that on property ownership suggested the contradiction between what Fijians resident in Suva do and what kind of people they feel they are.

Forms of movement: some conclusions

If migration is defined as a complete break of all connexions with one's area of origin, and if the existence of socioeconomic links, short-term visiting, and stated intentions to return are accepted collectively as indicators of circulation, then most Fijians interviewed in Suva circulate rather than migrate. To examine this conclusion further, all 400 heads of household are ranked on a scale to describe the nature of movement. First, they were divided into three groups according to the number and recency of visits made to places of origin, the degree of linkages maintained with those communities, and their statements of future residential intentions (Fig. 15.6a). Second, the separate scores were tallied for every mover to obtain a cumulative ranking (Table 15.5). A final score of three therefore means that a person had placed within the first group for each of the three criteria. The cumulative scores thus range from three to nine and form a scale of movement, on which three represents the highest intensity of origin-destination linkages, along with an intention to return ultimately to the village, and nine indicates no such contacts nor any such intention.

TABLE 15.5 A scale of movement, linkages, and future residential intentions of some Suva residents, 1977-78

	Fijians		Indo-Fijians		Total	
Cumulative score[a]	No.	%	No.	%	No.	%
3	42	21.1	0	0	42	10.5
4	30	15.1	0	0	30	7.5
5	70	35.2	13	6.5	83	20.8
6	23	11.6	33	16.4	56	14.0
7	24	12.1	144	71.6	168	42.0
8	4	2.0	5	2.5	9	2.3
9	6	3.0	6	3.0	12	3.0
Number of persons interviewed	199		201		400	

[a]Based upon amount of visiting, degree of socioeconomic linkages, and future residential intentions.

Viewed according to these criteria, most Fijian heads of household would have to be classified as circulators, although they range from high to moderate participants in circulation (Fig. 15.6b). About seventy per cent fall in the upper half of the cumulative ranks (3-5), eleven per cent in the middle rank (6), and seventeen per cent in the lower (7-9). Despite this conclusion, many Fijians have resided in Suva for more than half their lifetimes and even more for all their working careers. While many have visited their *koro* for short periods, few have remained away from the capital for more than a month or gone back intending to stay there. Many Fijians declared they would return ultimately to live in their villages, but also state this most likely would occur when they retire or when their children complete formal education: that is, at the end of their working lives. Many who now live in Suva are unsure of

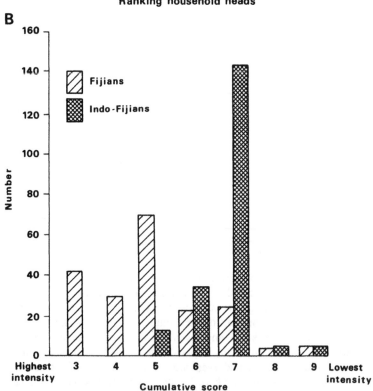

A

	GROUP 1 (1 point)	GROUP 2 (2 points)	GROUP 3 (3 points)
Visiting	Recent and more than one return to village	One visit, not recent	No visit since arrival in Suva
Linkages	Reciprocal help from family and village	Some help to family or village	No help to family or community of origin
Residential intentions	Intent to return to village	Unsure	No intent to return to village

Ranking household heads

B

Distribution of household heads by intensity of
origin-destination linkages

Source : Table 15·6

FIGURE 15.6 *A scale of movement for some Suva residents, 1977–78*

326

their residential intentions and others considered they would remain forever. If these facts are the deciding criteria, then most Fijian heads of household in this study would have to be classified as migrants. Their degree of commitment to urban residence, as well as the balance of other factors, indicates a fair degree of permanence, providing however that it is not defined as remaining in Suva for an entire lifetime.

This conclusion does not preclude, it must be emphasized, the possibility of some Fijians eventually returning to their villages, evidence about which is inconclusive. Nayacakalou (1975:98), a Fijian conversant with his people's thinking and lifestyle, believed 'the assumption that Fijians would return to their villages is largely unfounded' and that those living in Suva were destined to be permanently settled in Suva. Yet studies exist that document other Fijians who consider settling in one's village, even after long-term residence in Suva, a fact of life, together with some of the factors involved: availability of fertile, cultivable land and of some outlet for the sale of local produce to provide a small income for family necessities; and the strong emotional and cultural pull of the natal village that can impel even the highly educated to return, especially upon retirement from regular employment in urban places (cf. Racule, Tubuna, this volume). The difficulty of generalizing from such research is that it is silent about the proportion of village-born who do not go back, just as the present study was conducted in Suva and confined to those who had been resident for at least six months. Such constraints underline the observation of other field workers in Melanesia that 'when they were resident in villages circulation seemed more basic, but when in town migration appeared paramount' (Chapman 1978:563; cf. Bedford, this volume).

That a disproportionately large number of household heads had come to Suva from the small island provinces of Lau, Lomaiviti, and Kadavu (Fig. 15.2) must be underscored. Such Fijians not only stay longer but also fewer intend to go back, even though the links with their natal communities are just as strong (except for visiting) as those maintained by Fijians originating from other, less distant provinces. Could it be that the outer islands account for a higher-than-average ratio of long-term residents in Suva and that this explains why Fijians in town are more migrant than circulator in their mobility behaviour?

Conclusions about Indo-Fijian movement to Suva are less ambiguous since, by almost all criteria, such moves seem to be permanent. Most Indo-Fijians make brief visits to family and friends in their rural areas of origin but there are no other ties; very few remit cash and there is no custom of exchange, nor administrative encouragement to contribute to community projects. Grouped according to their visiting patterns, linkages, and residential intentions, more than three quarters of Indo-Fijian householders lie in the lower ranks (7–9) and consequently most would be classified as migrants (Table 15.5; Fig. 15.6b). Unlike the

Fijians, most stated a desire to remain permanently in Suva, while their median duration of stay and perceptions about advantages of city residence reveal a commitment that is as strong as that of their Fijian counterparts. The fact that Indo-Fijians have none of the security of communal land nor a durable village society upon which to depend in case of eventual return means that those interviewed in town are likely to remain throughout their lifetimes. Such a statement cannot be made with equivalent confidence for Fijian residents.

Circulation and migration as coexisting processes

Circulation and migration are not distinct or mutually exclusive processes; they are part of a mobility continuum which, in terms of different types, varies from intensely short-term circulation to permanent relocation. If any one, significant conclusion is to be reached, it is that this entire continuum is represented by the mobility of Fijians and Indo-Fijians between their rural areas of origin and Suva as an urban destination. Although different definitions of circulation and migration may result in varying conclusions about the dominant form of movement, it remains that both coexist, are to some extent contingent, and often substitute for each other (Chapman 1977:2).

It has been demonstrated that a certain complementarity exists between Suva and the rural areas by which the perceived needs of the total population are satisfied. Suva, as the capital city of a young independent country, fulfils many economic and social needs, whereas the rural communities meet most of the cultural and some of the social needs. In such a situation of territorial complementarity, the coexistence of circulation and migration is a logical outcome. Ullman (1956) has described the complementarity of areas and the resultant interaction between them of people as well as of goods and services. Although he was concerned largely with economic relationships and spoke in terms of demand in one area and supply from another, this notion also has been used to describe the movement of people: for example, by Baxter (1973) for the Orokaiva people of New Guinea and by Singhanetra-Renard (1981:Fig. 6.2) for the Mae Sa of Chiang Mai, northwest Thailand. According to Baxter (1973:114–15), movement for the Orokaiva 'has developed as complementary to village life. . . . The main reason that the village and urban areas have existed in a complementary fashion is that conditions have been so similar in each that movement between the two has been able to take place with few negative consequences for either the village or the individual'. He further suggests that the social networks of movers are based around two focal points—the village and the destination area—in the same fashion as Ryan (1970) previously had described for the Toaripi of New Guinea (cf. Ryan, this volume).

For Fijians, the complementarity between rural localities and Suva results more from differences in their social and economic condition and is perhaps emphasized, as for the Orokaiva and Toaripi, through the bilocal focus of social networks upon both kinds of places. This may be especially true for the large number of Lauans settled in Suva, but in general Fijian householders continue to make short-term rural visits despite lengthy residence in town. The complementary nature of mobility behaviour also varies according to ethnic group and place of origin. Rural areas offer less security for Indo-Fijians, so long periods are spent in the capital and the prevailing pattern of movement is more readily classified as migration. Similarly, the isolation from Suva of *koro* in the distant provinces of Lau, Lomaiviti, and Kadavu means that movement generally is of longer duration, fewer return visits are made, and the intent to return is less firm. Compared with Fijians from the main islands of Viti and Vanua Levu, those from the outer provinces engage dominantly in migration rather than in circulation. Yet, irrespective of place of origin, many had not come to Suva intending to stay and many had made several revisits before deciding finally to do so. The Fijians, especially, observed that their first journeys were for *gade* (short-term holiday) but most remained after one such trip and have now resided for many years in the capital. In short, an intendedly impermanent stay (circulation) evolved into fairly permanent residence (migration).

Finally, as forms of movement, circulation and migration may substitute for one another. Such a contingent relationship would be more evident had journeys to work been included in field inquiries. Among Suva residents, those originating from the provinces of Rewa and Naitasiri, within which the urban area is located, are under-represented. Fijians and Indo-Fijians from these two provinces can commute daily and weekly rather than engage in longer-term circulation or establish permanent residence in Suva. The frequency of contact with rural places was far higher for those household heads who came from communities within a few hours' bus journey of the urban area, and declined the greater the time and cost needed to reach less accessible parts of Viti Levu and other islands. For people born in the most distant provinces (Lau, Kadavu), long-term residence in Suva appears the only viable option, whereas those sufficiently fortunate to live nearby often substitute circulation for a prolonged or permanent stay in town.

Acknowledgments

Field research in Suva was made possible by a graduate scholarship awarded by the East-West Population Institute, Honolulu, Hawaii. In Fiji, particular thanks are due to Dr R. Thaman, who acted as field supervisor; and Abdul Sharif and Misaela Druibalavu, former students at the University of the South Pacific, upon whose shoulders fell much of the interviewing.

329

Part V
Melanesian circulation in broader context

16 Population movement in a small island periphery: the case of eastern Fiji

Richard Bedford

Throughout the western Pacific, small islands have become less and less attractive as places of residence for their indigenous populations. Whereas some of the most highly organized socio-political systems in Melanesia were once centred in such groups as southern Lau and Lomaiviti, eastern Fiji (Fig. 16.1), today the residents of these small islands are on the fringes of activity in the newly-independent nation states and last outposts of empire. Although colonial histories of Melanesian countries differ considerably, one common trend during the twentieth century has been the centralization of economies and societies. In each country the new centres of commerce and political power are on one or two of the larger islands, with the numerous small islands invariably on the periphery. Especially since the 1950s, real and perceived differences between places in life styles, economic opportunity, and the range of available services and facilities have increased. As one Melanesian remarked of the manifold changes that accompanied colonial intrusion: 'Our island world ceased to be. The world exploded and our island became a remote outpost . . . the last place in a country which has few centres and much remoteness' (Luana 1969:15).

Since the 'sixties, the course of population change in small island peripheries has been dominated by heavy net outmigration. In some areas, such as Lau province in eastern Fiji, there was an absolute decline in population during 1966–76, the decade between decennial censuses. Such cross-sectional information highlights an important redistribution of people. In most Melanesian countries, numbers in towns and cities are estimated to be increasing at more than five times the rate of those resident in rural peripheries. It is now widely accepted, however, that the rural-urban drift revealed in national censuses is part of a more complex process of mobility in which relocations of individuals and families are often temporary.

In this paper trends in the movement and redistribution of people

333

are examined for eastern Fiji, a remote region with few centres. This cluster of small, widely-dispersed islands comprises the administrative provinces of Lau and Lomaiviti and the district of Taveuni (Fig. 16.1). A brief description of the eastern islands provides background to recent demographic changes. Explanation of movement trends is then considered in two contexts: from the viewpoint of the 'necessary' conditions favouring aggregate flows of population from rural to urban areas; and with reference to a strategy of risk minimization which peasant producers, on the margins of the commercial system, adopt in the face of economic uncertainty.

An island periphery: eastern Fiji

In September 1976, the forty inhabited islands in Lau, Lomaiviti, and Taveuni had a population of 37,112 (Table 16.1). Whereas the great majority of residents were Melanesian Fijians (hereafter termed Fijians), the country's dominant ethnic group — the Indo-Fijians — comprise only ten per cent. Over half the islands had less than 400 residents, and forty per cent of them are less than 10 km² in area. Only one island, Taveuni, has an area more than 200 km² (almost the same size as Barbados in the West Indies) and a population greater than 5,000. Levuka, the early colonial capital of Fiji, is the only gazetted town in the region. Its urban and periurban population in 1976 was 2,764 — smaller than reported for the 1966 census. In only two other areas of the eastern islands are significant numbers employed in non-agricultural pursuits: a scattered complex of commercial enterprises, government offices, and educational establishments between Somosomo and Wairiki on the west coast of Taveuni; and headquarters of Lau province in Tubou village on Lakeba (Fig. 16.1). About 3,000 persons lived in these areas in September 1976 to give the region an urban population approaching 6,000, or 16 per cent of the total enumerated.

Most of the 31,000 rural residents are in small villages and agricultural settlements or on coconut plantations scattered around the coasts. Only ten of the inhabited islands have more than five concentrations of settlement, and almost all rural communities had less than twenty-five households in residence. Except for the plantations and some recent agricultural settlements on Taveuni, virtually all localities have at least one church, most have a co-operative store selling a limited range of imported consumer necessities, and the larger ones have primary schools and small medical centres. Secondary schools and major hospitals are concentrated on three islands: Taveuni, Ovalau, and Lakeba (Fig. 16.1). Most people in the eastern islands have reasonable access to basic social services and amenities, but for anyone wishing to acquire advanced education, specialized medical treatment, or professional services, residence outside the region is necessary, usually in Suva, the capital of Fiji.

TABLE 16.1 *Population growth and distribution in Fiji and the eastern islands, 1966–76*

	Population 1976		Intercensal growth 1966–76 (%)		% of population in Fiji	
	Total	Fijian	Total	Fijian	Total	Fijian
Eastern Fiji						
Lau province	14,397	14,132	−9.9	−9.2	2.4	5.4
Lomaiviti province	13,358	12,064	0.7	7.3	2.3	4.6
Taveuni district	9,357	7,165	6.9	14.7	1.6	2.7
Eastern islands	37,112	33,361	−2.3	0.9	6.3	12.8
Fiji						
Rural areas	339,574	180,618	7.0	17.3	57.7	69.5
Urban areas[a]	218,494	79,314	37.2	64.5	42.3	30.5
Suva[b]	117,827	48,303	46.8	82.8	20.0	18.6
Fiji	588,068	259,932	23.4	28.6	100.0	100.0

[a]The areas termed 'urban' in the census include the towns and associated periurban areas of Suva, Lautoka, Ba, Labasa, Levuka, Nadi, Savusavu, Sigatoka, and Nausori; and the 'unincorporated townships' of Korovou, Navua, Rakiraki, Tavua, and Vatukoula (Fig. 7.1).
[b]Suva city and the periurban area.
Sources of data: Volume 1, 1976 Census (Lodhia 1978); 1966 Census (Zwart 1968).

FIGURE 16.1 *The eastern islands of Fiji*

The agricultural economy of eastern Fiji is dominated by copra production; fifty-seven per cent of the total produced in the country in 1975 came from these islands. Cattle, fish, taro (a staple food crop), *yaqona*, and handicrafts are also important sources of cash. Communications between these islands and with the main centres of commerce and government on Viti Levu and Vanua Levu are generally poor, although three islands (Taveuni, Ovalau, Lakeba) have air services. On most of the islands four to six weeks may pass between shipping calls, and for some there is no regular transport link. The one wharf in the region is at Levuka (Fig. 16.1); in 1976 two others were under construction on Taveuni and Lakeba. While tourism is still of minimal importance, the

336

economies of some small Lau islands have been sustained in recent years by handicraft production for the tourist market in Suva.

Most islands, then, are small, relatively isolated, and their inhabitants dependent on one cash crop for the export market. Yet there is great ecological diversity in eastern Fiji despite only two basic geological elements: volcanic extrusions, and limestone of coralline and submarine origin. Traditional specialization in resource use led to systematic exchange of food from richer volcanic islands for canoes and handicrafts produced on limestone islands poor in garden land. With integration into the larger economic and political systems of Fiji and the northern hemisphere, a uniform export crop (coconuts) diffused and masked both this specialized production and the ecological diversity. While the 'coconut overlay' has served the eastern islands well in past times, today it is the fundamental cause of the depressed economy and deep dependence on the decisions and policies of central government in Viti Levu.

During the early 1970s, the extent of this dependency was demonstrated very clearly in the responses of both government and small-island residents to a combination of adverse market forces and natural disasters. Considerable fluctuations in prices for copra and handicrafts have been common since 1972; fluctuations which have been accompanied by rapid inflation for consumer goods and have had much greater impact on cash incomes of rural dwellers than on urban wage-earners. At the same time parts of the region have been devastated by hurricanes in successive years between 1973 and 1975. Government response to these crises has been to introduce a range of policies such as price stabilization, freight subsidy, and disaster relief, which are designed to cushion rural residents from the worst effects of fluctuating prices for cash crops and the devastation of assets in natural disasters. Local response to these crises has varied but two main ones can be identified: a marked acceleration of temporary and long-term movement out of the eastern islands; and, among those who have chosen to stay in the periphery, more emphasis upon welfare support from and more reliance upon decision-making by the centre.

Research into population movement in eastern Fiji

In 1975 and 1976, the small islands of eastern Fiji were the base for the first research project in UNESCO's Man and the Biosphere programme on the ecology and rational use of island ecosystems. This research was funded by the United Nations Fund for Population Activities (UNFPA). Some of the field inquiries into demographic change undertaken for this project are the subject of this paper; the wider study is described in UNESCO/UNFPA (1977) and the methods of its various components detailed in a collection edited by Brookfield (1980).

The examination of population movement was framed with the knowledge that there had been in progress, for some years, a major redistribution of Fijians born in the eastern islands. A central concern was to account for a trend towards absolute population decline in islands which could support far greater numbers at present levels of living, or alternatively higher production and thus improved incomes for current residents. Net outmigration to other parts of Fiji was perceived to be the most important feature of population movement in the region and inquiry was concentrated upon all movement which resulted in a change in place of residence for at least one month.

The arbitrary limit of one month in residence at the destination was chosen on partly conceptual and partly methodological grounds. Some lower limit along the mobility continuum has to be established in any study of population movement. In Fiji, primary concern was with movements which necessitated a substantial restructuring of the round of social and economic activities of those involved. Absences from home for a few hours or days were thus not considered relevant whereas a period of at least four weeks generally did result in a locational shift in the primary activities of an individual mover. Other temporal limits obviously could have been adopted, but the methods employed for much of the data collection imposed an additional constraint. Retrospective mobility histories were compiled for adult Fijians and it was found impractical for people to recall past moves which had taken them to other localities for a few days or even one or two weeks.

The decision to restrict inquiry to mobility involving a month or more of residence at place of destination meant the exclusion from the UNESCO project of numerous forms of circulation. Short-term social visits, hunting and fishing trips, participation in ceremonial activities in neighbouring communities, the cultivation of gardens, the journey to work, and visits to hospitals were not part of its concern. However in certain inquiries, especially those of Bayliss-Smith (1977, 1978) for Lomaiviti, some data on short-term circulation were collected.

To understand better the role of population movement as a regulator of demographic as well as social and economic change in eastern Fiji, inquiries were carried out at three scales: the region, the island, and the community. The following review of results refers particularly to five islands: Koro and Batiki in Lomaiviti province, Lakeba and Kabara in Lau Province, and Taveuni in the district of the same name (Fig. 16.1). All are strongly contrasted in area and population, geological constitution and ecological diversity, demonstrable isolation and marginality, and levels of economic development (Fig. 16.2). Despite this diversity, the recent history of the island region located east and southeast of Viti Levu was that of a social and political focus for all of Fiji, whereas today the area is entirely peripheral.

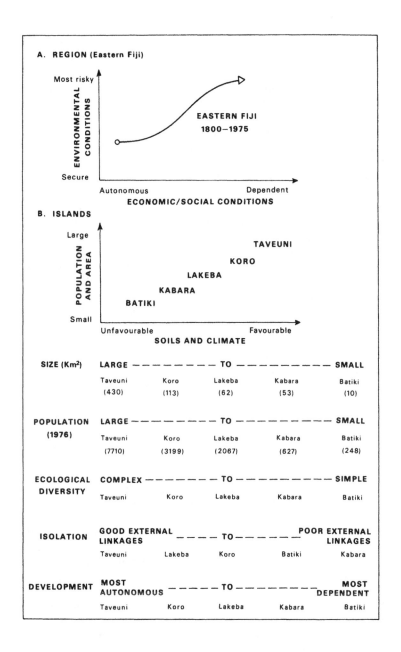

FIGURE 16.2 *Characteristics of eastern Fiji and selected islands*

339

The drift to towns

Decennial censuses have indicated that, since the mid-1950s, the populations of many small islands have been increasing at much slower rates than the national average. In 1956 islands in Lau and Lomaiviti provinces and Taveuni district contained eighteen per cent of the country's Fijians; by 1976 the proportion had declined to thirteen per cent. Over the intercensal period 1966–76, the divergence between national and regional trends became particularly marked: whereas the Fijian population increased by around twenty-eight per cent, numbers in eastern Fiji rose by less than one per cent. In Lau province, the main cluster of small islands, there was an absolute decline of nine per cent (Table 16.1).

Of the 39,000 Fijians born in Lau and Lomaiviti provinces, fifty-seven per cent were enumerated there in September 1976. If adults (fifteen years and over) are considered separately, only forty-four per cent were resident in their provinces of birth. Lau province had lost a greater proportion of its natal population (47%) than Lomaiviti (39%), but census statistics suggest that rates of movement away from both provinces have been closely parallel since independence in 1970. During the six years before the census, for example, one quarter of the Fijians in Lau and twenty-four per cent of those in Lomaiviti in 1970 had moved to another province. Unfortunately similar statistics are not available at the district level and it is only possible to estimate recent trends in the redistribution of population in Taveuni district.

To obtain a clearer indication of the magnitude and characteristics of net migration, the Fijian population reported in the 1966 census was projected forward by age group and sex, using survival ratios obtained from regional model life-tables (Coale and Demeny 1966). Expected populations, assuming no migration, were compared with those censussed in 1976 to derive estimates of numerical loss in different parts of the region. Although several limitations are associated with application of this technique in eastern Fiji (UNESCO/UNFPA 1977:137), the derived figures for net migration represent conservative estimates given the assumptions made about mortality rates.

Over the decade 1966–76, eastern Fiji lost the equivalent of slightly more than a quarter of its residents through net outmigration (Table 16.2). Losses were heaviest in Lau province (32%) and lowest in Taveuni district (10%), the largest and most diverse of the island areas under study. Even more important is the age-sex composition of these flows. Proportions of children and women were much higher than expected, given an initial hypothesis that population movement was essentially short-term and circular. Family or household relocation has been very common in recent years and this is likely to indicate long-term rather than temporary movement from the region.

TABLE 16.2 Net migration of Fijians by age group, eastern Fiji, 1966-76

Demographic	Age group in 1966[a]			Total Fijians
characteristic	0-9	10-29	30-49	aged 0-49
Numbers lost 1966-76				
Lau	1,920	2,100	530	4,550
Lomaiviti	730	1,050	170	1,950
Taveuni	220	250	100	570
Eastern Fiji	2,870	3,400	800	7,070
Sex ratio[b]				
Lau	102	91	71	93
Lomaiviti	135	102	55	107
Taveuni	175	176	43	138
Eastern Fiji	114	99	63	99
Losses as percentage of residents in 1966[c]				
Lau	31.9	39.6	19.1	32.3
Lomaiviti	18.9	25.4	8.2	19.4
Taveuni	10.5	10.9	8.2	10.2
Eastern Fiji	24.0	29.0	13.2	26.4

[a]Net migration statistics refer only to Fijians aged 0–49 years in 1966 and 10–59 years in 1976.

[b]Males per 100 females in the net migration flows.

[c]Net migration losses as a percentage of the resident population in 1966.

Source of data: UNESCO/UNFPA (1977:138); Volume 1, 1976 Census (Lodhia 1978).

Place of residence of landowners offers another perspective on population losses from the standpoint of the island and the local community. The Native Lands Trust Board maintains registers of all Fijians who have customary rights to property in primary land-owning units (*mataqali*). Even though these registers are incomplete, they provide further indication of the extent to which *de facto* populations in eastern Fiji are only a fraction of the number who have rights to land. On Kabara in southern Lau, for example, less than one third of the registered landowners were in residence in October 1975 (Bedford and McLean 1978:51). In 1974, a government-sponsored survey on Batiki, Lomaiviti province, found an even smaller proportion present (23%; Bayliss-Smith 1978:87).

The main destinations for those who left rural communities in eastern Fiji are towns on Viti Levu and Vanua Levu (Fig. 7.1). Between 1966 and 1976, for the country as a whole, the Fijian population enumerated in urban areas increased by sixty-five per cent; in Suva by eighty-three per cent (Table 16.1). Boundary changes since the 1966 census are responsible for some of this change but the drift of Fijians to towns, and especially to Suva, is a major characteristic since the mid-1960s.

Over half of those born in Lau and Lomaiviti, but resident in 1976 outside that province, were enumerated in the capital city. More than eighty per cent of absentees among the Lau and Lomaiviti-born, as well as most of those who moved during the six years before the census, were living in provinces where Fiji's main towns are located. Not all these migrants were in urban areas, but the far higher rates of growth for town-based than for rural-resident populations suggest that a drift to towns, rather than intrarural relocation, is the most important dimension of contemporary internal migration in Fiji. There are counter streams: in 1976, residents of Lau and Lomaiviti included Fijians born in every province, but flows into these small islands replaced only a fraction of their outmigrants. Whereas over 16,000 had left, only 4,000 Fijians born in other provinces were enumerated in Lau and Lomaiviti and a quarter of these inmigrants had come from other parts of the eastern region.

Reviewing trends in Fijian population movement in the late 1950s, Ward (1961:265) commented:

> For many years it has been regarded as almost traditional for young Fijian men to leave home, to go to the gold mines or some other employment centre for a period of a few months to one or two years, and then return to their villages. . . . This type of short-term employment, however, has become less and less typical. As the contrast between urban and rural living has increased (particularly in the post-war era), and as the number of Fijians in the towns has grown (thus making it easier for the newcomer to find friends) the migrant's desire to return home has grown progressively less. . . . Once a family, or part of a family, is established in Suva or some other urban centre they frequently find themselves supporting kinsfolk who have followed them to town. Many of these are school children sent in by parents unable or unwilling to move themselves yet anxious to have their children educated in the city. Another section of this secondary wave of migrants consists of elderly people whose younger kinsfolk on whom they must depend in their old age have migrated earlier. . . . The second wave of elderly migrants helps increase the problems of crowded housing conditions in the towns. It also marks a further step in the break from the land.

Ward's argument, that the pattern of short-duration circular mobility from a base in the rural area was shifting towards one where movement involved long-term residence in urban areas, had the support of an eminent Fijian anthropologist, the late Rusiate Nayacakalou. Around the same time, Nayacakalou (1963:34) stressed: 'There is little doubt that the great majority of Fijians living in Suva are destined to be fairly permanently settled there'.

Evidence from studies undertaken during the 1970s by staff and students from the University of the South Pacific (Suva) highlights this trend towards stabilization in urban residence amongst migrant Fijians (for example, Harré 1973; Racule 1974; Walsh 1976, 1977). To see if the same held true for Lauan migrants in Suva, a sample survey was conducted in the periurban settlement of Qauia. Of the eighty-one resident in January 1976 and born on a Lau island, over forty per cent had spent most of their lives in town (Table 16.3). Both men and women aged more than twenty-five had been just as stable in urban residence as their younger kin; indeed over seventy per cent of the older men had arrived in Suva before 1965. Almost eighty per cent of the Lau-born were living in Qauia in 1971 and the majority had moved there from another part of Suva. Intraurban and interurban residential shifts dominated the mobility histories of these eastern islanders.

TABLE 16.3 Aspects of the mobility of Lau-born Fijians in Qauia, 1976 (%)

| Aspect | Males | | Females | | |
	Under 25 (21)	25+ (22)	Under 25 (17)	25+ (21)	Total (81)
Spent most of life in towns	42.9	45.5	47.1	33.3	41.9
Came to Suva before 1965	33.3	72.7	29.3	52.4	50.6
Only lived in Qauia in Suva	57.1	40.9	64.7	42.9	50.6
Resident in Qauia/Suva January 1971	76.2	81.9	76.5	80.9	79.0
Have not visited home village since 1970	57.1	68.2	47.1	47.6	54.4
No plans to visit home village	90.5	86.4	82.4	90.5	87.6

Source of data: Qauia mobility survey, January 1976.

There was little evidence of frequent circular movement between town and village among Lauans in Qauia. More than half the sample claimed that, during the previous six years, they had not been back to their home villages for even a short visit, let alone a month (Table 16.3). The effective base for their social and economic activities was Suva, even though most of the older men and women claimed they would eventually return to the village. Whether this will occur is debatable — witness the following comment by a Lakeba woman who was living in Qauia in 1973 and has been away from her village for more than thirty years:

If it wasn't for Ritova [the woman's husband, also from Lakeba] we'd be back now in Lakeba. . . . Ritova and I are getting old. We can't expect the children to look after us in old age. I mean it wouldn't be right our imposing on them. They have their own families and everything around here, everything now costs money. . . . But Ritova just won't listen to me. He feels that here the children at least have a chance of making a living, there's opportunity for jobs. Although he's the eldest in the family and therefore has the right to the land over his younger brothers, he feels that he cannot claim the land selfishly for our own use. We're already here in Suva; the others are in Lakeba: let them use the land and make a living from it. Here there are more ways open to us for making money: those on the land depend only on copra to get them money. I quite agree with him but all the same it would be nice to go back and die in Lakeba. I mean, it is our island! (Racule 1974:14–5).

Although many may not go back to retire, Lauans in Qauia refuse to cut ties with former rural homes. Evidence of limited circulation between town and village does not mean an absence of interest in events or of communication with people in the eastern islands. Urban residents send money, clothing, and food to relatives in the village, and provide accommodation for rural kin when they come to Suva. However, face to face interaction between members of the town-based and village-based components of eastern Fiji's population is fostered mainly by rural residents, who periodically seek short-term employment or medical treatment, sell handicrafts or root crops, purchase supplies, or call on friends and relatives. A short visit to Suva often becomes a much longer stay, especially for unmarried men and women with no immediate family commitments. Over time, the loss of younger, more dynamic members of small island populations continues as numbers in urban settlements grow at the expense of the villages.

Rural perspectives

Inquiry into the movement behaviour of villagers tends to produce rather different perspectives on the pattern and process of internal migration than those obtained from study of urban residents. Research in rural Melanesia highlights the circular nature of much population movement and considerable emphasis is laid on the continuing import- ance of the village as the primary centre for economic and social activity. In eastern Fiji, field investigations concentrated on rural com- munities because of the project's focus on ecological, socio-economic, and demographic characteristics of small islands. A rural orientation to field research, however, did not result in failure to appreciate the funda- mental redistribution of people which has been occurring for some decades. Circular forms of movement from a rural base were found to dominate mobility histories of village residents, but both *de facto* and *de jure* enumerations revealed that the numerical base is dwindling, especially on very small islands.

Village populations are in a constant state of flux. When examining changes in the demography of Batiki, Bayliss-Smth (1978:87) found that the resident population had declined from 253 to 227 between two *de facto* censuses in September 1974 and September 1975. During that year, however, the total number of Batiki-born or Batiki-related persons who had spent at least one night on the island was 288 − higher than the *de facto* population at either census. In his analysis of mobility to and from the island, Bayliss-Smith used a *de jure* base comprising all those born on, or related to people born on Batiki, who had been in the four villages over the two years before the 1975 enumeration. This population totalled 338 and since 1973 the great majority of adults (66%) had made at least one move to and from the island. An examination of recent mobility experiences for a similar *de jure* population of Nacamaki village, on the larger island of Koro (Fig. 16.1), revealed much the same pattern (Bayliss-Smith 1977:29–31). Circulation between village and other rural as well as urban communities outside the island is now part of the lifestyle of all except the very young.

Rarely is it possible to know the degree to which the circulation of people between different parts of a country has intensified over time. For Nacamaki village, Bayliss-Smith was fortunate to have data available on inter-island movement in 1958. In a study of this village in the late 1950s Watters (1969:139) found that eighty-seven per cent of 344 residents had never travelled outside Koro island. Sixteen years later Bayliss-Smith (1977:29) reported that seventy-five per cent had been to other islands and lack of travel experience was confined almost entirely to people aged less than fifteen. The presence of large numbers of Koro islanders living elsewhere in Fiji encourages such experience. A desire to maintain contact with widely-dispersed kin who belong to a particular place, in the sense of having customary rights to land, is reflected in reasons given by Batiki and Nacamaki residents for their outward journeys between 1973 and 1975. Holidays, often quite prolonged, with relatives in Suva or Levuka (Fig. 16.1) emerged as the most important stimulus for temporary absence (Table 16.4). Next came movements associated with employment − either to find a job, to return to one or, in the case of sixteen men employed on two commercial fishing boats operating out of Batiki, to pursue a particular occupation − followed closely by the outward trips of teenagers to secondary school.

Considering how different the two islands are in size, resource endowment, self-sufficiency, and overall prosperity, Bayliss-Smith was surprised to find such striking similarity in mobility patterns for the villagers of Nacamaki (Koro) and Batiki. He commented 'these two Lomaiviti islands have very similar requirements for contact with the outside world which they manage to satisfy in very similar ways' (Bayliss-Smith 1978:89–90). While a valid conclusion on the basis of stated reasons, there is a major difference in the net effect of that recent mobility upon the longer-term course of population change for each. Between 1966 and 1976 the resident population on Batiki declined by twenty-three per cent; for Koro, it increased by thirteen per cent. These growth rates highlight the contrast between larger and very small islands in eastern Fiji. Batiki has lost a higher proportion of individuals with land rights to long-term residence in other parts of the country. Between 1973 and 1975, Bayliss-Smith points out, only thirty per cent of 1,100 people who might have spent some time on Batiki did so. Although comparable information is not available for Koro, it is unlikely that such a large proportion of Fijians with rights to property there were 'without direct economic interest in their home island during this period' (Bayliss-Smith 1978:90). The populations of Batiki and Koro may have similar requirements for contact with other parts of Fiji, but it is questionable whether they are managing to satisfy these in the same way.

TABLE 16.4 Reasons for movement from Batiki and Nacamaki, 1973–75

| Reason | Stated reason for outward trip (%) | |
	Batiki	Nacamaki
Social	37	49
Employment	28	21
Schooling	23	17
Buying/selling produce	5	6
Hospital	5	4
Marriage	2	3
All trips	100	100

Source: Bayliss-Smith (1978:89).

The population history of Taveuni, the largest island investigated by the project, differs greatly from that of other parts of eastern Fiji. A classic plantation economy, sizeable non-Fijian minorities, and scattered commercial/administrative facilities on the west coast of the island have favoured net inmigration at various stages over the past 100 years. Since the second world war, the Fijian population has increased by 150 per cent — growth far in excess of the modest sixty per cent for the whole region. In recent years, however, population growth has tended to be much slower, especially for the village-resident component. Between 1966 and 1976 Fijians living in the nucleated settlements officially termed villages *(koro)* increased by a mere five per cent, or less than half the rate for the total Fijian population (Bedford 1978:153).

The percentage of Melanesians living in villages had declined from seventy, in 1946, to thirty-eight in 1976; the majority now reside in dispersed clusters of houses on subdivided *mataqali* land, labour lines on commercial plantations, or the 'urban' area around Somosomo-Waiyevo (Fig. 16.1).

A detailed account of demographic change in Taveuni has been published in Bedford (1978). The following summary outlines some findings for one village, Qeleni, where cross-sectional and longitudinal data on population movement are comparable to that available for villages on a southern Lau island, Kabara (Fig. 16.1). It must be emphasized that Qeleni is not representative of Taveuni in the sense that Bayliss-Smith was able to argue that trends in Nacamaki were indicative of Koro as a whole.

That much population movement has been and remains of the type classed as circular is clearly demonstrated when longitudinal data on mobility are collected. Movement histories for adult males in Qeleni village and Kabara revealed that, during the five years before interview, considerable proportions had resided for at least a month in other parts of Fiji. In Qeleni just over forty per cent of those aged fifteen years and over had made more than one such move since 1971. But whereas almost half the villagers stated they had lived in a town, only nineteen per cent obtained paid employment while there. As in Lomaiviti, the majority had been visiting relatives and friends. Rural wage employment at some stage over the past five years had kept marginally more men (20%) away from the village for at least a month, and plantations on Vanua Levu and neighbouring Laucala (Fig. 16.1) were the main destinations. Even though some men had been quite mobile, eight-four per cent claimed to have been resident in Qeleni in 1971, and an even higher proportion (89%) had spent more than three of the five years in the village.

In terms of number of absences for at least one month, and time in residence or employment in towns, Kabaran household heads gave evidence of more extensive mobility since 1971 (Table 16.5). A slightly higher proportion of Qeleni males had had rural employment away from the village, which is hardly surprising given the proximity of plantations offering jobs on Taveuni and neighbouring islands. While proportionately fewer had been absent from Kabara than from Qeleni in the six months before the survey, forty-eight per cent had been living elsewhere for at least a month between October 1974 and October 1975. Main destinations and reasons for last absence were quite different. Fifty-nine per cent of Qeleni household heads cited destinations in other parts of Taveuni and on Vanua Levu whereas Suva was specified by over two-thirds of those on Kabara (Table 16.5). Employment for wages was given as the objective of the last move by fourteen per cent of the Qeleni men, compared with fifty per cent on Kabara.

TABLE 16.5 *Movement experience between 1971 and 1975 of house-hold heads in Qeleni and four villages on Kabara (%)*

Characteristic of mobility	Household heads	
	Qeleni (29)	Kabara (44)
Absent 1+ month more than once	51.8	63.6
Lived in town	58.6	77.3
Worked in town	17.2	40.9
Spent 6+ months in urban employment	10.3	31.9
Worked for wage in rural area	37.9	22.7
Away within 6 months of survey	48.2	15.9
Destination last absence: Suva	31.0	68.2
Reason for last move: work for wages	13.8	52.3

Source: Bedford (1978:176).

It is not surprising there are some major contrasts in the recent movement experience of adult males living in a Taveuni *koro* and those on an island battered by two severe hurricanes between 1973 and 1975 (Campbell 1977). Although Qeleni villagers have been affected in recent years by the same fluctuations in price which profoundly affected the copra industry throughout Fiji, the resultant impact of declining cash returns has been far less than on Kabara. Coconut groves on Taveuni were not devastated by hurricanes in December 1973 and January 1975, which effectively ended copra production as an income-earning activity on Kabara until the late 1970s. Qeleni farmers have also been able to turn to the production of root crops (taro, *yaqona*) which are in demand in urban markets but cannot be cultivated in Kabara's harsh environment. Demand for wooden handicrafts, the alternative source of cash on the latter island, was also severely depressed in the mid-1970s with recession in the tourist industry. During 1975, incomes generated locally were low and Kabaran households were heavily dependent on wages earned off the island (Bedford and McLean 1978:39). In October, for example, heads of ten out of thirty-three households in Naikeleyaga, the largest village on Kabara, were either employed or seeking work in Suva. By contrast, only one household head was absent from Qeleni in August 1975. For Qeleni village and on Kabara island, natural disasters and adverse market forces largely explain the different patterns of recent mobility and urban employment among adult males.

The nature of recent net outmigration is also different. Using records from the Fiji National Archives, it was possible to compare names of residents enumerated in these two areas with those present nine years earlier. In both Qeleni and Kabara, less than half (44%) of those resident in 1966 still lived in the same communities at the time of the 1975 surveys. While levels of net outmigration were similar, the demographic composition of the flows and destinations of those who left were not. In Qeleni, the major loss was of males aged fifteen to twenty-nine years. The main destinations, for over sixty per cent, were other communities on Taveuni or neighbouring islands, and many had revisited Qeleni since departure. Whole families rather than male members had left from Kabara, so that adult women and young children accounted for over half the net loss to the island's population. Most had moved to Suva and return visits appeared very infrequent. Out of 179 absent relatives of resident household heads, three quarters had left before 1966, just under half had not returned to the island since departure, and for a further thirty per cent no one could recall when they had last visited. Just as Bayliss-Smith described for Batiki, most of Kabara's *de jure* population were not exercising their right to use land on a small, relatively-isolated island.

Retrospective movement histories compiled for village populations invariably reveal that 'return migration rather than linear migration is the dominant kind of mobility' (Chapman and Prothero, this volume). Inquiry into population mobility at a range of scales in eastern Fiji, using a combination of cross-sectional and longitudinal methods, clearly demonstrated that over the past century an intensification of circular forms of movement has been accompanied by a fundamental redistribution in population. Evidence from rural and urban surveys in Fiji supports a claim made by Ward (1971:103) for rural-urban drift in Papua New Guinea: 'to talk of temporary urban dwellers with the implication that they will go home in due course is largely wishful thinking'. This is not to argue that Fijians now living in Suva have severed their connections with former communities of residence, or that they feel permanently committed to remaining in the particular suburb, squatter settlement, or town in which they were enumerated. It is merely to recognize that many islanders who have left eastern Fiji to work or stay in Suva do not return to live out their years in former rural homes.The base for their vast range of temporary excursions has shifted from village to town.

Conditions favouring rural-urban drift from island peripheries

In a brief but perceptive review of the current status of migration theory, Janet Abu-Lughod (1975) has noted that the reassuringly simple explanations of mobility provided by the mathematical gravity flow, push-pull, and psychic cost-adaptation models have proven much more time-, place-, and culture-specific than originally anticipated. Despite their limitations, some hypotheses contained within these reassuringly simple explanations offer useful starting points to analyze what Mitchell (1959, 1969a) calls the 'necessary' conditions for population movement. Necessary conditions, Mitchell argues, are situations and events important as underlying causes of movement and largely account for aggregate flows between places. However, they are not 'sufficient' conditions to explain why an individual chooses to move at a particular time from his rural home. Economic factors such as pressure of population on limited land resources, income disparities between rural and urban areas, rates of job creation and unemployment in town, prices for cash crops, and the changing educational status of the labour force, are critical necessary conditions for rural-urban drift in most third world countries, but by themselves are not sufficient to explain when and why a particular migrant leaves or returns to the village.

The necessary conditions for population movement from small islands in eastern Fiji to towns on Viti Levu and Vanua Levu were created by processes which converted diversified, integrated and autonomous economies based upon subsistence into disadvantaged and dependent units on the margins of a wider market-exchange economy. There are two dimensions to this shift. The first was the spread of an export crop, coconuts, which transformed highly-specialized systems of resource exploitation and degraded a range of indigenous skills and practices. As noted earlier, the 'coconut overlay' has served residents in small islands well at different times over the past century, whereas marked fluctuation in prices received by copra producers are a source of considerable dissatisfaction. Not only has extensive planting of coconuts failed to generate the desired income, but also the Melanesians' concept of land as a resource changed radically with increasing involvement in the market economy. Competition for productive tracts has intensified and, on islands where small size limits their availability, coconut planting is partly responsible for perceived shortages of land. Within and between groups in the rural population, furthermore, cash cropping has generated considerable inequalities in ownership of and access to land.

The second dimension relates to radical changes in people's expectations as to what constitutes a satisfactory standard of living, a desirable occupation, a stimulating residential environment, a suitable mix of accessible services and facilities. Reliance on imported foods and goods has intensified with rising aspirations for the material wealth and life styles of the white man. Inhabitants of small islands were often the first to experience protracted contact with agents of colonialism — traders, missionaries, recruiters, settlers, administrators — for they were more accessible than Fijians who lived inland of the inhospitable coastal fringes of many larger islands. The climate and terrain of small coral and volcanic islands also were more attractive to foreigners. Despite exceptions, culture contact initially was most intensive and disruptive in small islands. Melanesians in Lau and Lomaiviti, for example, were the first in Fiji to become involved in cash cropping and wage employment, to be heavily exposed to missionary teachings and education, and to experience the excitement of urban living. As foci of alien commercial interests shifted to the larger islands of Viti Levu and Vanua Levu, eastern islanders were also the first to experience frustration and disenchantment as their aspirations ran ahead of any possible attainment within a small-island world.

Although competition for productive land has risen in recent years, excessive pressure for numbers on limited local resources is not the primary reason for frustrated aspirations. Population densities have increased since the 1930s and residents on small islands will argue persuasively that, without the safety valve of outmigration, there would have been considerable problems over land. But shortage of land, by itself, did not motivate all those now living elsewhere to leave; many villagers with access to extensive tracts have been as mobile as the land-hungry. What is important to appreciate is that, in small-island peripheries, local resources have not provided the sole source of livelihood for some considerable time. One of the benefits of integration into larger economic and political systems under colonialism has been an increase in the range of opportunities for satisfying needs in different places. Even in small islands where productive land is not in short supply, the perceived returns from using local resources more intensively have gradually become less attractive as compared with benefits accruing from employment and residence elsewhere. In short, rapid population growth coincident with rising demand for land for cash cropping is a 'necessary' condition for the drift to larger islands, but for many of those who have left eastern Fiji since the 1960s it is not by itself 'sufficient'.

A more powerful explanation for net outmigration exists in the increasingly imbalanced distribution of employment opportunities which more and more Fijians regard as socially and economically desirable. Since the 'sixties, as in most third world countries, differences between agricultural incomes and those obtained from wage employment in towns have widened. Although the subsistence component in Melanesian village agriculture makes comparison difficult, some indication can be given of such rural-urban disparity. In 1974 industrial workers in Suva earned each year on average around $F1400 ($F0.75 to £1.00; $F1.20 to $A1.00; UNDAT 1974). Based on both cash-earning activities and an imputed monetary value for subsistence production, less than thirty per cent of Taveuni's village households received a comparable income in 1975 (UNESCO/UNFPA 1977:220; Nankivell 1978:258). When placed beside other parts of Fiji, incomes in the eastern islands were low and exhibited considerable inequality. Indeed, there was more inequality in village incomes in Taveuni district in 1975 (gini coefficient 0.469) than for Fiji's urban households in 1972 (gini coefficient 0.423).

With financial rewards from coconut production proving so unreliable and small-island producers suffering a number of hinterland penalties (rising freight costs, rapid inflation in imported goods, deteriorating shipping services), it is hardly surprising that most younger Fijians envisaged a more profitable future in the towns. Moreover, this perception has remained even though the rate of urban unemployment (percentage of the labour force without work and seeking it) has been rising steadily since the mid-1960s. While such trends support Todaro's (1969, 1976) hypothesis of 'expected incomes', to explain recent rural-urban drift in purely economic terms is too simplistic and omits the important dimension of social status. Throughout the Pacific, agriculture as an occupation has been losing prestige; one of the most significant obstacles to agricultural development is the fact that working the land is now held in low esteem. Sevele (1973:16), for example, has remarked that 'in Tonga one often hears parents expressing the wish that their children "work at something better than agriculture", even though they themselves are farmers. This "something better" invariably refers to white-collar jobs which carry with them a lot of prestige'. In eastern Fiji, similar attitudes were expressed in many rural households and a secondary school education, especially for sons, rated highly on the list of priorities.

A survey in 1975 of occupational aspirations among secondary pupils in two Taveuni schools demonstrated clearly the strong preference for employment or higher education in town. Whereas only thirteen per cent of Fijian students gave farming as their first job preference, on leaving school sixty-seven per cent hoped to learn an industrial trade or enter a centre of tertiary training; sixty-seven per cent specified a town as their chosen place of work; and over three quarters listed 'the government' as the preferred employer (Bedford 1978:235). There is nothing surprising about these preferences; they express the occupational aspirations of secondary school pupils and their parents throughout the country. In the words of one Fijian:

> And so we went to class rooms, grew up to put on ties and sweat in jackets, to pull white socks over our black skins. We looked around and took stock. We could tell those still with some way to go for they walked barefoot. . . . We leave the old and infirm in the villages and rush to the towns. Some are employed, how lucky they are! Many more keep searching, looking for the jobs they may never find. . . . Their diploma of achievement is spelt in capital letters – FRUSTRATION (Vusoniwailala 1978:11).

Consequently the decision to move from a village in eastern Fiji to a town on Viti Levu may reflect many reasons other than the immediate desire for urban employment. For at least a decade most village residents have had a close relative living in Suva. In detailed migration histories compiled for young adult males, visiting kin in town was given frequently as the main reason for departure. A job there may have followed, but the motivation to leave — the sufficient conditions for movement — were not necessarily economic.

It cannot be denied, however, that over the past twenty years the most important necessary conditions for the drift of Fijians from the eastern islands has been changing occupational aspirations. Opportunities for non-agricultural employment in the periphery are few. Movement to Viti Levu has been necessary to obtain the desired jobs; that this process has been occurring for some considerable time is reflected in the high number of eastern islanders who occupy senior positions in both the public and private sectors of Fiji's main towns. Yet despite the relocation of substantial proportions of small-island residents, circular movement remains an important component of population mobility. To understand why demands focus upon continuity rather than change; upon elements of the indigenous socio-economic system that have been deliberately retained.

Explaining circular movement

Unlike the drift to towns in Melanesia, circular forms of mobility were not initiated by colonial intrusion. In Fiji, movement involving both short-term and lengthy absences from places of usual residence was an essential element in customary social and economic systems. At the time of European contact extensive networks linked various groups on the small eastern islands and were sustained by reciprocal exchanges, marriage ties, and warfare. The most common form of circulation occurred within and between islands, generally involved participants in return journeys of less than one week, and was associated with gardening, fishing, feasting, and fighting. Related to this were movements of longer duration often motivated by payment or collection of tribute, exchange of specialized products only able to be acquired in certain areas, and negotiation of alliances with rival groups.

In addition to circular forms of mobility, there was also long-term relocation. Two common reasons for individuals leaving their communities of birth were marriage and residence with maternal kin. Intermarriage of chiefly families on different islands was part of deliberate policy to strengthen alliances and weaken rivals. Customarily, Fijians can claim special privileges in the community where their mother was born; it is a place of refuge in times of political unrest. Finally, there was group migration associated with the division and relocation of villages. Some groups in the population were much more mobile than others and the spatial domains over which movement occurred often were quite limited. But there was nothing locationally static about traditional Fijian society.

Both continuity and change characterize Fijian mobility systems

353

over the past 130 years. The short-term, oscillatory movements linked to the daily round of subsistence and social activities have persisted, to which have been added some new nodes, in particular trading stores, schools, and churches. Migration following marriage has continued, especially for women and, while Fijian villages have become more fixed in certain locations, the process of community division and reformation has not ceased. There have also been obvious changes. The large-scale and long-distance voyages to collect tribute, engage in war, or even trade local produce have disappeared from the mobility scene. Perhaps the most dramatic, however, came with demands for labour to work on plantations and later in towns. Circular labour movement not only served to strengthen some traditional circuits but also added a new dimension to social and economic activities throughout Melanesia.

Introduction of a capitalist mode of production required cheap labour to work foreign-owned plantations, exploit mineral deposits, develop communications networks, and construct towns and ports. Where force was not used to obtain the necessary manpower, inducements in the form of trade goods and head taxes were introduced to encourage men to seek monetary work outside their rural communities. It was generally assumed that those seeking wages had limited pecuniary wants and would return to their villages or farms once they obtained 'target' incomes (cf. Elkan 1960, 1967 for African examples). Consequently colonial administrations and major employers often adopted policies of paying low wages, to ensure some stability in labour forces, and providing minimum accommodation for temporary male workers.

This convenient rationalization of the host population's response to wage employment persisted for many years in the western Pacific. The result was that Melanesians could guarantee social and economic security more readily through participation in both rural-based activities, subsistence as well as cash crop, and periodic employment for cash. The circular movement of wage labour was thus a mechanism — in some places voluntary, in others institutionalized and sustained by government policies — whereby members of the indigenous population satisfied a dual dependency on two economic systems within the colonial state. Wishing, or being compelled to retain the security of traditional institutions associated with residence in rural communities, while also obtaining some benefits from involvement in non-indigenous economic activity, they circulated between village, hamlet, or farm and the centres of wage employment: plantations, mining settlements, and towns.

The circular movement of wage-labourers is generally regarded as a transitional or compromise form associated with an early stage of economic development (Moore 1951, 1967; Moore and Feldman 1960; Kerr 1960). With the transition from subsistence-based to market-exchange economies and the changing aspirations among people as activities become commercialized, ties to particular localities are assumed to weaken. Improvements in transport and communications facilitate mobility, and with increasing diversity as industrial and urban centres evolve, there emerge the necessary conditions for more extensive movements involving long-term relocation. In a dynamic economy, it is in this context that internal migration is viewed as an equilibrating mechanism where 'new opportunities are continually created in places to which workers must be drawn, and old enterprises are ruthlessly abandoned when they are no longer profitable' (Lee 1966:52).

In some areas, however, the dual dependency created by colonial intrusion has proven remarkably durable, even despite substantial modification of the institutional framework within which the circulation of labour evolved. Government controls introduced early in the colonial period to regularize movement of Melanesians to town have been withdrawn and people nowadays are free to leave rural communities to seek employment in urban areas. Local response to this more open labour market has not been 'ruthless abandonment' of old enterprises once perceived to be less profitable. A relatively stable compromise emerged whereby Melanesians actively retain access in rural areas to productive resources and socio-political rights and privileges while also participating in the wage economy. Circulation thus facilitates involvement in both the traditional socio-economic system and the market-exchange economy introduced by colonialism.

The pattern of behaviour evident in circular movement can be explained within the premise that such decisions are intendedly rational. However, a behavioural model based on the assumption that individuals attempt to maximize monetary· gains through movement, such as proposed by Todaro (1969), is inadequate. Only if it is assumed that participation in economic activities in both villages and other areas is a necessity can circulation be said to maximize returns from time and labour. For most Melanesians such an assumption is no longer valid. Options for satisfying a need for money can be met either by cash cropping in the village or through wage employment in town; participation in both is not essential and attempts to do so usually result in sub-optimal returns, in a strictly economic sense. More relevant to understanding circular movement is the notion of risk minimization. As Brookfield (1970) has argued, in the game against an uncertain world Melanesians retain security in the traditional system while making selective use of opportunities for gaining access to some perceived benefits of the foreign commercial system. In this way they can minimize risk as disadvantaged operators on the fringes of that commercial system, while utilizing a growing range of options for economic activity in both village and town.

Most Melanesians wish to retain access to land for subsistence gardens and cash crops, as well as rights to participate in local social activities. They have deep psychological attachment to the rural communities in which they or their parents were born and where many have resided for a significant proportion of their lives. To satisfy social and economic needs at different times, Melanesians adopt strategies which ensure that a range of activities can be pursued in both villages and other locations. For most adult males, subsistence gardening and short-term wage employment represent viable options. Others may include cash cropping, sale of handicrafts, various types of entrepreneurial activity, as well as temporary withdrawal from the monetary economy and thus utilization of savings and traditional forms of wealth in the local social system (Brookfield with Hart 1971:262). In a certain locality, circumstances may combine to favour concentration of inputs in a particular activity, but most strive to maintain their freedom of choice and minimize risk rather than maximize incomes, by balancing the security of their known world against the uncertainties associated with full participation in the market economy. Where access to land still depends on membership of a social group and continuing interest in village affairs, physical involvement in rural-based activities remains important, since long-term absences can result in loss of social prestige and even rights of land.

Evidence presented earlier on the contemporary movement of Fijians born in the eastern islands does not greatly support this kind of explanation. Less than half the adults from the Lau and Lomaiviti provinces were in residence in 1976. Recent movement histories for some urban-resident Lauans contained few references to visits to their natal villages. Those living in villages on such islands as Kabara suggested that most absent kinsmen had not returned since departing for Viti Levu. It is apparent, based on these findings, that a significant proportion of Fijians from the eastern islands do not endeavour to utilize a wide range of options for economic activity in both rural and urban areas.

An absence of direct participation in village affairs is not evidence, however, of total disinterest in their rural homes. Some long-term urban residents stressed they would prefer to revisit the village more frequently but that this was not feasible given their inaccessibility. As demonstrated elsewhere, communities on most small islands have experienced deterioration in both the frequency and regularity of shipping services linking them to Viti Levu (UNESCO/UNFPA 1977: 261-3; Brookfield 1977). For those with salaried employment in the civil service or private sector, protracted absences are impossible if they wish to retain their positions. The risk of being stranded in the village while awaiting transport back to town is all too real (cf. Nair, this volume). Where better transport links exist between centre and periphery, as for the three islands with airfields and Batiki with its commercial fishing boats, there is far more short-term movement between town and village.

The system of land tenure is another partial explanation for the less intensive involvement of urban Fijians in rural activities than might be expected from other parts of Melanesia. The colonial administration codified Fijian land tenure in great detail and made more rigid a flexible indigenous system. Legally, every Fijian has rights to property in his father's *mataqali*. Whether the father or child has never lived where the land is located is irrelevant, since rights are ensured by legislation. Unlike other parts of the western Pacific, Fijians do not have to visit their villages periodically to ensure their birthright. Detailed village studies in Taveuni and Batiki show this situation to be changing (Bayliss-Smith 1978; Bedford 1978). In addition, many born outside the small islands where they have legal rights to land seem unconcerned about ensuring a future place for themselves in rural society, by taking 'active steps during the whole of their working lives to contribute to its continuance' (Elkan 1976:706).

Movement histories for some village residents nevertheless indicated that a small minority of Fijians born and raised in town, or who spent significant proportions of their working lives outside the islands where they had access to land, did return eventually. It was surprising to an outsider from a different cultural background, conscious of the limited natural resources of Kabara, its isolation, and small range of amenities, to encounter young men who had spent most of their formative years in Suva and achieved good secondary educations. Most cited reasons of kinship for their return, such as caring for elderly parents, providing for a widowed mother and younger siblings, or helping rebuild houses destroyed in the devastating hurricanes of 1973 and 1975.

Despite a traumatic period of readjustment to the demanding physical work of gardening, fishing, and repairing damaged buildings, as well as to restrictions imposed by village elders on certain social activities popular in town (especially western-style dancing and drinking alcohol), these young men claimed to find village residence deeply satisfying and had no plans to return and live in town. They repeatedly contrasted virtues of freedom and variety in village life with a more regimented, clock-bound, and costly life style in Suva. The opportunity to choose if and when to go fishing, visit gardens, make handicrafts, or sit around and gossip was highly valued by those who had experienced the routine of labouring, office work, or factory employment. A comment made to a European years ago by a resident of Rotuma, one of the most remote of Fiji's small islands, admirably summarizes their view: 'I feel sorry for you white men; you have only one way of life to choose from. We not only have your's but our own as well' (Howard 1963:77).

It is a declining proportion of Fijians from the eastern islands, however, who choose to exercise the option to participate in two ways of life by circulating between village and town. The village is still a spiritual home to most in that it represents their roots (Racule, this volume). But, it is no longer true that rural communities constitute the place of

357

permanent residence for most absentees from small islands. Among eastern Fijians, dynamic centres of social, economic and cultural activity are found nowadays in migrant communities on Viti Levu.

A concluding comment

Adequate explanation of population mobility in any societal context requires consideration of hypotheses relating to the movement process at both the macro (aggregate) and micro (individual) levels. The focus of much recent inquiry in Melanesia has been at the micro-level, since few countries have the statistical data required for adequate treatment of the macro-dimensions of population movement. Detailed investigation of selected communities has generated a wealth of information on many circular forms of mobility, especially where the behaviour of rural residents has been the focus. In part, the documented prevalence of circulation has been an inevitable consequence of definitions adopted for 'movement': absences of twenty-four hours, a week, a month. The concern with circularity also reflects a deliberate attempt to derive more relevant theories of movement behaviour for different cultural contexts and as a reaction to global statements advanced to explain what Abu-Lughod (1975:204) has termed the 'central type of mover' in western literature: 'the rural migrant who left his village to live in a large city quite removed geographically, culturally, and in terms of easy communication from his place of origin'.

In eastern Fiji, there was an intuitive assumption that mobility included two dimensions: net migration from small islands to towns, identifiable from national censuses; and an intensive system of circulation which served to maintain the central focus of Fijian society in rural areas. Internal migration was deliberately defined to ensure that much short-term movement was not automatically excluded. The method employed to collect field data, retrospective histories, tends to highlight circularity in movement behaviour. Despite this inherent bias towards identification of temporary mobility, evidence from eastern Fiji suggests that circulation from a rural base has been replaced by a massive exodus from the villages.

The consequences of this population movement for the size and structure of resident groups in the periphery have not been discussed here. As the Graves (1976) have noted for an outer island of the Cook group, there is an important discrepancy in rural communities between the real and perceived demographic implications of net migration. The roots of dissatisfaction with life in the periphery are deep and varied but not yet essentially demographic. Since the early 1960s accelerating net out-migration coupled with declining fertility levels, have reversed a previous trend of progressive numerical growth but, except on the smallest islands of eastern Fiji, the age-sex structures have not been so severely distorted by the number of dependants as to create new problems for village residents. In 1976 in Lau province, for example, more young men and women aged between fifteen and twenty-four were in residence than in 1966. Over the economically productive age range, fifteen to fifty-nine years, there has been a decline of only three per cent while in Lomaiviti the productive age-group actually increased by nine per cent.

Compared with the demographic consequences of net outmigration, the transformations in life style accompanying greater involvement in the cash sector, increased dietary reliance on store-purchased foods, recent devastating hurricanes, deteriorating shipping services, and decline of interest in group economic and social activities (UNESCO/UNFPA 1977) have much greater relevance for the hardships of living in the periphery in the mid-1970s. None the less, the smaller the island population, the greater the potential for structural imbalance caused by slight changes in the processes of movement, fertility, or mortality.

More than a decade ago Ward (1967:96) remarked that although such a prospect was sad, 'it certainly seems that many of the smaller islands will cease to be viable socio-economic units as present trends in culture change continue'. By 1976 some of the smallest in eastern Fiji had been reduced to the status of absentee farms, used occasionally as sources of copra by land-owners who now live elsewhere. Such a process will continue as resident populations decline in numbers relative to absentees and as a new generation of islanders grows up with no more than sentimental connections with the birthplace of their fathers or grandfathers. A fundamental problem is how, indeed whether to try and reorient the direction of demographic change in areas such as eastern Fiji. Whatever the academics, advisers, or planners decide, the real decisions will be made by those 'communities of stalwart natives . . . who are meeting and solving difficult problems in ingenious ways' (Thompson 1940:v), in islands which many former residents as well as some of Fiji's present policy makers describe as 'beautiful, but no place to live'.

Acknowledgment

This paper utilizes material collected during the course of the UNESCO/ UNFPA Population and Environment Project in the eastern islands of Fiji. The views expressed are the author's and do not necessarily represent those of UNFPA, UNESCO, or the project as a whole. For help with data collection, special acknowledgment is due Vaseva Baba, Latileta Kiti, Joeli Walesi, Lasarusa Moce, and John Campbell. Harold Brookfield, formerly Chief Technical Adviser on the Fiji project, is the source of many of the ideas on social and economic change in eastern Fiji and which form an essential part of any understanding of population movement.

17 The disconcerting tie: attitudes of Hagen migrants towards 'home'

Marilyn Strathern

In Papua New Guinea, studies of population movement between countryside and town have a fundamentally economic bias. The very phrase 'labour migration' identifies people in terms of employment and job opportunities. To subsist in an urban environment, migrants usually need to seek paid work and may even explain their departure from the local community as a quest for money. Certainly a demographic shift from rural to urban areas has economic consequences, affecting the development of commerce and industry, as well as the earning capacities of individuals. Yet we should not necessarily take consequences for causes nor people's economic-style explanations ('We are in town to earn money') for personal motives.

This paper raises a theoretical point about the kinds of inferences to be drawn from migrant statements about motivation. It is prompted by a study undertaken in Port Moresby over seven months in 1970-72, the period to which these observations refer. The research was concerned with men from the former Hagen subdistrict of what was the Western Highlands district (Fig. 14.1). They numbered 360, of whom about two-thirds were in unskilled occupations, from a rural population of 90,000. Compared with other groups in town, they were unusual for their small numbers in Port Moresby (Fig. 2.1) proportional to the district population; in coming from an area where people, sometimes to a quite passionate degree, discourage the departure of the unskilled; and because availability of land meant a lack of compulsion to leave, plus the fact that a sojourn in the capital city is regarded as neither educative nor prestigeful. My concern is not with the formally educated, who are seen as embarking on a 'road' valid in its own right, but with unskilled migrants, whose wage-earning fares poorly when rural people make comparisons with their own economic enterprises.

Since Hagen mobility patterns are not typical of Papua New Guinea, the rather unusual circumstances of Hageners in town help pinpoint certain methodological factors which hold for any investigation into rural-urban movements. The basic proposition is that people's statements about their actions must be related to the context in which they are made. Thus although a number who leave say they are escaping from local constraints, each knows very well what rural people think and rationales given for their behaviour must be understood in this light. Similarly, attitudes towards Hagen are best understood in terms of the kind of urban society that migrants make for themselves.

The argument falls into two parts. The first takes certain ideas expressed by migrants about the local society as reflected in communications with folk there. The act of maintaining links can place Hageners in town in something of a dilemma, and the second part considers the idioms in which this dilemma is voiced. Two significant sets of statements made by many of the unskilled are that: first, they came to Port Moresby to 'find' money – and the corollary – they cannot leave because they have not saved enough; secondly, they intend to return, at a date precisely or vaguely defined, but often put a few months from the time the statement is made. Although the idiom employed is one of personal intention and motive, in some cases appearing to correspond with past or future behaviour, the full force of such remarks cannot be understood without realizing their ideological or mythical dimension. Coming to town to earn money may be one among several motives but its verbal prominence should be viewed in relation to the kinds of things talked about as much as to what is in people's minds.

Much of the quantitative and anecdotal material on unskilled Hageners is given elsewhere (Strathern 1975). The explanation of their behaviour in terms of money, as much as the stated intention to go back, takes rural society as a reference point. We are thus dealing with attitudes towards 'home', although some are less consciously formulated than others, so that this paper is a companion to an earlier one concerned with 'home' attitudes towards the migrants themselves (Strathern 1972).

When alternatives become dilemmas

Communication

Hageners have been coming to Port Moresby since the late 1950s, although in numbers only since the mid 1960s. Many return, but there is a growing core of men who plan at least a few years more of town life although most aver they will go back at some stage. At the moment of departure the move is envisaged as temporary, a short respite from the responsibilities which adulthood will bring and which the migrant will later assume. Few at that point are men of low status, since most depart before such a judgment could be made: migrants are mostly young (late teens and early twenties), male, either newly married or more often on the verge of being, and with good prospects to which they see themselves as shortly returning. It is their staying away, almost invariably longer than intended, which alters these. Among those town dwellers who forever postpone this return are some who could still resume a viable life in rural Hagen and others who by their prolonged absence have jeopardized what were formerly reasonable prospects.

Evidence has already been given (Strathern 1972) that the general Hagen disapproval of moving excludes the well educated. A low evaluation of unskilled wage-earning must be understood in reference to the high value put on local business enterprise ('bisnis'). One element is the role of money in a subsistence economy, a valuable to be used in capital and social investment, by contrast with a wage-based economy where it has to be spent on subsistence – a simple, rural-based model that poorly reflects migrants' actual use of money. Social interaction between village and town comprises visits to Port Moresby; return visits by those away; migrants who go back; repercussions of politically-significant events in either area; sending back small gifts; and the conveyance of messages, which is my main concern here.

Material goods (gifts, money) are usually remitted with either returning workers or visitors, a social situation which breeds its own complications (status of the gift, whether it is to be reciprocated later, for whom it is intended). Migrants do not regularly entrust others with presents for their parents or other kin, with whom the most direct and frequent communication is by either letter or verbal message, where delivery is more reliable. There is enormous variation in the amount of correspondence. Both at Hagen and in town it is moderately easy to find persons who can write pidgin but it is the literate, including the self-taught, who are the most prolific correspondents. An energetic and prominent man, who also relays messages on behalf of other migrants, might despatch three or four letters a month. A notion of reciprocity is involved, so that a slight is suspected if a letter is not answered quite promptly. Lapsed correspondents may complain that they stopped writing because they never received any replies or because all they get are requests for money.

In Hagen, there is a quite strong notion that immediate kin have a duty to keep absentees informed of demographic events: birth, marriage, death. Marriage is crucial, for the transactions involved in the bride-wealths of brothers and sisters affect the migrant's potential position. Wage workers complain quite bitterly they are being cut off by kin when they hear not directly but in a roundabout way of the marriage of a close relative. Such evidence of neglect may also be used to bolster one's independence. In one conversation between a migrant and a visitor, evasive replies were given about the possibility of return. When the visitor began pushing for a positive promise, this was successfully deflected by the migrant saying that apparently his brother-in-law's daughter had been married but he had not been informed directly. The visitor, admitting this was true, was now on the defensive, apologizing that he thought others had conveyed that message. By managing to insinuate neglect by affines, the migrant's refusal to promise a return had become unanswerable.

A Hagener in town is suspicious of any indication that kin, in forgetting him, have somehow sprung a trap. When he hears that his father or brothers have given away all their pigs in *moka* (ceremonial exchange), it can seem as if previous promises ('Come home soon; we have our pigs ready to get you a wife') were more devices to persuade return than genuine expressions of support. Indeed, migrants seem very sensitive to the idea of neglect or discrimination by relatives, which in turn reflects the freedom this generates for themselves ('Oh well, if they are not thinking about me at home, I shall not think about them'). Yet equally, local pressures are distrusted. While a worker may be upset about the rumour of a sister's marriage he does not know of, he is likely to ignore any message that attaches pressure to the information. Parents do inform their sons of what is happening, but cannot resist the added plea to come back and enjoy the benefits of the bridewealth. Sometimes a sister sends word she is waiting for her brother before marrying, but he is as likely to react to the coercion as to the sentiment of the message. In sum, migrants want to think of themselves as having a place, perhaps a marriage, to which to return, but resent others using this attitude to influence the timing of their decision.

Many young men with good local prospects consequently are subjected to considerable psychological pressure from kin to go back. Such can become a threat if the migrant has left a wife behind ('Your wife will divorce you if you do not come home at once'), yet there is a communications gap between village and town. Relatives will send the strongest pleas they can: if the migrant does not come soon his wife will run away; the bridewealth they have collected will get dissipated; now is the time to obtain a wife and they are ready to start negotiations. But a strong bachelor ethos in town insulates the migrant from such pleading. Some give as reasons for coming in the first place that they are not ready to marry and these may still hold good; others look to town friends to find an consort, Hagen or not, so that some aspects of marriage are enjoyed without incurring the responsibilities entailed by returning to marry. Migrants may also regard local persuasion as quite unfair, without rejecting the legitimacy of claims made by relatives.

Rather less than half the unskilled workers apparently have pigs at home, which in the case of single men would be cared for by parents. Where the person looking after the migrant's pigs is not from among his agnatic kin and thus under no special obligation to do so, he may be particularly well fed and entertained should he visit Port Moresby, for the relationship must be kept alive by solicitous transactions. Thus when one pig-minder came to town, the owner presented him the most substantial gifts of any Hagen migrant and more than twice those made by the visitor's own younger brother. On a later visit, that same person tried to pressure the younger brother, noting that kin were ready to obtain him a wife, that in any case his pigs were being a great nuisance and breaking into food gardens, and that he should return to sort out his affairs. The migrant's reaction was to say he would come home later, not yet, for he regarded the news about his pigs as an attempt to influence him.

Occasionally pressure dissolves into fiasco. One well established migrant, living with his wife in the capital city, was visited by her brother, who previously had come by arrangement and been given money in acknowledgment of their relationship through marriage. This time he arrived unannounced. One of the messages brought was from the wife's father, repeating a request already sent by letter that the older man hoped to purchase a pig and wanted monetary help. Then the wife's mother arrived by the next plane, because she did not want to see all the wage-worker's savings going to the men (her son and husband). The daughter was disappointed, since she had been putting aside various items for when she returned later and now her mother had come and as it were snatched them before they could be given. Her husband was in a quandary. He could send a little money to his father-in-law, buy the return air ticket for his mother-in-law, but not both. This became translated into a general complaint that people in Port Moresby could not be expected to be constantly sending money and material items if visitors came so frequently. Other migrant relatives were annoyed, because not only was the situation so embarrassing but also they must pay for the old woman's return, and she knew it. There was much talk about rural Hageners not understanding what wage work was like and some quite close relatives simply avoided the visitors.

Reactions

The incidence of communication between rural area and town can be indicated by four examples spanning a period of three months, including messages conveyed by me. Roderick, foreman of a garden team whose wife had gone back to have a child, received six letters and messages from at least two carriers other than myself. These were from his wife, mother-in-law, father, mother, a married sister, two brothers-in-law, and two clansmen. Simon, a driver whose wife in the village had virtually divorced him, heard through me from his parents, married sister, and a clansman, and his father's message was later repeated by another carrier. A domestic servant Brian, unmarried but with wealthy father and brothers, had messages from his father, an elder brother, and a lineage brother, conveyed by three carriers including myself, with the father's message repeated three times. An office worker Francis, also single but sponsored by his mother's brothers, heard by letter and through another carrier as well as myself from both his mother's brothers and their mother (who had raised him), plus a clan brother.

When in 1971 I went to Mount Hagen (Fig. 14.1), I was thus given a number of messages, notably by relatives of unskilled workers to carry to Port Moresby. In total, fifty-five were sent through me, from fifty-eight people to forty-two migrants from two tribes, Tipuka and Kawelka. Some communications derived from several persons (perhaps both the migrant's parents and a father's brother) while for others one individual sent news to many absentees. The messages themselves can be decomposed indefinitely into their constituent items − a plea, a request, a piece of news − but the number of fifty-five refers to the total information received by one migrant from one source. Communications from rural kin overwhelmingly made a plea for return; reminded the migrant of transactions of which he might be part (bride-wealth, *moka*: ceremonial exchange) and of his parent's health (a covert threat at what might happen if return was slow); and contained specific promises to find a wife or set aside gardenland for cash crops, as well as frequent admonitions to save the money earned in town. Requests for material goods were few. What were the migrants' reactions to such messages, recorded for thirty-seven of the fifty-five?

Parts of several messages contained straightforward descriptions of local events that were received with interest but little comment. Sometimes the news was stale, as in one instance of a migrant hearing that a former courting partner was married. Several were moved at the thought that I had seen their parents and the reminder apparently put a few into an intolerable position. One man refused to listen, saying he had heard it all before; another burst into tears at the sight of his mother's photograph and friends said he was *popokl* (angry, upset, resentful, pressured) because he had not seen her for so long.

The reception of actual demands or pleas for return, a component of most messages, were of two kinds. Some men listened attentively and responded sympathetically; others rejected the sentiments expressed by kin. The typical response to the first category was that the migrant was indeed intending to go back but had no money yet, hence the delay. One said he was on the verge of getting a job promotion and had no 'bisnis' in Hagen, so what would he do there? Another that it was entirely his own inclination when he went ('laik bilong mi': it's up to me), since people cannot save money in towns as easily as in rural areas. The second group responded negatively, in some cases to almost identical news. When one man heard that his parents were becoming old and sick, he said that indeed he knew this and would return to care for them, especially since his other brothers were not doing so. Conversely his friend and clansman, on learning that his two wives were ill, dismissed the issue angrily: the insinuation was that this resulted from his absence and consequently not completing certain funeral rites for a child who had died. These illnesses, reacted the migrant, had nothing to do with him because women always get sick when angry about pigs and other troubles in the local community. Three years later, both men were still in Port Moresby.

Where pleas to return were accompanied by promises of transactions – a debt would be repaid, bridewealth raised – those responding negatively poured derision on the intentions of kin. Several migrants said the latter were lying and their words were mere devices to entice them back. Information that a wife would take another if the absent worker did not return soon became a 'trick,' while suggestions that kin wished to find him a consort might be met with a laugh and the short comment that he was not yet thinking about women. One wage worker, whose village wife was on the verge of leaving him, noted that although he wanted her to come to Port Moresby he had no house – a stock answer for many years without any attempt to secure accommodation. Yet questions of future marriage could be met sympathetically, as in the case of one renowned saver, who had made very specific and large investments in visitors. He said he knew that once he returned his brothers would select a wife, to which he was looking forward.

Reactions to the suggestion that a particular kinsman might come for a town visit were almost unanimously negative, a mixture of surprise and chagrin. This was often linked directly to the tacit purpose, which would be to raise money to take back. Migrants pointed to their own poverty; they had no alternative but to provide, entertain, and give, but how to do this properly? The only proposed visit contemplated with equanimity was from a mother's sister's son, who said he would bring pork from an impending feast and had not forgotten how much he had been helped in town last time. There were some outright requests for cash, mostly received rather impatiently ('All they can talk about is money!'), with one or two migrants observing that rural people had their much vaunted 'bisnis' to produce money. The few other requests, for a gun or furs from the market, were met with much less heat.

A final class of messages were those concerned with ongoing rural transactions, to which the response was often positive. Thus the migrant who was so negative at the news that his two wives were ill and who regarded further information that his father-in-law might withdraw one of them as a malicious threat, reacted very favourably towards details about various *moka* (ceremonial exchange). He admitted to certain pig debts and sent his wife instructions how to handle one of these; he was interested about the stage which *moka* plans had reached, because he had been a previous donor and was now ready to receive. He detailed quite precise intentions of what would be done with the pigs reciprocated, declaring he would go back when time for prestation grew near. His reactions thus contained two distinct elements: resentment when rural Hageners tried to tell him what to do and interest in local events, which indicated continuing involvement. His desire was to participate without responsibility, to profit from local connections without having his freedom diminished by family expectations. But the general interest to hear about *moka* plans and ceremonial did not extend to returning to swell the line of dancers.

Some emphases can be extracted from the tone of migrant reactions. Pressures to return are very strong and may be met with sympathy when interpreted as an expression of sentiment, but also with resentment should rural kin show authoritarianism which threatens freedom of action. Although a migrant may be anxious about being ignored by kin, promises of future marriage payments are likely to be received with indifference. There is some interest in maintaining exchange transactions with partners in Hagen, though direct requests for money are begrudged. Finally, most migrants are generally receptive to news and information from local communities, even if they object to the influence particular individuals try to exert. There is thus a kind of elasticity in how far townsmen see themselves as attached to or free from rural society. This can be viewed as a product of the ambiguous social situation in which migrants placed themselves by coming to Port Moresby, so that they oscillate from one viewpoint to another in judging their personal situation; or, more positively, as a product of the way in which links with rural Hagen become relevant for the definition of the urban society. In the statement 'I want to go home but cannot', we may be dealing with both a personal dilemma and a set of ideas about the relationship between village and town.

The migrant's dilemma

The evolution of an urban Hagen community could almost be described as a case of population mobility without motivation. Departure from the rural area is discouraged and seems to occur for trivial reasons. Since most migrants phrase their aims in terms of seeking money, they say they plan to go to Port Moresby for a while, acquire some cash, and then return. But what in Hagen seemed an alternative, an enlarged scope for money-raising activities and of a person's social field, in town becomes an alternative of a different kind. Through involvement in an urban society, the migrant has not enlarged his contacts as much as temporarily eclipsed the rural ones.

This is true in several ways. If money is given as a reason for leaving, it also prevents one's return. From the perspective of the local community, it might have seemed that the young man, departing before becoming involved in 'bisnis', is simply maximizing his chances of doing well at a stage in life which contemporaries spend largely in self-amusement. But the longer the absence the more is jeopardized the possibility of being able to return and act as before. While in terms of a subsistence economy wages are 'extra' and can be used to further prestige, in town the migrant finds that he has to spend what is earned not only on food and clothing but also on maintaining his urban network. The genuineness of a migrant's desires to maintain rural contacts is linked in the eyes of other Hageners in town to the vigour with which he maintains strong relations with them. But investment in the urban network, a declaration of continuing Hagenness, costs money. The dilemma is that the migrant cannot draw upon differentiated assets to behave in one way to rural folk and in another to urban associates, because money is the sole resource at his disposal. And money is very much a limited good; expenditure in one direction must be at the expense of others.

Those intending to return do not abandon town acquaintances (fellow Hageners), for this would also imply disavowing rural connections (being a Hagener). They also take a lively interest in local politics, which may become another factor inhibiting return. When one unskilled worker said he had decided to stay a little longer because the town was safer than Hagen, which had experienced a recent resurgence of tribal fighting, then escape becomes a trap. Migrants learn that the longer they stay away the more vulnerable they become upon return, because their local knowledge is outdated, inadequate, and they are unable to detect local enmities. Moreover, accidents in town are taken seriously by rural folk and the few individuals associated with the deaths of fellow Hageners would find it virtually impossible to travel back with impunity.

In the Salisburys' (1970) terms, Hagen mobility looks at first like a rural strategy. The local society remains a reference point and people do not readily admit to any long-term intention other than to go back. But what begins as an alternative for a career, money earning, and politics, becomes a dilemma for those who stay. Increasingly, the cost of the decision to leave Hagen cumulates: the cost of 'bisnis' opportunities forfeited, of the restricted ends to which money can be used, of political security. Consider now the various elements summarized here in terms of the individual migrant from another perspective, that of Hagen urban society.

When intentions become ideologies

Attitudes

While my aim is to contrast the attitudes of rural Hageners with those of urban migrants, the comparison is not between two world views of comparable order. First, there is an imbalance between the receptivity of each side. The migrant workers, being from a common origin are more familiar with the viewpoints of village people than they are with theirs. Rural society is very relevant to the urban, but Port Moresby life impinges little on it except over certain events, such as homicide, with all its political repercussions. Second, rural Hageners are extremely articulate about their criticisms of those away, drawing on dogmas and widely-agreed values to support their position, whereas migrants lack their own coherent ideology. Urban Hageners make the most precise definitions of identity in opposition to rural society, a device which gives negative values to local ideas. Knowing full well that kinsmen want them to return as soon as possible, and consciously adopting a manner that will be found uncongenial if not distressing, they defiantly subscribe to the ideal of personal autonomy ('I shall do what I like when I like') and avoid questions of social responsibility ('We are all on our own here; no one is going to tell me what to do; if my wife at home wants a divorce that is up to her; I know my mother may be ill, yet I can't go home, I don't have any money'). But because rural views are the more articulate, migrants paradoxically may choose to see things in these terms. They may disparage the town in precisely the same way as do visitors (hot, expensive, without good food); when they deter someone from visiting because there is no money, they are recognizing their consequent obligations. If a migrant says he cannot come back yet because not enough money has been saved, he is acknowledging the subsequent rural demands and agreeing that they are at least partially legitimate.

Migrants' attitudes are thus a mixture of rural views and opposition to them. Lacking any positive evaluation of town dwelling, it is hard to appreciate in which terms they see it. There is little dogma about the idea of urban living and no image of townsmen as mature, whole persons analagous with rural adults. Rather than categorizing themselves as townsmen with specific attributes, migrants seem to formulate their position in highly individualistic terms, which are given positive value as an aspect of autonomy ('I shall do what I want'). Consequently, if people are asked about living in town, the reply usually takes the form of self-opportunism ('I'll stay in this job until I find a better paid one') or personal intention ('I'm pretty fed up and plan to go home at Christmas').

In short, the attitudes of migrants towards rural Hagen society are not the urban counterparts of local values nor formulated as coherently. In the absence of a counter ideology, there are only the stated intentions of individuals, which constitute a social style (cf. Douglas 1970). To a degree, rural values are shared: openly accepted, the return is planned with greater or lesser enthusiasm, or the migrant says they do not affect him at this particular moment rather than that he will never go back. The constant stream of visitors not only reinforces the rural evaluation of money but also considerably irritates the migrant. What is revealed at such times is the divergence from what is expected (thinks of kin and saves money for their use) and what he does (spends money and defers going back).

What people say about motives and preferences cannot be placed as private intentions in contrast to public attitudes when the latter hardly exist. Perhaps we should ask if these personal statements do not embody tacit values about urban life and the identity of town dwellers. They may be couched in an individualistic idiom, but the regularity with which Hagen migrants make the same observations about coming to town and not going back is itself a phenomenon and amounts, meagre though it is, to a set of symbols about urban society.

The repercussions of disapproval

Unskilled workers are quite aware of the disapproval bestowed on their staying in town, which in turn colours their attitudes towards rural society. It is not unusual for a migrant to speak of leaving Hagen as an act of escape. Selecting any particular reason from a multitude of factors will reflect bias; one gloss which recurs is the desire to avoid marriage. Such an explanation also has considerable symbolic value for the migrant's status in town. Marriage stands for the general assumption of domestic responsibilities, since entry into adulthood is marked by no other ceremony or form of initiation. Marriage brings both responsibility and the chance to embark on a career: taking part in public affairs, becoming active in ceremonial exchange. First marriages are generally arranged by a youth's elders, so that the migrant's protest is not so much against such an arrangement or particular matches as against the idea of 'settling down'. Although youths intend to return, the timing of their departure strongly suggests an avoidance of paternal authority, which is now becoming more direct. The commonly expressed disapproval of going away consequently reinforces the migrant's notion of escaping from irksome conditions and leaving becomes an assertion of autonomy. Wage workers may lack an ideology directed towards urban life-styles but they do have a migrant ideology, whose central component is personal independence and which greatly influences behaviour in town. There seems an obvious connection between the urban migrant putting a high value on personal freedom and formulating the reasons for departure in terms of escape.

A related aspect, not admitted in so many words, is that by prolonging bachelorhood young males are also extending adolescence. Most unskilled migrants arrive in Port Moresby when just eligible for marriage, and may not return until beyond the optimal age of the early twenties. The ethos of adolescence, derived from rural society, is highly adaptable to town conditions. Its styles include thinking mainly of oneself; joining peer groups which emphasize equality and discount leadership; not taking the adult world too seriously; spending a great deal of time on amusement (card playing, drinking); experiencing a freedom from domestic obligations; feeling diffident about participation in public affairs that involve speech-making but being enthusiastic about dancing and courting sessions.

In rural society there is considerable adult tolerance of such behaviour, since marriage stands for future responsibilities rather than an immediate enforcement of duty, so that young married men often continue for a time in their old ways. Apart from participation in dancing and courting sessions, urban migrants can pursue a life style which is strongly adolescent. The shift from a rural to an urban context seems further to mean that the adult world and its obligations become largely identified with the former, from which the migrant has 'escaped', so that those in Port Moresby who do marry and perhaps have a child are able to perpetuate many aspects of bachelorhood. There is a certain equation between remaining in town and not becoming mature, though I infer this from people's behaviour rather than their statements. To many, being in Port Moresby seems a phase, as adolescence is a phase, out of which they will grow. Those who stay into middle age may still talk of themselves, a little impishly, as insouciant children. Absent is any ethic of maturity. If migrants in local eyes are 'absentee businessmen' (Strathern 1972), then rural folk in the migrants' eyes are 'absentee adults'.

Male maturity in rural Hagen is marked by assuming responsibility for a wife's gardens, obtaining livestock for her management, deploying valuables in transactions, and gradually becoming prominent in public affairs. Whereas the adult world is competitive, the adolescent is egalitarian. Status distinctions between those who are or are becoming *nyim* (of significance) and those who never achieve this (*korpa*: rubbish men) are highly prevalent; the ethic of maturity incorporates using one's capacities to the fullest. Conversely, migrants avoid talking about people in status terms. They stress equality among themselves ('We are all the same') or with reference to the claimed superiority of rural people, their common low status ('We are all *korpa* here'); are reluctant as individuals to be seen as prominent; and have no real leaders. They organize exchanges and public occasions but make no formal speeches. In disputes, issues and relationships are reduced to their simplest rather than inflated to their most complex form. This contrast is summarized in the remark of an urban Hagener at a beer distribution: 'We are all brothers here, there is nothing to be said'; in rural Hagen its analogue would be: 'We are all brothers here, so let's use the opportunity to say things . . .'.

The known attitudes of rural men again affect the migrant's position. Visitors to Port Moresby are frequently leaders and persons of substance; articulate and sometimes heavy-handed; often appalled at the migrants' lack of political sense, 'ignorance,' and inability to speak in public. They come with involved stories of local events, from which every drop of political meaning is squeezed. This frightens the migrants, who see politics in the rural world as dangerous and ramifying. They are enough of it to appreciate the insinuations produced by and manufactured out of particular incidents (a car accident, a theft, a 'big man' falling sick); but they have been away too long or are too far removed to influence the repercussions. News travels frequently between village and town but migrants can only know the general outlines, not the little details upon which decisions rest. There is sufficient congruence between rural and urban events for crises to be mutually relevant, but too much distance for the migrant to grasp enough information to make rational judgments. The likely reaction is withdrawal, to say he is not affected since he belongs to town.

Money

When visitors arrive, they are entertained as guests and generally refrain from explicit expressions of disapproval. But their very presence and demeanour is an imposition, particularly given the expectation that they will carry back the migrant's savings. One uncomfortable reminder of rural life is the guest's air of prosperity; some arrive with considerable amounts of money to spend and both parties know it was not obtained from wage labour. Money, in fact, is a bridge between rural values and town behaviour. Migrants say repeatedly that they heard Port Moresby was a place where money could be found, which in turn somewhat justifies their absence in the eyes of village people. Here, statements about money are as much a matter of ideology, relating behaviour to certain values, as of economics. Although Harris (1974) has considered the role of noneconomic factors in population mobility in Papua New Guinea, even statements of an 'economic' kind ('I came to Moresby to earn money') may have the opposite implications.

In phrasing presence in town as a matter of economic motivation and in using an idiom that will elicit a little approval ('At least he is earning money'), the migrant becomes vulnerable to an exploitation from which it is almost impossible to extricate himself. Earning money at unskilled work is justified only if substantial amounts are saved and remitted to kin for investment in rural social relationships. What relatives find difficult to understand is that money is used differently in Port Moresby. Hagen townsmen complain strenuously about the cost of subsistence — shelter, food, electricity, transport — but their budgets suggest the highest expenditures are on, first, urban luxuries and entertainment (cars, cards, beer); second, social transactions in town (loans, contributions towards car purchases, parties); and third, gifts to visitors and those returning. All three types of spending help maintain urban ties, since most friends are fellow Hageners. But while visitors approve of the third deployment, anything else is 'waste': luxuries are an extravagance and expenditures on urban social relationships are not noticed.

The identity of being 'Hagen' in urban society is thus partly maintained by expenditures quite unappreciated by rural people. The fact that visitors find Hageners in contact with one another, that friends of his own contacts will help, and that he may be entertained by far more persons than his immediate relatives reflects relationships which migrants have set up amongst themselves and are given value through financial transactions (contributions, loans) or costly entertainment (beer, cards). The friend of a friend who contributes a small sum towards the guest's return air fare does so out of urban friendship as much as to express rural links. Many visitors arrive back in rural Hagen with both transportation paid and cash, raised generally from several men: a close brother, a clansman, a tribesman of others. But often small monetary gifts are received from migrants from groups with whom there is little contact in local society. The urban contributors are helping one another; later, should their close brother come to Port Moresby, they may reasonably hope for some reciprocation. Interestingly, then, the presence of visitors does stimulate migrants into giving money; perhaps not as much as was wanted, but if nothing has been saved then at least the next fortnight's wages. Given guests from rural Hagen, the obligation to send back money is partly an obligation to the urban host.

Why does the migrant bother to respond and why is behaviour explained in terms of economics? There are several answers, reflecting Hagen perceptions of social change and modernity, but it must be stressed that the financial transactions of migrants are invariably two-sided. Contributing to the return fare of a friend's brother both recognizes the necessity of wage-earners sending back some money and strengthens intraurban ties between associates. Spending on friendships among fellow Hageners in town reinforces intraHagen bonds there, which in turn makes it more likely that at times the migrant will think of rural society. In using the idiom of money, in drawing attention to how much has or has not been saved and spent, to what makes return difficult just now, the Hagen migrant is also reminding his village acquaintances that he belongs to an alternative society. Nevertheless the option of returning to the rural place must be kept open and the small gift, handed over with apology that it is not larger, expresses this perfectly.

Rural orientation and urban society

Most migrants state they intend to return but, even when dated to

373

within a few months hence, in their time-scale it is long term. With this long-term expectation goes a readiness to entertain visitors, listen to news about relatives, and keep abreast of broad political developments. Many out-of-work relationships involve other Hageners and it is to them that the actual returnee looks for support. Fellow migrants customarily raise a sum to cover and even exceed the cost of an air fare (about $A50 [$A2.05 to £1.00]). Most migrants also have some savings and many, paradoxically, could gather from both their own resources and other financial transactions amounts of $A100–200. Consequently they need make no solid effort at private saving, despite contrary statements that such keeps them in town ('We cannot save much and so cannot come home quickly').

In addition to recalling specific debts, many also hope for gifts from friends. As in rural Hagen, friendship follows paths defined in trans-actional terms. Contributions will be most forthcoming to those men who, over the years, have participated actively in the urban Hagen net-work, invested in fellows, and spent on urban luxuries and social relationships by participating in parties, judicious social drinking, and the organization of large-scale events (funerals, feasts, buying cars). The occasional solitary person cannot expect much help from others. In short, to raise a sum of money appropriate to going back, the migrant has had to manage his urban relationships, which in turn may mean a lack of conspicuous saving. Some migrants do return exclusively on their savings but, interestingly, someone not broadcasting the intention to leave may be criticized for removing the chance of others to con-tribute. What does this mean? A man who participates actively in Hagen social life in Port Moresby is ensuring the probability of return. Con-versely urban Hageners, in identifying themselves with this ultimate intention – however many times the actual event is deferred and how-ever personal and individualistic the sentiment may appear – are defin-ing their identity of being Hagen. Reference to rural society thus has an ideological significance.

The intention to go back is found even among those who appear most settled and entrenched in Port Moresby, a number which is likely to increase (cf. Ryan, this volume). This partly is a product of the sojourn in town being regarded as a stage in life, during which a little 'home' is constructed so that through personal associates the migrant can continue being a Hagener, although no longer present in rural society. The idea of returning is two-sided. Although usually cast as a short-term prognostication, it indicates a long-term eventuality, and while rural oriented it has significance for how migrants think about their urban social life. Many unskilled workers have nonHagen contacts and friends but the most easily activated friendships come from fellow Hageners in town, and urban migrants regard themselves as an ethnic body. When they talk about links with rural kin they are emphasizing common cultural origins, one rationale for town-based relationships, and also how mutual interests differentiate them from members of rural society. People back in Hagen do not realize how 'impossible' it is to save money in town, make large issues out of trivial events, and think that everyone is a 'big man'. Rural society thus provides a reference point for conceiving the elements of being migrant and some bases for common identity. The most prominent element is appreciating the financial predicament of town dwellers when set against rural expectations. The man who says he has no money for a return ticket is also identifying himself with fellow migrants; he may even add: 'I am not coming home yet because I am waiting for my friend X. When he is ready to come, I will too'.

If the ideological significance of statements of personal intent are not grasped, then the migrant's position can be misjudged. It is not unusual to find a Hagener who asserts very firmly that he will be leaving shortly, who is heavily involved in intraHagen relationships in town, but who may also be thinking of obtaining a Papuan wife or buying a truck (both indicating little likelihood of prompt departure). In this, the migrant is both differentiating himself from rural society and relating himself to it. Indeed the most energetic individuals, those interested in making social capital out of their relationships, often face in all directions – to be the most emphatic about going back and to be the most involved in town (cf. Oeser 1969). This parallels the perceptive observation of Harris (1974:175) that lack of involvement with other ethnic groups cannot be taken as indicating a lack of urban commitment when it is a matter of style.

Conclusion: the disconcerting tie

The concluding title is to draw attention to similarities between some points discussed here and Silverman's (1971) description of the Banaban, who had to transfer their society from one location in Kiribati to another in Fiji (see end map). In considering the ways Banaban think about their situation, Silverman uses the contrast between identity, the idiom in terms of which relatedness is perceived, and code, the rules which affect the conduct of relationships.

When Hagen migrants think about rural society they dwell on their personal situation, the actual ties they maintain or would prefer to let lapse, and attempt to imprint on that situation rules for its conduct. Thus many subscribe to the idea that if money is remitted to kin, then kin in turn will be obliged to help when they decide to go back. Treating visitors hospitably has rules too, for customarily they return with presents. Migrants refer explicitly to the 'lo' (rules, customs) followed in dealing with the occasional despatch of corpses. But thinking involves both cognition and perception. In addition to the way particular relationships are perceived for their implications for behaviour ('Yes, my father is old; I shan't stay here for ever'), they incorporate identity ('Yes, I drink beer with that man because I am related to him through my father'). The second part of this paper argues that one set of relationships (ties with rural society) can be viewed as an idiom in terms of which another set (intraHagen ties in town) receive identity. The significance of ethnic origins for urban associations is an old one in movement studies. Among Hagen migrants, ethnicity is of little relevance for urban strategies against other groups but is most significant for the individual's self-definition. Consequently, it is not surprising that people's remarks are couched in the personal idiom of preferences and intentions ('Oh, I'll go when I want to'). What is clear is that the degree of stated commitment to rural kin cannot be used as an indicator of dissociation from urban life where it also serves an ideological purpose.

Hageners indeed find their ties with rural society disconcerting: both necessary and irksome, appearing now as unalterable givens, now subject to a novel degree of choice. They are also disconcerting to the investigator. If expressions and explanations of personal intent are partly a matter of social and group identity, their link to individual motivation may be highly obscure.

Acknowledgments

My thanks to the Hageners in Port Moresby for letting me participate a little in their society; to Andrew Strathern and Christine Nolan for discussion and comments; and to the former New Guinea Research Unit, Australian National University, for practical assistance. This chapter is an edited version of one that appeared under the same title in R. J. May (ed.) 1977, *Change and Movement: Readings on Internal Migration in Papua New Guinea*, Canberra: Australian National University Press and Port Moresby: Institute of Applied Social and Economic Research.

18 Circulation in southeast Asia

Sidney Goldstein

Rapid population increase in southeast Asia has led to vigorous efforts in a number of nations to reduce its magnitude (Stamper 1973; Keeney 1973; Whitney 1976). In contrast, much less attention has been given population redistribution, its effects on urban and rural growth, its volume and character, or its implications for development. Lack of concern with such issues stems partly from paucity of data, compounded by serious questions about the appropriateness of concepts employed to measure people's mobility.

All too often, when data are collected and analyzed for southeast Asia (Fig. 18.1), concepts based upon research in the western world have been employed uncritically. As a result, the value of available data is seriously limited by the definitions used, the politico-geographical units by which movement is measured, and the range of mobility encompassed. Restricting the definition of movement to permanent transfers across boundaries, which generally delimit such large areas as provinces, ensures the virtual exclusion of both short-distance and temporary movement. Yet, numerically, such mobility may be extremely important and have significant implications for both the persons involved and their places of origin and destination.

Throughout southeast Asia, the rising importance of movement in population dynamics and the pressing need for its incorporation into development planning demands a critical examination of existing data sources, their use, and how they might be exploited more fully, as well as innovative attempts at collecting new types of information. With these needs in mind, this paper assesses the strengths and limitations of illustrative data sets on southeast Asian movement; reviews some recent studies that were designed to evaluate the full range of migration and circulation, including commuting; and evaluates the insights they provide on mobility and the development process, their implications for policy, and the needs they identify in research and data collection.

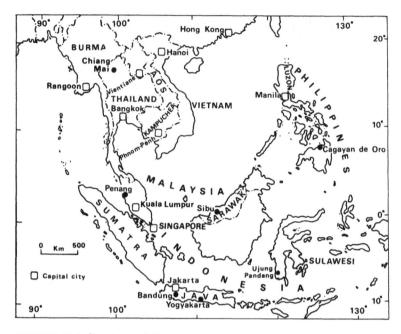

FIGURE 18.1 *Southeast Asia*

Limitations of data and concepts

In southeast Asia, as in other third world regions, much of our knowl-edge about the role of population redistribution and of urban and rural growth relies heavily on indirect information — either estimates of net migration or comparative growth rates of urban and rural places (United Nations 1970). During the 1970s, in attempts to rectify this situation, census planners considered the incorporation of and appropri-ate wording for questions about population movement. In several countries, including the Philippines, Thailand, and Indonesia (Fig. 18.1), national surveys have been undertaken that give greater attention to mobility, and still others are planned (UN Economic and Social Com-mission for Asia and the Pacific 1980). Most usually, however, the processes of rural-urban movement and urbanization must be gleaned from the more limited information contained in official censuses.

Conceptual difficulties also exist, in that migration and circulation usually are distinguished as moves that are permanent and those that are not. To overcome problems arising from the absence of a universally-accepted definition of permanence, Gould and Prothero (1975b:42) suggest that 'If there is a specific desire on the part of the individual or group of individuals who are moving to return to their place of origin, and when before leaving in the first place this intention is clear, then the movement may be considered as circulation rather than migration'. Their suggestion, that permanence be defined in terms of desires and/or intentions to return to place of origin expressed before the actual move is made, requires information rarely available in censuses or surveys, except on a retrospective and often distorted basis. More seriously, there can be no assurance that moves initially viewed as temporary do not in fact become permanent, nor that the reverse may occur when disillusionment sets in because of unachieved goals. At least in the Asian setting, the use of desire or intent is a dangerous basis upon which to distinguish migration from other forms of population movement.

Despite clear evidence of the importance of recurrent forms of mobility in Africa and Melanesia (e.g., Gould and Prothero 1975b; Chapman 1976), most demographers studying southeast Asian societies have confined their attention to permanent, often lifetime migration, as defined in national censuses and other sources of official data. This focus reflects the evolution of data collection in the region and the comparatively recent inclusion of questions on population movement in census or registry systems.

Indonesia, to take one example, has had a question on place of birth in several of its censuses, thus providing the basis to measure lifetime migration. Not until 1971 was a second question included to determine whether respondents had lived previously in a province different from their present residence and/or birth and, if so, which was the province of most recent residence and how long had they lived in the present one. Consequently it became possible to identify individuals who had returned to their province of birth. Yet the Indonesian data suffer from all the limitations of adopting provinces as units of analysis and of ignoring additional moves made in the period between birth and most recent change of residence.

The situation in Thailand is similar. A question on birthplace was first asked in the 1947 census but this information was not tabulated (Prachuabmoh and Tirasawat 1974:19). Since 1910, Thailand has had a system of population registration which records many details about migrants. Published registration data, however, only contain statistics on the number and gender of those moving into and out of each district, the official reports of the population registry aggregate them for the province, and the accuracy of this record for temporary movement is particularly questionable. As Prachuabmoh and Tirasawat (1974:15) point out, the registry data provide incomplete coverage due to the indecision of migrants about whether their new residence will be permanent, the tendency of rural-urban movers to maintain registration in the home district in case of return, and the frequent delays in registering until one's entire family has relocated.

The inherent problems of registries have been further documented by Chamratrithirong's (1979) research on migrants in Bangkok metropolis. Of 498 interviewed, only one in five reported their move as permanent, over half thought it temporary, and another eighteen per cent were unsure. Reflecting the high percentage of moves considered temporary or open to change, only one-fifth of the migrants had registered their new residence in Bangkok. More than three quarters also continued to list on their identity cards their place of origin as the permanent residence. Yet there is an inconsistency of response in these data. Although many individuals considered their move temporary, almost four-fifths had never thought about leaving Bangkok. It may be that this ambiguity reflects the phrasing of the question or a reluctance to answer affirmatively unless a departure was being actively considered (Chamratrithirong 1979:29–30).

Not until 1954, when the National Demographic and Economic Survey was undertaken, were questions on places of both birth and previous residence asked for Thailand as a whole. The first comprehensive set of migration data became available with the 1960 census, which sought information on places of not only birth but also of residence five years prior to census date. Migration, in both instances, was defined as a provincial change in residence.

The 1970 census asked basically the same questions, except that the five-year question was altered to enquire about length of residence in village or municipal area of enumeration. Consequently the place of origin for those who moved during the previous five years could be identified as urban or rural, while the wording of the question allowed determination of whether the transfer was within or between provinces. No attempt was made in published data for the 1960 and 1970 censuses, however, to ascertain the extent of repeat or return movement by using the questions on either place of birth or of past residence. Availability of 1970 data on tape has facilitated such cross tabulation and these have begun to be exploited for this purpose (e.g. Arnold and Boonpratuang 1976; Goldstein and Goldstein 1979).

As in Africa and Melanesia, surveys will most likely provide the full range of information needed to assess the extent of migration and circulation, including journeys to work, and the interrelations among them. Yet to date, very few such surveys have been undertaken in southeast Asia. Based on these and on the exploitation of census and registry data, what do we know about the extent of circulation and other forms of temporary movement like commuting, and how in turn do these relate to migration? What broad patterns emerge from the selective field evidence about people's mobility, what research problems characterize these studies, and how far do they suggest future avenues of enquiry?

Thailand

In the 1950s concern arose in Thailand about the many people from the northeastern provinces who were moving into or passing through the capital city. Textor's (1956) study of some 12,000 pedicab drivers working in Bangkok (Fig. 18.1) confirmed that a majority came from the northeast, with most of the rest originating from other parts of the nation. This research documented that circulation was common in Thailand; pedicab drivers moved back-and-forth between the northeast and Bangkok rather than settling permanently in the latter.

Stimulated to move first by the depressed conditions in the dominantly rural northeast and second by considerations of kinship ties, adventure, and the prestige achieved by going to the south, migrants nevertheless had great difficulty adjusting to Bangkok. Few wives and children accompanied them, since most moves were considered temporary. Many travelled together and maintained friendship ties in the city for social and psychological security, for sharing of dialect, food, and housing, and to exchange experiences arising from a common occupation. Apart from basic needs, the income obtained in Bangkok was destined for the northeast: as consumption goods, to build a new house, to extend an existing structure, or for brideprice.

Although engaged as pedicab drivers, migrants viewed themselves as farmers and consequently learned few new trades or skills of later use in their village. Drivers nonetheless represented a potential source of social change, returning with new food tastes, clothing styles, items of technological efficiency, and more hygienic habits. Since Textor's conclusions are based on limited data, their major value lies in documenting to what extent pedicab drivers in Bangkok were temporary residents and in raising the possibility that similar forms of circular movement characterized many other migrants to the capital.

Soon after Textor's study, Marian Meinkoth (1962) initiated another, not restricted to pedicab drivers, on movement from the northeast to Bangkok and the coastal provinces in the central region. Undertaken in July 1957, her research tried to capitalize upon the fact that almost all such migrants travelled by train to Bangkok. Teams of interviewers several times met arriving trains, questioned as many disembarking people as possible, and some teams later surveyed the migrants' northeastern provinces of origin (Meinkoth 1962:4-5).

Of 537 migrants interviewed, sixty-two per cent planned to stay in Bangkok one or more weeks and a further twenty-two per cent to continue on to other urban centres. Only two per cent of the former thought they would become permanent residents, a finding quite different from evidence available from the Central Bureau of Statistics, in which many northeasterners in Bangkok qualified as 'permanent' residents. This contradiction most likely reflects different study designs and problems of concept and measurement rather than that many sojourners subsequently changed their plans (cf. Chamratrithirong 1979). Meinkoth's study also stressed the temporary character of much movement: thirty per cent recalled previous periods in Bangkok, a pattern parallel to that found by Textor. Limited opportunities for economic expansion in the northeast and the generally low standard of living were the main stimuli, especially during slack periods of agricultural activity. Yet most movers clearly preferred northeastern villages as their permanent residence.

The 1960 census, as noted earlier, provided the first comprehensive opportunity to assess migration in Thailand. Analysis of questions about province of birth and of residence five years earlier documented high levels of movement into metropolitan Bangkok: as many as one-third of all residents were born in a different province and about eight per cent reported having relocated within the previous five years. By contrast, only seven per cent living in rural, agricultural households said they had been born in a different province and less than two per cent that they had resided in their current province for less than five years (Goldstein 1977:101). From this it was concluded that people in rural Thailand were highly stable, yet the census results are clearly influenced by both the emphasis upon residential change and the use of province as the unit of measurement. This perhaps explains why Bangkok registered a comparatively low level of immigration during the previous five years, despite a very high rate of population growth.

The 1970 Thai census, which used not only the province but also the village or municipality as the basis for determining migration, found that 19.2 per cent of those aged five or more years who lived in a municipal area had changed residence during the previous five years, compared with 10.4 per cent in rural areas (Arnold and Boonpratuang 1976:9). The much higher level of migration recorded in 1970 than in 1960 seems to reflect some actual increase rather than simply the change of areal unit by which the data were collected. In the absence of details about temporary mobility, repeat and return migration could be identified by relating places of birth to those of past and current residence.

Since birth place information is available only for provinces, the results of such analysis suggest a locationally stable population. Just over four fifths of all those aged at least fifteen had continued to reside in their province of birth, while a further ten per cent lived in the same province during 1965 and 1970 even though not their province of birth. Nonetheless, as many as a quarter of those who changed their province of residence between 1965 and 1970 are repeat or return movers, who play an important role in population redistribution. Predominantly, such persons are aged between twenty-five and thirty-four, have more than a primary education, and work in white-collar occupations. Clear relationships between type of migration (repeat, return) and socioeconomic attributes do not always obtain, however, indicating that a multiplicity of factors are involved. Repeat and return movement permeates all educational and occupational segments of the population but tends to be higher in lesser urban places and rural areas than in Bangkok (Goldstein and Goldstein 1985; cf. Arnold and Boonpratuang 1976).

In 1968, the Institute of Population Studies at Chulalongkorn University (Bangkok) initiated its Longitudinal Study of Social, Economic, and Demographic Change in Thailand. Detailed residential histories were collected, especially of household heads, and anyone absent for longer than one month was classified a 'permanent migrant'. On this basis, only 2.9 per cent of the rural population and 1.7 per cent of the urban were temporarily absent; similarly, one per cent of those enumerated in rural households and 2.9 per cent of those in urban were temporarily present. While these results indicate only a limited amount of short-term mobility, obviously they also reflect both the minimum period that defines 'permanent migration' and the phrasing of the survey question.

The data on population movement in the Longitudinal Study were neither collected nor coded with a view to examining circulation; they therefore yield only limited details about temporary residents in town and city locations. During the course of their lifetime, forty-nine per cent of the Bangkok inmigrants and forty-six per cent of those in lesser urban places had lived in three or more provinces. When different places of urban residence became the focus, one quarter of the Bangkok migrants and thirty-one per cent of those in lesser towns and cities reported having lived in at least three. Repetitive movement, in both cases, was far greater among those of urban than of rural origin. Furthermore, the number of times that migrants had lived in Bangkok during their lifetime provides some idea of the amount of return in such mobility. Of those currently resident in the capital, twenty-seven per cent said they had resided there at least twice, compared with only seven per cent of those then domiciled in other urban centres.

Those migrants with a history of multiple movement characteristically had higher education, suggesting a critical role in the economic, educational, and administrative functions of urban places. The disproportionate number of government officials indicates a career associated with frequent relocation, while the high ratio of professional, administrative, and government personnel who were repeat migrants in smaller towns and cities reflects the growing importance of these places in the urban hierarchy.

The reliance on census data to study population movement in Thailand and the limited use of coded information from the Longitudinal Survey to document recurrent mobility makes difficult any comparative assessment of circulation and migration, especially for the rural population. In the mid 1970s, this need was addressed by two research projects: one by Lauro (1979a, 1979b) of a central plains community and the other by Singhanetra-Renard (1981, 1982) in the northwest. Focusing in depth upon a rice-growing village, Lauro used a range of field techniques: the life-history matrix, reconstructed genealogical records, and participant observation. The life-history matrix, first developed by Balan *et al.* (1969) in Mexico, is a concisely formulated instrument for ordering, stimulating, and cross-checking an individual's recall of personal life events, which can encompass information on a wide range of demographic events. In Lauro's (1979b) study, the 819 life histories of current residents included data on fertility, mortality, mobility, occupation, and the ownership and use of land. Also compiled were 4,971 genealogies for individuals who had lived in the village during the past 131 years, thereby providing the basis to reconstruct its demographic history as well as to identify the population at risk to such particular events as movement.

Lauro recognizes that, for analysis of population movement, the genealogical approach presents some serious problems. Reliance on survivors for information means that outmigrants who left no family members behind may be omitted, while the shallowness of Thai lineages places serious constraints on historical reconstruction. Still another difficulty of village research in Thailand, as elsewhere, is that despite the great advantages arising from the intensive study of a single community, the particular choice – in this case of a remote one with reasonably detailed and accurate written records – inevitably raises questions about its typicality. Despite such methodological concerns, these field data provide many insights about people's mobility within the context of changing social and economic conditions in a rice-growing community.

Lauro identified a range of circular movements, from daily journeys to work to occasional return visits by those who had taken up relatively permanent residence elsewhere. He focused particularly on seasonal wage labour and found it to have increased significantly since 1960, along with expanding job opportunities in construction and agriculture. As many as two-fifths of the men had been seasonal workers, although the frequency varied inversely with the amount of land owned. Despite the alleviation of economic stress afforded by outside employment, little difference was discerned in the behaviour or experience of those villagers who had gone away to wage labour and those who had not, mainly because of their very limited contact with the societies in which their work sites were located. Circulation clearly was acceptable behaviour among central plains villagers and was seen by Lauro as having its roots in the Thai traditions of military service, corvee labour, and the monkhood.

Singhanetra-Renard (1981, 1982) also found circular mobility to be very common among the peasant communities she studied around Chiang Mai (Fig. 18.1), the major urban centre of the north. Legitimately critical of the heavy reliance of Thai migration research on models developed in the western world, she argues that census data fail to identify much movement because of its temporary nature. As a native of Chiang Mai, Singhanetra-Renard was impressed with the striking similarities in patterns of circulation noted for the Pacific region and therefore sought to examine the complete range of short-term, repetitive, and cyclic movement. Included in these aims was assessment of how villagers conceive different mobility experiences, broad comparison of several villages in terms of migration and circulation, and testing the applicability of Zelinsky's (1971) stage-type model of the mobility transition.

The key village in this research is Mae Sa, some thirteen kilometres from Chiang Mai, in which a wide range of data were collected: a detailed census, a register of all moves of at least six hours' duration made into and out of the village, life histories from 150 residents, and observations and interviews by key informants. Movers also were studied in Chiang Mai, in *miang* (tea) villages, and in the foothill gardens where many of them worked. To ensure comprehensive coverage, a move was defined as an absence of six or more hours from the place of reference and circulation as any movement that originated and ended there.

The underlying hypothesis is that circular mobility results from the perceived complementarity between places of both origin and destination. The former provides the individual villager rights, privileges, and security offered by land, family and shared beliefs; the latter wage jobs, formal education, agricultural and forest land, medical and other services; and circulation maximizes the benefits of both. As a result, circulation is infrequent where economic and other needs can be met in the origin community; circulation occurs in the absence of alternative local opportunities; and as movers become involved in a set of social relationships at point of destination and as they acquire prestige and security there, so they gradually withdraw from the village and become permanent outmigrants.

Singhanetra-Renard distinguishes between commuters, persons still part of ongoing village life; circulators, those temporarily absent from the village who consider it their home; and migrants, who have left and do not plan to return. These three categories are not discrete, since commuters may travel daily, periodically, or seasonally; circulation varies from as little as one week to as long as fourteen years; and migration, depending on intent, from a similarly short period to one's lifetime. This typology articulates movement sets in local terms, to which the conventional coordinates of time, space, and purpose subsequently

385

were fitted, and their overlap documents that Mae Sa villagers do not conceive mobility in discrete categories of distance or elapsed time. Such a conclusion illustrates the challenge in evolving concepts to clarify the ambiguities inherent in the study of movement behaviour while also capturing its meaning for the society involved.

Although recognizing the value of such a typology, especially in an exploratory study, the considerable overlap of time and distance creates special problems for comparative analysis among different populations as well as placing undue burden on the researcher, if not the village mover, to identify the appropriate category for observed behaviour. The complexity of data needed relegates its collection to intensive research of the kind conducted by Singhanetra-Renard but which larger-scale surveys would find difficult to replicate. In future studies, simplification of concepts, procedures, and related questions consequently should take high priority.

Noteworthy in Singhanetra-Renard's enquiry of Mae Sa, regardless of the technical problems noted, is the high degree of mobility and the presence of all three types during the past century. Since 1960, circulation to *miang* (tea) villages has declined but has been replaced by that to factories in Bangkok (Fig. 18.1). Migration has persisted, tied predominantly to traditional forms of marriage, and a dramatic rise in commuting reflects the combined factors of land shortage from population increase, new desires and aspirations for material goods and formal education, and the easier availability of motor transport. Indicative of the widespread occurrence of mobility is the observation that it is not a question of who goes but of when, since at one time or another practically every villager travels to the city for work, education, shopping, or some other activity.

In sum, Singhanetra-Renard's study has broken new ground by indicating how much the mobility of the rural Thai is masked by official statistics and census-based definitions. The insistence on permanence as a key concept in delimiting movement misses its key role in the lives of both individual villagers and whole communities as well as in maintaining social ties and raising living standards. Short-term mobility influences social and economic change through the impact that movers have on both their local communities and their various places of destination. Since many urban and rural employers rely upon villagers for their labour force, these locations do not have to provide housing, sanitation, education, and other facilities as they would for permanent residents (cf. Nelson 1976).

Further evidence about return movement from Bangkok is provided by Chamratrithirong (1979), who over six months monitored all relocations made by persons arriving in the metropolis during the previous two years. A subsample of 498 migrants aged between fifteen and forty-four was randomly selected from the 1977 Survey of Migration in

Bangkok (Thailand National Statistical Office 1978), as were 300 non-migrants. Following initial interview, questionnaires were mailed each month, so as to identify further mobility. At the end of six months, efforts were made to reinterview all those migrants who, according to mailed questionnaires, had not moved again. Where direct interviews proved impossible, contact was made with neighbours and relatives so that, in the end, information was obtained about the mobility behaviour for ninety per cent of the migrant sample (498).

After six months, almost half the migrants resided at the same address and another twelve per cent had moved within Bangkok. Just over one-fifth of the men and almost one-third the women had returned to their community of origin, most often while they were still in the younger ages (men 20-24; women 15-24). Migrants with relatives in the city departed less often than those without or those who were single upon arrival. Return was more common among persons originating from rural rather than municipal areas, as well as for those with less than five years schooling. Although somewhat surprising, the fact that migrants who had a job at first interview were more likely to leave than those looking for or without work reflects their dominantly agricultural background, with the implication that urban residence was expected to be temporary.

The recent focus on temporary mobility in Thailand augurs well for its more accurate assessment and for testing different concepts and measures of population redistribution. In the next decade, many of the issues raised should be easier to pursue because of a question on journey-to-work included in the 1980 Thai census, the efforts of ESCAP to initiate a series of country migration surveys (including one in Thailand), and because of continuing efforts by the Institute of Population Studies at Chulalongkorn University to monitor patterns of population redistribution through both national surveys and studies of particular groups.

The Philippines

Although several comprehensive studies of population movement have been undertaken in the Philippines, most reflect census-based definitions and standard measures of migration (e.g. Hendershot 1971; Pernia 1975). As demonstrated for Thailand, however, census data may indicate the existence of circular or repetitive movement. Using 1970 tabulations, del Rosario, Lourdes, and Kim (1977) have constructed a typology to assess differences in the social, demographic, and labour force characteristics of movers and stayers.

From census questions, the location of household heads was obtained at four points in time: birthplace, and residence in 1960, 1965, and 1970. On this basis, fifteen mover-stayer patterns were identified and then collapsed into six major types: non-migrant, primary migrant, secondary, tertiary, return, and circular migrant — a classification elaborated from that initially developed with United States data by Hope Eldridge (1965). 'Return migrants' are those who, by 1970, had gone back to their birthplace and 'circular migrants' those who, having left their birthplace at any reference period (1960, 1965, 1970), had by 1970 returned to any previous place of residence other than that of birth. According to these four reference points, only one fifth of all household heads had moved more than once. Of these, just under half had once returned to their place of birth or previous residence, and circular movers outnumbered returnees by four to one. Even at the provincial level, there consequently exists clear evidence of much return and circular movement throughout the Philippines, which would constitute an even higher proportion of total mobility were data available for more time periods and more refined politico-geographical units of analysis.

That circulation is more extensive than census data would indicate is confirmed by van den Muijzenberg's (1973, 1975) research in central Luzon (Fig. 18.1) during 1968–69. This stresses the need to consider 'circommuters', temporary workers in urban places, even though they are difficult to quantify. Because of low levels of land ownership in the village studied, a number of persons take jobs in Manila to both meet basic needs and expand local resources. Mostly adolescent and adult males, the 'circommuters' work part of the year in the city, returning to the village for planting and harvesting of rice as well as at weekends once or twice a month. Thus strong links are maintained with the village and in fact may have been strengthened by increased demands for rural labour as a result of the Green Revolution (van den Muijzenberg 1975).

As the distance that villagers travel to Manila increases, according to van den Muijzenberg, so circulation will become less feasible and competition will rise for the jobs undertaken by 'circommuters'. Nonetheless, the high cost of living in the capital, as in other third world cities, will force temporary workers to leave families behind and total circulation will increase. Places of both origin and destination benefit from 'circommuting': the village, by having employment brought within its resource field; and urban areas, by requiring less investment in education or most other social services needed by permanent residents. Interaction between city and village is similarly a two-way process. Teachers sent to rural communities represent the national centre and, in so doing, help integrate the provincial periphery – an aspect of circulation that has received too little attention.

Stretton's (1978, 1981) study of the building industry in Greater Manila lends considerable support to van den Muijzenberg's conclusions. In construction work, circulation is the important means by which labourers cope with uncertain employment and irregular income. Skilled tradesmen only moved to Greater Manila when work was available; once their contract ended, they returned to their home towns or *barrios* to await the next opportunity. Circulation thus permitted construction workers to improve living standards, as income was earned from intermittent but quite regular employment and living costs at work sites were reduced by families remaining in the local community. By retaining close links to place of origin, labourers were also assured of acceptance whenever a job had been completed. To become a permanent migrant, without regular employment in Manila, would have offered far less security and involved the threat of a marginal job, with lesser income and higher costs, especially if the entire family relocated.

Circulation, Stretton emphasizes, benefits the metropolitan building industry even more. It depends on an efficient network of contacts maintained by independent foremen, but is able to function with minimum bottlenecks of labour since redundant workers usually return to their home province. Temporarily resident while employed, skilled labourers also make few demands of urban facilities. On the other hand, when the building industry is without contracts, rural areas come under strain as they are expected to not only provide food and shelter for the unemployed but also absorb the resultant drop in remittance income. The causes and consequences of circulation therefore reflect the tight links between urban and rural conditions.

A different perspective is provided by Sutlive's (1977) comparative research on urban migration in Sarawak and the Philippines (Fig. 18.1). Conducted in 1975, the Philippine segment focused on Cagayan de Oro, northern Mindanao, which had grown from 68,000 in 1960 to an estimated 164,000. Interviews with captains of twenty-four out of forty rural *barangay* (neighbourhoods) encircling the city indicated that desire for regular income most commonly stimulated movement, as a result of landlessness, diminished soil fertility, deforestation, and difficulties in obtaining clear title to land. But Cebuanos have little prospect of returning to rural areas unlike the Sarawak population or the village communities described by Stretton and van den Muijzenberg. As Sutlive (1977:368) phrases it, 'once "in" they find that there is no "out" '. How far circulation in the Philippines is a viable alternative to relocation thus varies considerably, depending on opportunities available in places of both destination and origin.

Malaysia

Despite censuses dating from 1881, there is little national data on population movement in Malaysia. Nagata (1974), in assessing the situation, notes that comparative analysis of places of birth with current residence are bedevilled by changes in definitions of urban places, urban boundaries, and ethnic identity. The 1970 census is more promising, and offers some prospect to assess return movement, in that not only was country of birth asked but also years of residence in present locality and the place and name of previous domicile (Cho 1976). In 1967–68, an intercensal survey of west Malaysia asked about places of residence and of birth (for those born before 1957). Within each of the five zones, additional details were elicited on both current and 1957 residence, but these ten-year data cannot be related to that of birth to determine place of residence for three different periods as a means to gain insights into the extent of repeat and circular movement.

Given such limitations, Nagata's analysis of these survey materials inevitably shows great locational stability for the Malaysian population. At least ninety-two per cent of those questioned in four out of five zones were still living in their zone of birth. The ten-year data indicate that most movement occurred within the same zone, so that the intercensal survey mainly documents more permanent and longer-distance mobility.

These general results were both elaborated and questioned during fieldwork in two cities and a village (*kampong*) of west Malaysia. Much movement, in fact, does not follow a unidirectional pattern towards either rural or urban places but rather alternates 'back and forth between city and country, and often also between zones' in what Nagata (1974: 317) terms 'oscillation'. This recurrent pattern reflects several factors typical of Malay life, which in turn duplicate many of the underlying forces reported for circulation in Black Africa and Melanesia. Since formal education is a key mechanism for social mobility and high quality instruction is available only in urban areas, rural-born persons and sometimes entire families will move temporarily to the city to obtain schooling for either themselves or their children. The fact that many teachers in rural communities maintain a residence in the nearby town and return there frequently adds to these circular flows. This entire process underpins Nagata's contention that the distinction between urban and rural in west Malaysia is unclear. Rural homes are perceived as security against crises and failures confronted in the city, so that ownership of land or at least a homestead makes far simpler the decision to leave for a time and try to improve one's economic lot. In fact, many move to town in the hope that rural conditions will improve during their absence, at which point they will return.

Inheritance also plays a key role in oscillation. Under Islamic law, all children inherit some property, and the various socio-economic strategies followed by urban dwellers almost always result in rural-urban flows. At one extreme are those who return to settle and work the village land; at the other are absentee landlords, whose major rural involvement is through investment to improve their property. The exchange of money, ideas, and goods resulting from absentee ownership is significant and may help explain why, 'socially, economically, and ideologically, as well as demographically, the rural-urban distinction for Malays is somewhat blurred' (Nagata 1974:319). Closely related is the common practice by which Malays, having moved to urban places, plan for social and financial reasons to retire to the *kampong*. Government policy, as in Thailand, also contributes to circulation as civil servants, who constitute a significant proportion of those employed outside agriculture, are transferred from one location to another.

Circulation is intensified by situations associated with the life cycle or social and religious norms. As in India, many women in Malaysia prefer to bear their children at the parental home, while individuals return in times of sickness or for religious instruction. In summary, Nagata (1974:323) suggests that oscillation between rural and urban places may entail little change in life styles or basic values and attitudes — but considerably more research is needed, preferably longitudinal in character, to ascertain the effects of such intense mobility upon both individuals and communities.

These observations about circulation in Malaysia are supplemented in field research by Strauch (1977, 1980, 1984), who also argues that the emphasis of western models upon residential change is inappropriate for many third world societies, particularly since departure from and return to rural places not only occurs but also demands considerable adaptation for those involved. Thus Strauch, unlike Nagata, is convinced that the circulation of wage workers can have profound social and political impact on their home communities. Primarily focussing her attention on Chinese in Hong Kong and Malaysia, she notes that their movement,

> both traditionally and today, occurs typically in two distinct forms: 1) the back and forth shuttle migration within a relatively bounded regional system, showing considerable internal variation, and 2) the long-distance and long-term overseas migration. . . . In both types, the intentions of the migrants are typically those of the sojourner, whether short or long term. The goal is to return home successful, to retire with honor in the bosom of the native place (Strauch 1977:2; cf. Watson 1975).

Although the regional circulation of Chinese has received little attention, comparatively high levels were found for a Chinese-Malaysian village studied by Strauch in 1971–72. About thirty-eight per cent of the working population of Sanchun, a pseudonym, were 'shuttle migrants'; one such migrant was found in forty per cent of all households, and one quarter of them included a spouse who continued to reside in the village. Most absent labourers were engaged in logging, construction, or tin mining; their frequency of return visits ranged from once a week to once yearly; and the distances travelled varied from a few to several hundred miles. Some 'shuttle migrants' even worked or studied in other countries.

That circulation is of positive benefit is reflected in wage labourers having gained self-confidence through being away, as well as a broadened outlook on national issues, especially the role of Chinese. These changes, concurrent with the maintenance of close ties to their home communities, saw many migrants providing leadership among younger members and supporting quite radical action that could lead to socioeconomic change in the village and its greater incorporation into the national polity (Strauch 1977:3).

A comparable study of wage labourers in Hong Kong, undertaken in 1978, demonstrates that kinship links remain strong between a coastal valley in the new territories and a distant metropolis (Strauch 1980, 1984). This valley community is less cohesive than the Malaysian one: whereas the locally born shuttle constantly back and forth, more recent members immigrant from mainland China gradually transfer their allegiance to urban Hong Kong. The cultural and economic incentives to become an urban resident thus reflect not only the absorption of ever present material values but also whether the labourer was born of or immigrant to the origin community. In Strauch's (1980:19) words: 'Patterns of population mobility are determined by a dialectic comprising individual needs and calculated choices on the one hand and overarching frameworks of political and economic systems on the other. Such systems present some options but preclude others, thereby structuring the range of [mobility] choices open to the individual'.

As previously noted, Sutlive's (1977) research focused on both the Philippines and Sarawak (Fig. 18.1). The Malaysian segment considered the movement of Iban into Sibu, a city in the Rejang valley of Sarawak whose population of 70,000 contains about 3,500 Iban. Their high mobility reflects a complex of natural and cultural factors including rather infertile soils, swidden cultivation as a way of life, an opportunistic ethic, and a general desire for economic advancement. Interviews in Sibu with a random sample of 200 Iban revealed considerable range in the intended length of stay, depending on their degree of job security and such benefits as guaranteed housing. Least likely to leave town, despite statements to the contrary, were government or church employees, for whom housing was often provided. Service workers and day labourers, by contrast, averaged one or two years in town since, as Sutlive (1977:360) puts it, 'After they have saved enough money and brought enough goods to demonstrate their success . . . they return in triumph to their longhouses'. Unlike the Cebuanos, for whom a return was unlikely, absent Iban retain viable options to be part of their local communities.

Iban who do not return maintain frequent contacts with rural kin. Of the sample, more than three quarters regularly helped their parents with money, food, clothing, or their labour, and were in turn supplied with rice or other rural foodstuffs. The close links that persist between the rural and urban members of a kin group consequently have significant implications for both the adjustment of migrants in town and the ability of rural stayers to share the benefits from mobility.

In 'Village Paths and City Roots', preliminary results from another community study, Strange (1976) identifies three categories of women born in a coastal village of northeast Malaysia. Village-oriented women engage in subsistence, cash-oriented, or wage-paying activities, or claim to be only housewives, a status of high prestige. Of the town-oriented women who now live in east coast towns, some hold wage or salaried positions while others are housewives. The third group, dual-oriented women, live in the village and work in town. Dualist influences also affect both village- and town-oriented women: for the former, through children's education, kinship ties, and retail facilities in town; for women who are the latter, as a result of family and relatives, life-cycle events like birth, and festivities specific to the village.

Strange, like Nagata, observes that in recent years more women have been going to town, which leads to greater understanding of urban ways and makes it easier for others to visit. Women learn that education and work opportunities exist outside the local community, exposure to which creates impatience with the village and contributes to it being viewed as dull and restrictive. Slowly and steadily, the reality of urban alternatives becomes apparent to the villagers (Strange 1976:12). Rural and urban distinctions are thus blurred by circulation and Strange, unlike Nagata, sees this as an important force in social change and the development process.

The different implications of rural-urban complementarity noted by Nagata and Strange call for more attention to local contexts and situations. Penang (Fig. 18.1), the place of Nagata's research, is one of Malaysia's largest and most industrialized urban concentrations, whereas the east coast, site of Strange's fieldwork, is quite underdeveloped even by local standards. Socioeconomic changes observed to be occurring rapidly in the latter probably took two centuries in Penang. Any assessment of the impact of circulation must therefore consider an area's socioeconomic condition, its levels of urbanization, and prior links between rural and urban communities.

Indonesia

To date, the most comprehensive assessment of circulation in southeast Asia is for Indonesia. The work of Graeme Hugo (1978) on west Java constitutes a milestone, both for its innovative character and the many insights about the role of temporary mobility in people's behaviour. Undoubtedly this research will serve as a model for other's conducted within the region and already it has influenced Mantra's (1981) study of wet-rice communities in central Java.

As elsewhere, the Indonesian investigators are concerned that

393

western ideas and values too greatly colour the assessment of population mobility. A key goal is to distinguish between migration and circulation, based on the criterion of permanence, especially since Indonesians have long embodied such distinctions in their language. *Pindah* refers to permanent migration whereas *merantau* implies that, for the mover, the destination is not a goal in itself but rather a means to improve or stabilize one's position in the place of origin. Although such a move may become permanent, the orientation and behaviour of people involved reflects an intent to return.

Mantra (1981:166) argues that the linguistic distinctions for types of movement extend far beyond this broad dichotomy. In Javanese, he points out, *merantau* refers only to those who go to another island for a relatively long period before returning to the origin community. By contrast, Javanese use *nglaju* for those who travel to a place but come back the same day, *nginep* if the absence from the home community is for several days, and *mondok* if movers remain at their destination for several months or years. The very existence of such words strongly supports the need for more research into the great complexity of people's movement.

Comparison of birthplace data in the 1961 and 1971 censuses of Indonesia shows a decline in lifetime migration to the capital of Jakarta (Fig. 18.1), the dominant destination in the country. Hugo (in press) argues that such apparent change reflects nothing more than arbitrary census-based criteria of time and space revealing neither the full range of movement nor the substitution of one form for another. If the question on province of most recent residence asked in the 1971 census is compared with province of both current residence and birth, then 1.53 million persons are found to have returned to their birthplace. These account for 7.7 per cent of all migrants, compared with only 6.4 per cent when birthplace statistics alone are used. Census data also suggest two interesting patterns: first, the highest levels of return movement characterise provinces with greatest rates of outmigration; and second, metropolitan Jakarta is a major origin of return to provinces of birth. Consequently these data, for all their limitations, point to a high degree of circulation in Indonesia.

Hugo's (1978) research is particularly effective in providing evidence that many kinds of mobility have occurred throughout much of Indonesian history. His fieldwork was undertaken in 1973 in west Java, where a quarter of the country's population live on a minute portion of its total land area. The region is also a major source of migrants to Jakarta and Bandung (Fig. 18.1), two of Indonesia's largest cities. Fourteen villages were surveyed because they supplied outmigrants; studies also were conducted in the two cities; and documentary materials were gathered.

In the village surveys, information on mobility was collected from all current residents as well as, by a method of family reconstitution, about all those living in the same households during the previous ten years. These movement histories contained details of migration and circulation, which was divided into 'circular migration' and 'commuting'. 'Commuters' were defined as villagers who regularly departed, although not necessarily each day, to work or attend an educational institution but returned most nights, while 'circular migrants' remained at their destination for continuous periods of up to six months.

Most rural commuters travelled less than ten kilometres from the village, compared with up to fifty for urban commuters. Whereas about four fifths of the latter depend upon buses, microbuses (*oplet*), or rail, the former rely heavily on walking and the bicycle. Most commuting occurred on working days, suggesting that the major occupation of villagers was at their destination. At the same time, most regular travellers worked in the village whenever possible, especially during planting and harvesting seasons for rice, either at weekends or by not commuting for a few days. This strategy also is followed by villagers in northern Thailand.

Although commuting among urban residents is heavily tied to modern modes of transport, for most west Javanese villagers daily journeys to work are not new. Helping with rice planting and harvesting in neighbouring villages causes intense seasonal mobility and there has always been frequent travel by itinerant traders. What has changed is the increased opportunity for commuting to a wider range of urban centres as a result of more available forms of transport. Commuting nonetheless continues to be limited by constraints of space and time and where these become restrictive, greater reliance is placed on 'circular migration', with patterns remarkably similar to those observed by Singhanetra-Renard for northwest Thailand.

Conceptually, 'circular migration' is a temporary move in which return to village birthplace will occur within a consciously planned period. Recognizing the difficulty of precise measurement, Hugo adopted six months as the maximum period a villager could remain away continuously, and still be considered a 'circular migrant'; this fitted with both census criteria and whether a person was inscribed in the official village register as having departed temporarily or permanently. Yet many villagers were absent for longer periods and, intending to return, qualified as 'circular migrants'. Adoption of an absolute time frame therefore proved a source of frustration for Hugo and, as for Singhanetra-Renard (1982), underlined problems inherent in evolving unambiguous definitions of circular mobility.

Among reconstituted households, in nine of the fourteen villages, 'circular migration' was the most common rural-urban strategy. Two thirds went to urban destinations, with Jakarta the most preferred. Commuting was the dominant form of movement in two villages, both understandably closer to Jakarta or Bandung than the others, while in only three did migration prevail. 'Circular migration', although not new, has expanded considerably throughout Indonesia since independence, as a result of improvements in public transportation, growing pressures upon agricultural resources, and substantial unemployment in rice-growing areas due to the seasonal nature of labour demands. Also important was the desire to supplement village incomes and raise living standards by working in urban places. Circulation that occurred within rural areas mainly reflected seasonal differences in the agricultural cycle, especially the timing of land preparation, seeding, and harvest, while other villagers found such nonfarm activities as carpentry, hulling rice, and basket making.

Particularly valuable in Hugo's analysis is what various forms of movement mean for villagers and their impact upon the local community. Links with previous destinations are especially important in acquiring jobs, arranging housing, and providing a sense of security in environments. Moving into the city is facilitated through urban sub-communities defined by village-based relationships and reflected in not only residential but also occupational clustering. The resultant savings are critical if much of the income earned from commuting or circular migration is to be sent to the village. All west Javanese movers remitted money and four out of five brought back goods. Such remittances accounted for as much as sixty per cent of the total income in commuter households and about half in those of circular migrants, thus helping to meet daily needs as well as long-term plans for education and purchasing material items.

Among movers and stayers, at least three out of four felt that money was the most positive local contribution of people's mobility. In both groups, most regarded the experience gained by movers in town as beneficial and a considerable number also mentioned the new ideas with which villagers returned. Some concerns, however, were expressed about the breakdown of traditional values, different styles of dress, and changes in behaviour patterns. Nor was the flow of remittances in one direction. Villagers often subsidize the travel costs and initial city expenses for both circular movers and permanent migrants – and after visits home there is commonly a reverse flow of foodstuffs. Although strong rural-urban ties exist in both directions, the economies of all fourteen villages would be further depressed if the flow of remittances generated by circular mobility were to stop.

Beyond documenting the rate and significance of circular movement for both individuals and entire communities, Hugo's study addresses both policy and theoretical questions. Confronted with the massive flow of rural people to cities, especially Jakarta, the Indonesian government has attempted to discourage such practice by requiring both registration and the deposit of return fares. In fact, such regulations encourage circulation since temporary movers can obtain permits in the village as well as avoid registering in the city. Given the significance of remittances to the village and of cities as places where the number of jobs continued to expand, circular mobility can be viewed as a redistributive mechanism. Similarly, less demand is made on urban services, housing, and financial resources than in the case of migration. This view is consistent with Elkan's (1967:589) for east Africa, which argues that circulation should not be regarded as an evil but as a process lowering the cost of development. Although further evidence is needed perhaps circulation, more than migration, has greater potential for spreading to rural areas the knowledge, ideas, and attitudes that contribute to socio-economic change. On this basis, Hugo concludes that southeast Asian governments ought to encourage various forms of circular mobility.

On a theoretical level, Hugo's (1983) investigation provides a careful test of Zelinsky's (1971) hypothesis of the mobility transition. Little evidence was found that mobility patterns in Indonesia conformed to this evolutionary sequence, while in west Java the major historical force was colonialism rather than modernization, as argued by Zelinsky. During the eighteenth and nineteenth centuries, a history of forced labour on government plantations, seasonal movement, and coolie labour all contributed to a tradition receptive to seeking part-time occupations beyond the village. In Java, colonial policy kept industry decentralized to create rural employment. Even after independence, rural population pressure was countered through such mechanisms as deliberate confinement of industry, agricultural intensification, and the enduring circulation both within rural areas and to a hierarchy of cities. Despite increased urbanization, much of the colonial social and economic structure remained to influence the processes of industrialism and modernization. *In situ* adaptation continued in the villages and, with improved transport, circular mobility flourished. In general, third world and western countries therefore differ in the pace at which particular stages of the mobility transition are reached and Hugo advises: 'The experience on which it has been formulated is bound by a specific time, culture and political economy' (Hugo 1983:103).

Reacting to the evidence from field research such as Hugo's, Zelinsky (1979) has recognized limitations in the mobility transition hypothesis. In a revised statement, he posits that 'mobility characteristics associated with underdevelopment' may be 'quite distinctive', that circulation itself may be one such symptom, and that it 'promises to endure . . . as long as underdevelopment persists' (Zelinsky 1979:187). This provocative suggestion undoubtedly will stimulate further rejoinder from students of third world mobility, as well as reconsideration about the role of circulation in the complex links between people's movement and the development process.

Whether circulation represents a mobility phase eventually replaced by migration is addressed by Maude (1980), in a study of 976 movers from 325 households located in eleven villages of west Sumatra (Fig. 18.1). Despite a long tradition of return to the home village, in most, short-distance circulation gradually had evolved to long-distance migration in response to socioeconomic changes in both rural and urban places. Fewer resources are thus likely to flow back to the village. Even so Maude, like Hugo and Stretton, stresses the positive contribution of circulation to rural communities and the need to acknowledge this in any attempts to curtail such recurrent movement.

Mantra's (1978, 1981) study of wet-rice communities in central Java was inspired by the same frustration as Hugo's: the fact that Indonesians in general and Javanese in particular were viewed as locationally stable reflected the focus of prior research on residential change. At the same time, the sizeable region of west Java made it difficult for Hugo to evaluate in detail the processes underlying mobility behaviour. Mantra's criticism points to a major dilemma confronting those concerned with population movement for which census and survey data do not exist. To obtain such details, with appropriate information on motivation and impact, necessarily restricts enquiry to a few communities. This limitation, in turn, inevitably raises the question of how far those communities typify the more general patterns characterizing a region or country, given the unique ecological, economic, and social conditions probably found in most villages. The ideal is a combination of approaches, as Hugo attempted, although villages ought to be chosen so that results can be combined to assess the overall relations among various forms of movement and diverse socioeconomic influences.

To overcome the constraints of secondary data and of a more broadly-designed regional study, Mantra investigated two wet-rice *dukuh* (hamlets), Kadirojo and Piring, in Yogyakarta Special Region (Fig. 18.1). By examining the behaviour of hamlet residents and the processes whereby they obtain information and make decisions to move or to stay, he attempts to encompass the full range of mobility and to understand the varied conditions under which it occurs. Reflecting such a detailed approach, a move is defined to have occurred whenever a person crosses the *dukuh* boundary and stays within or beyond the hamlet for at least six hours. Given this minimum time and the fact that a hamlet (*dukuh*) is one constituent unit of a village, the research design identifies the maximum mobility without reducing time or areal unit to the point of absurdity. Nonetheless, the challenge faced in such an inventory is evidenced in Mantra's (1981:39) observation:

> The identification of all moves at or close to the time they occurred was made more difficult by the surprisingly large number of *dukuh* people who were involved. . . . People find such movements hard to

remember, because often they happen spontaneously, but it is very rare to find people who are absent from their village for longer than one month.

How much more serious would this problem be if such data were obtained retrospectively following a longer interval?

Migration is defined as an intentional shift of residence out of the *dukuh* for at least one year and circulation as an absence of more than one day but less than a year. Commuting consists of absences spanning more than six hours but less than one day. Compared with Gould and Prothero's (1975b:42–43) framework, return to the community after one or more years is not considered circulation. For nine months during 1975–76, a total of 17,407 moves were recorded for the 738 people studied in Kadirojo and Piring. Of these, ninety-two per cent were commuting, eight per cent circulation, and far less than one per cent migration, although migrants kept strong ties with the village and often planned to return. The extent to which ecological context can affect mobility is seen in the ratio of 7.8 commuters to every circulator in Piring, in contrast to 1.6 to 1.0 for Kadirojo, a difference that highlights the inability of villagers in the former to cultivate rice during the dry season.

These results document one problem in measuring different forms of mobility. If a move qualifies as migration, then the number of both moves and migrants will be identical. By definition, however, the number of moves per commuter or circulator is far greater than of persons involved, so that counting the moves made rather than participants inevitably inflates the data in favour of commuting and circulation. This bias occurs frequently in statistics of population movement based on registration data (Goldstein 1964) and underscores the need to present field counts for both moves and movers. Recognizing this problem, Mantra also classifies the study populations by movement status, from which it is found that only twelve and sixteen per cent of those in Piring and Kadirojo did not commute and/or circulate during the period of research.

Several marked differences emerge in the purposes of and destinations for commuting and circulation. In Kadirojo, half of all commuting was for wage work, another twenty-eight per cent for trade, and a further fourteen per cent for education. In Piring, by contrast, earning money accounted for thirty-nine per cent, trade another fifteen per cent, and schooling about twenty-four per cent. Whereas about a quarter of Kadirojo commuters went to an urban destination, thirty-two per cent of those from Piring did so – and even more if trading or schooling were involved. Clearly the purposes of commuting, as well as of circulation, are much affected by differential opportunities in the village and the accessibility of rural and urban places. In both *dukuh*, at least four fifths of those circulating had returned within a week of departure and all but two or three per cent within one month. Neighbouring hamlets and other rural places were the dominant destinations, since major urban centres were far distant and lesser cities offered few job opportunities. This, together with the importance of rural destinations for commuting, further emphasizes the need to include agricultural areas in any assessment of people's mobility.

For central Java, as generally for southeast Asia, both economic and social factors underpin circular mobility as a compromise between leaving the community permanently or not moving at all. The distinct advantages of maintaining close social ties, continuing to work the land, and the lesser cost of *dukuh* living complement those of taking jobs in either city or rural locations or of travelling there for education. As in Thailand, most of the *dukuh* born who work outside eventually return at retirement. Again, lower living costs make the village more attractive than the city, especially on a reduced or eliminated income, while close ties to kin and community as well as the high value placed on living and dying in the ancestral home exercise a strong pull on those who have been absent for long periods.

In contrast to detailed surveys in villages of origin and/or urban destination, Forbes (1978) and Jellinek (1978) consider particular occupations. Forbes (1978) studied trishaw drivers in Ujung Pandang (Fig. 18.1), who constitute about ten per cent of a work force of 108,000. Nine out of ten were born in rural areas and fifty-seven per cent had lived in the city less than a decade. Many drivers maintained close links with their local communities: four fifths returned at least once a year and thirty-six per cent five or more times. Seasonal labour demands in rice production were an important reason, and sixty per cent returned for planting and over two thirds for harvesting. Money remitted by at least one third of the drivers during the previous year was essential to agriculture and used to purchase land or livestock, in house construction, and for ceremonial purposes. As elsewhere, few trishaw drivers were accompanied to Ujung Pandang by their families, who remained behind to work the rice fields.

Overall, Forbes' research stresses the interdependent ways in which circulation links urban and rural areas. Entrance into both economies was relatively easy, the flexibility in labour requirements of both sectors permitted the absorption of more migrants and greater sharing of workload and profits among drivers, and seasonal differences in the agricultural cycle among regions enabled alternating patterns of movement between city and village. The fact that circulation is a mechanism to cope with the marginal livelihood provided by rural environments means that any success in controlling trishaw drivers would lead to greater unemployment and overcrowding in the informal sector. Economically, permanent return to the village is not a viable option.

The study by Jellinek (1978) focuses on ice-cream traders in Jakarta, who circulate between villages in central Java and the *pondok*: a facility shared with fellow workers on which they rely for lodging, meals, equipment, and supplies. All see Jakarta as little more than a place to earn income, for rural life is what really matters. This system is mutually beneficial for both the *pondok* owner and the village-born trader, as well as to their communities of destination and origin. Thus the increase in circulation is not only one of volume but also greater substitution for permanent migration, given deteriorating village conditions and the realities of city life: shortage of housing, high living costs, and reliance on contacts for local jobs. Like Hugo and Forbes, Jellinek argues that ice-cream traders, because they lack education, contacts, and funds for bribes, find it difficult to shift into the formal sector, even if this is seen as desirable. For these reasons, government efforts to curb traders would markedly diminish rural incomes, while technological innovations (for example, in soft-drink production and distribution) would destroy limited opportunities for villagers to earn money and greatly add to their hardship.

Research at the village level, of urban destinations and of selected occupations, reinforces Mantra's (1981:169) conclusions: 'Economic needs . . . underlie population movement but do not fully explain it. Social and kinship ties, the desire for continued education, and the perception of opportunities at other destinations are often an integral part of the decision-making process'. When economic conditions do lead to a move, it is usually to a nearby place so that continued contacts, if not actual residence, can be maintained in the village. There is a high premium placed on social ties as well as, wherever possible, the chance to earn rural incomes. For the Indonesian hamlets and villages, as for others in southeast Asia, Africa, and Melanesia, a major characteristic is their bi- or even multi-local residence patterns. Thus, as Chapman (1977:3) notes, local communities remain comparatively stable in their demographic composition but consist disproportionately of individuals in fairly constant motion between the village and other rural and urban locations.

Analogies with Melanesia

Most attempts at cross-cultural analysis are plagued by questions about comparability of concepts, units of analysis, availability and measurement of data, scale, and socioeconomic condition. These difficulties are magnified when the issues being investigated are novel and, furthermore, are being pursued from various perspectives by social and behavioural scientists of several disciplines. Yet, as attested by the papers in this volume, the rich body of field data available for Melanesia has permitted comparative assessments of population movement, yielded valuable insights about its dynamics, and suggested research models for use in other third world regions. In fact, some of the earliest studies on Melanesian circulation stimulated research in southeast Asia, especially in Indonesia and Thailand (cf. Ward 1980:128).

Both the Melanesian and southeast Asian enquiries document not only the extent but also the theoretical and practical significance of circulation in the third world. The opportunity to examine comparatively some of the assumptions and conventions of movement research should lead to more realistic assessments of the role of mobility in demographic change and how it facilitates diverse strategies in the adjustment of resources to socioeconomic needs. It also should help elucidate the extent to which policies of redistribution can build upon the variety in people's mobility so as to ensure more equitable conditions in rural and urban communities.

Given a literature that suggests the opposite, what is striking about circulation is its pervasiveness throughout southeast Asia and Melanesia. Despite substantial differences in population size, geographic area, natural environment, prior political history, socio-economic structure, and settlement systems, levels of mobility in both regions are substantially higher than those revealed by census data, and circulation accounts for a substantial proportion. In ranging from daily commuting on the one hand to residential change on the other, movement may well function as an adjustment factor in population dynamics, by which individuals acknowledge the combined advantage of urban places of temporary destination and rural places of more enduring residence. Upon returning to local communities, short- and long-term circulators may spread urban benefits, income, and knowledge, and thus have greater impact than if a group of villagers were to relocate in an urban centre (Lipton 1977:227).

Throughout southeast Asia, as throughout Melanesia, circulation has long been an important element in people's mobility. What has changed in both regions, and what undoubtedly varies from one to another, is the particular combinations of movement and the processes that account for its constant change. One cannot automatically assume that circulation patterns for southeast Asia replicate those for Melanesia and Black Africa. Much refinement in concepts and methodology is needed before effective cross-cultural comparisons are possible between the form and/or extent of movement and the kinds of interactions between movers and their places of both origin and destination. Is there a point in urban growth or in demographic pressure on rural resources at which circulation is replaced by more massive rural exodus, as has happened in eastern Fiji (Bedford, this volume)? Or, on the contrary, does the condition of urban places result in diminution of migration and greater resort to circulation? From the standpoint of both successive generations and the persons involved, do different populations

402

follow parallel patterns in declining ties with places of origin (cf. Frazer, this volume), and how far do such ties explain the persistence of circulation?

Evidence from Fiji (Bedford, this volume) suggests that the smaller the island population, the greater the structural imbalance caused by slight changes in movement, births and deaths. Do such relationships hold when the resource structure and socioeconomic opportunities for return are different, as in many southeast Asian countries? Under what specific circumstances is circulation mutually beneficial to places of both origin and destination? What processes determine when circulation becomes a viable alternative to outmigration, as in so many Melanesian societies, and when it fails to serve such a positive function – as suggested by Sutlive (1977) for the Philippines? In short, both the southeast Asian and the Melanesian data stress the need for careful diagnosis of the complex links between circulation and socioeconomic change.

Field research in both Melanesia and southeast Asia similarly confirms the need for prospective studies of sufficient duration to document whether transitions do occur from temporary to permanent mobility and by what process apparently permanent transfers are reversed. Studies of individual villages have used longitudinal designs to assess mobility behaviour, particularly if examining the full range from residential change to frequent journeys for work, schooling, commerce, medicine, and entertainment. In Chapman's (1975) research on south Guadalcanal and Mantra's (1981) of central Javanese hamlets, registers were maintained for five months in the former and nine in the latter. Records were kept of individuals who left the study communities for more than a specified number of hours (south Guadalcanal: 24, central Java: 6), as well as about their planned destinations, the routes taken, and the reasons for departure or return. Although such longitudinal designs can only be implemented in villages where the control of the exits is simple and the total population being observed comparatively small, they provide challenging insights about the diverse nature of people's mobility.

A variant of prospective registers was implemented by Conroy and his associates in a ten-year study of school leavers in Papua New Guinea (Conroy 1972, 1975, 1977; Conroy and Curtain 1973; Curtain 1975). In 1968, details were obtained about employment aspirations and intended place of residence from 1,200 students in their first year at selected primary, secondary, and vocational schools. Former primary students were contacted by mail in 1969–70 and March 1971, by which time half of them had left their villages. In June 1972, fifty of those who had moved to Port Moresby (Fig. 2.1) were interviewed and yet another effort, in January 1974, located most of the original sample. Although concerned with a select segment of the population, this research monitored how far mobility intentions were put into practice and thus presented field evidence, rather than informed opinion, about a basic distinction between circulation and migration.

Research priorities and policy implications

This review has focussed on voluntary movements stimulated by the wide range of factors that operate within the multivariate context of local situations. In southeast Asia, as in other parts of the third world, involuntary moves also are important. Hugo (1983) notes that most studies of refugees consider international flows and overlook the counterpart streams within countries, such as Malaysia, the Philippines, Indonesia and, more recently, Vietnam and Kampuchea. Given that these are forced moves reflecting war or scarcity, many refugees probably intend to return to their places of origin. How many have done so or did not, what contacts those not returning maintained with natal places, and the role of government policies in affecting the rate of both return and onward movement are all questions requiring intensive research.

As suggested by the studies in southeast Asia, greater theoretical attention must be paid to the circumstances under which populations resort to various forms of mobility. Zelinsky's (1979) reassessment of the mobility transition notes that complex and diverse patterns exist within a single region, and even within its countries, as a product of history and of fluctuating interactions within and between different places. Building upon field results available for southeast Asia, Africa, and the island Pacific, there is need to specify the historical and contemporary contexts within which circular mobility occurs in traditional societies and how far they are linked with the modernization process.

As Hugo (1983:86) argues, 'there is clearly a pressing need to conceptualize rural-urban mobility in such a way that *all* movers, not just a selected subset of them, can be distinguished from stayers [and] meaningful distinctions ... made between the various types of movement'. Degrees of commitment to places of both origin and destination is one approach, measures of which in west Java would include whether the family of procreation accompanies the mover; whether land, a house, or some other property remains owned in the village; if money or goods are remitted, and as what proportion of total income; what political and social roles the mover plays in the home village; and, finally, how frequently absentees return to it (ibid:87-9). If the distinction between temporary and permanent mobility is to have meaning for comparative research, then attention must focus upon such issues as well as those referring to time and space. Questions ought to be included in censuses and national surveys to provide the context for indepth community studies, whose typicality otherwise is open to challenge, and to facilitate comparative analyses of population movement with respect to differences of scale and socioeconomic condition.

Use of registers for mobility analysis is another possibility, particularly in those southeast Asian countries where some form already exists. In comparative research, assessments must be made of what kinds of movement are legally included and to what extent people report them. Coverage may be reduced markedly if some segments of the population view a registry system as an effort at administrative control. Whenever major efforts to assess population dynamics provide the opportunity to introduce such systems on national or sample bases, then efforts should ensure that equivalent attention is given movement in addition to births and deaths.

From the standpoint of the policy maker, many questions may be raised. To what extent are rural pressures relieved by various forms of movement, alone or in combination, and how far does short- or long-term circulation reduce problems that cities like Bangkok, Manila, and Jakarta would face were all movers to become permanently resident? Does the interchange between urban and rural places, particularly that of returning villagers, confer positive benefits in the form of new ideas and behaviour and through money and goods remitted? How crucial, in fact, are such remittances for the basic needs of rural communities and to what extent do they result in the more equitable distribution of urban incomes? Does the circulation of such elites as teachers and government officials enhance the spread of urban values or create a disdain of rural ways, thereby encouraging either the modernization of rural populations or the exodus of villagers into the cities as proletariat or into the jungles as dissidents?

Should southeast Asian governments, as Elkan (1976) and Hugo (1985) have proposed, foster circular mobility so as to lessen demands on urban services and contribute more visibly to the socioeconomic development of rural places? Would such encouragement provide the dual advantage of maintaining strong social ties with rural communities and deriving economic gains from jobs available in towns and cities? How do planning efforts, as for improved systems of transport and education, affect levels of migration and circulation and what is their impact upon socioeconomic growth? Should southeast Asian governments reach beyond the gradually evolving patterns of urban change and population movement to adopt policies designed to create growth centres within their national boundaries? Such centres, if established, might alleviate the pressures upon existing city and village communities, as well as create even further opportunities for interchange between rural and urban places.

The heavy reliance upon circulation by rural people throughout southeast Asia signifies the intense interaction between urban and rural places. It further indicates the need for integrated development planning: integrated in the sense that it attempts to take concurrent account of the aspirations of both rural and urban populations and to stress the

405

linkages rather than the differences between them. Because of bi- or even multi-focal residence patterns, many southeast Asians are neither exclusively rural nor exclusively urban. The interests and contributions of such people, as well as of the rural or urban communities of which they are part, are best met by policies that address the needs of not only rural and urban locations but also those who move between them for varying amounts of time.

Acknowledgments

Research for this paper was facilitated by a Ford Foundation grant to the Population Studies and Training Center at Brown University. A different version appeared as Goldstein (1978), which includes a fuller assessment of most studies reviewed here and on which the comments of Judith Strauch proved most helpful.

19 The context of circulation in West Africa

R. Mansell Prothero

Although numerous, studies of mobility in tropical Africa are almost inevitably selective (Gould 1977a). They concentrate upon particular types of mobility associated with particular groups of people, in limited parts of the continent and for limited spans of time. This is perfectly understandable, since time and resources are limited yet the possibilities for investigation endless. Such selectivity, however, had led to bias and imbalance, as illustrated by two examples.

First, there has been preoccupation with rural-urban movements to the neglect of mobility in rural areas. While towns and cities of tropical Africa rightly demand concern for their present problems and future prospects, greater attention should be given to the countryside from which their migrant populations originate. Most peoples in tropical Africa continue to live in rural places, few of which are yet experiencing major depopulation, consequently the mobility occurring within them is significant for social and economic change. In a study of inter-regional mobility in Uganda, Masser and Gould (1975:71) state:

> It is precisely this important element in overall mobility that is identified in the birthplace data, for the four major flows — Kigezi to Ankole and Toro, West Nile to Bunyoro and Bukedi to Busoga — and several of the other large flows may be held to contain *large elements of spontaneous rural/rural migrants.* (Italics mine.)

Little has been done to identify, let alone study such mobility.

Apart from the near obsession of economists with rural-urban migration, a more urban than rural emphasis is found in two important works on mobility in west Africa (Amin 1974) and east and central Africa (Parkin 1975), both of which include essays from a range of social science disciplines. In contrast, a seminar convened about the same time on internal migration in Nigeria (Fig. 19.1), which took place in that country and largely considered papers by Nigerians, was balanced equally between rural and urban considerations (Adepoju 1976); while in November 1977 a discussion at the Afrika Studiecentrum, Leiden, focused on migration and rural development in tropical Africa (van Binsbergen and Meilink 1978).

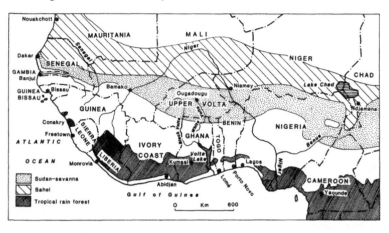

FIGURE 19.1 *West Africa*

The second bias has been in the attention given migration at the expense of circulation, though in some instances the former term has been used to describe the latter phenomenon: as in mobility research in northwest Nigeria (Prothero 1957, 1959) and elsewhere in west Africa (e.g. Rouch 1957, 1960; Berg 1961, 1965; Skinner 1960, 1965). Overall, there has been far too little concern with circulatory movements that involve no permanent change in place of residence. Where such have been considered, the focus has been on the circulation of labour, particularly in the context of socioeconomic relationships that exist within and between source and destination areas in the plural societies and dual economies of south-central and southern Africa (e.g. Mitchell 1959, 1961b, 1969; Garbett 1975; Garbett and Kapferer 1970).

Typologies of mobility

Thirty years' experience has underlined the critical need to recognize the immense range of mobility in tropical Africa to comprehend both its totality and the ways in which its different forms are related to one another. For example, where the movement of people is a factor in the transmission of disease and in planning public health programmes for its control and eradication, any form – and often many forms – of mobility may be relevant (Prothero 1961, 1964, 1965, 1977). Daily journeys over short distances may bring those involved into contact with disease vectors; seasonal movements over short or long distances, from one set of ecological conditions into or through others, may result in differential exposure to infection; longer-term and often long-distance transfers may present physical and mental problems of adjustment in new psycho-socio-economic environments. Frequently of crucial importance in these situations are details about the size, spatial and temporal character of such movements, and how they vary among groups of people and in different areas.

Such practical requirements have influenced wider thinking on mobility and how its various forms can be categorized and integrated. Most generally, a three-fold distinction of movements has been proposed: those of the past that, while no longer practised, somehow influence contemporary patterns of population distribution, density, and population-resource relationships; traditional forms that continue to the present; and more contemporary ones which have evolved during the present century (Prothero 1964, 1968). Revised thinking has led to questioning the validity of this trilogy, in particular the need to relate movements in each of these categories to one another, to identify analogies in present mobility behaviour with the past, and to investigate continuities from one period to another.

Two detailed typologies of contemporary mobility have been prepared. The first was a simple attempt to identify and describe the various forms associated with different groups of people and to assign those forms to two categories of primary motivation: economic and noneconomic (Fig. 19.2A). Even in a descriptive typology, however, it would be naive to overlook the highly complex nature of causes of mobility which more sophisticated attempts at explanation have failed to comprehend. Previously, this descriptive typology has been published only in the works of others: verbally by Hance (1970:163ff) and in diagrammatic form by Mabogunje (1972:41). It has many limitations. Other than through the activities of groups involved, there are no realistic parameters for distinguishing between different forms of mobility, nor are circulatory ones separated from the migratory even though this is fundamental to understanding the movement process in both tropical Africa and many parts of the third world.

The outlines of a second typology (Fig. 19.2B) were set down during the early 1970s preparatory to a wide-ranging survey of mobility in tropical Africa that aimed to systematize existing knowledge. The typology was an initial and essential step in this direction. Parameters of space and time were chosen to provide the basis for classification (see Gould and Prothero 1975a), which was subsequently applied to groups involved in different kinds of movement (Gould and Prothero 1975b) and to their exposure to various health hazards (Prothero 1977). Spatially the typology distinguishes between rural and urban, while recognizing that this is not clear cut in the real world and that the relationships between the two are often close. Temporally the major distinction is between circulation and migration.

Moves that involve no permanent change in place of residence (migration) are either regular, made deliberately without immediate coercion or compulsion although perhaps not entirely voluntarily; or irregular, occurring involuntarily as a result of natural or human factors

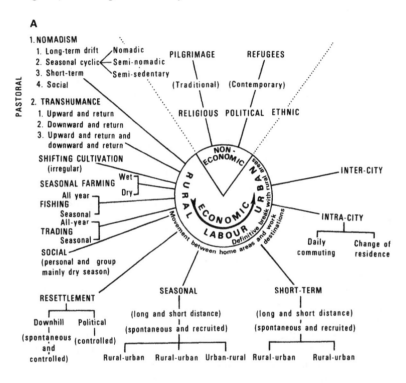

FIGURE 19.2 *Typologies of population mobility in Africa*

separately or in combination. In the former the migrant can make a decision concerning change in residence place but in the latter cannot.

The concept of circulation in the typology extended far beyond that of labour circulation discussed notably by Mitchell, amongst others, for southern Africa (Mitchell 1959, 1961b, 1969; Garbett 1975; Prothero 1974). It evolved in accord with Zelinsky's ideas on circulation, which in turn were developed independently of those of Mitchell and his associates. Zelinsky's (1971:225-6) definition of circulation has been widely quoted; according to him, the term refers to 'a great variety of movements, usually short-term, repetitive, or cyclic in nature, but all having in common the lack of any declared intentions of a permanent or long-lasting change of residence' (cf. Chapman 1970:181-3).

It is recognized that there are many problems inherent in defining 'any declared intention', both ante- or post-facto (cf. Ward 1980; Goldstein this volume). But from direct experience in Africa, it is quite clear that people leave their home places for varying periods of time, from less than twenty-four hours to many years, and then return to them. Intent to return is of course less equivocal for shorter periods of absence. Possibly daily circulation should be designated commuting (cf. Hugo 1975), though the latter has the connotation of travelling to work in modern sectors of socioeconomic activity (the office, factory, or school) and thus ignores the many traditional kinds which are equally relevant (cf. Singhanetra-Renard 1981:158-9). 'Declared intention' presents no special problem for periodic circulation, involving absence for more than twenty-four hours but less than a year, or seasonal circulation, which is periodic in nature but can be separately distinguished.

When it comes to long-term circulation, where people are away for at least one year, intent to return or not is more difficult to determine. Again, a great variety of evidence from all parts of tropical Africa indicates that links are maintained with home areas by those away in towns and elsewhere, who might well have been expected by outsiders to have severed their connections. Nabilla (personal communication), who studied the mobility of his own people in northeast Ghana, writes that some Ghanaians are designated 'NGBs' (Never Go Backs): those who consider they have severed connexions with their local communities, yet who 'will in the end go back, either their bodies for burial, or even without this return home would be recognized by those who would perform funeral celebrations in places of birth whether or not their bodies were returned'. In tropical Africa, total severance of anyone from the natal place is very difficult and this inevitably makes the distinction between circulation and migration more problematic.

The typology is sufficiently flexible to accommodate a great variety and range of mobility. It can accommodate movements that are widely recognized (labour circulation) as well as help identify the less well

411

known (short-term and short-distance journeys to work, to schooling, or to rural health services), to which greater attention should be paid in rational planning. The basic concern with forms of mobility also permits them to be related to different purposes and consequences. Thus the sixteen categories of circulation identified can be related to health hazards (Table 19.1) and reveal 'different effects in the exposure of population to disease, in the transmission of disease, and in the development of programmes for the improvement of public health' (Prothero 1977:264-5). Although basically descriptive, the typology consequently has some analytical value with respect to the influence of space and time but limited explanatory power. From its academic reception, however, it has assisted other scholars working in Africa and, for comparative purposes, those undertaking research in third world countries of Asia and Latin America (eg. Hugo 1975; Skeldon 1977; Mantra 1978).

TABLE 19.1 Typology of circulation in tropical Africa, with examples of associated activities and health hazards

Space	Time			
	Daily	*Periodic (24 hours- 12 months)*	*Seasonal (One or more)*	*Long-term (12 months- several years)*
Rural-rural	Cultivating* Collecting (firewood, water) (1)	Hunting (1)	Pastoralism (1) (3)	Labouring (1) (3)
Rural-urban	Commuting (1)	Pilgrimage (1) (2) (3) (4)	Labouring (1)	Labouring (1) (2) (3) (4)
Urban-rural	Cultivating (1)	Trading (2)	Labouring (1)	Trading (2) (3)
Urban-urban	Intraurban Commuting (3)	Pilgrimage (2) (4)	Trading (2)	Official/ commercial transfer (4)

Some examples of health hazards

(1) Exposure to diseases from movements through different ecological zones (e.g. malaria, trypanosomiasis, schistosomiasis, onchocerciasis)

(2) Exposure to diseases from movements involving contacts between different groups of people (e.g. smallpox, poliomyelitis)

(3) Physical stress (e.g. fatigue, undernutrition/malnutrition)

(4) Psychological stress-problems of adjustment

* Examples of activities/groups associated with different categories of circulation.

Source: Prothero (1977:265).

General statements on West African mobility

During the last decade, Mabogunje (1972) and Amin (1974) have made general statements about mobility in West Africa that stand in marked contrast, although both are somewhat limited in what they say about circulation. Mabogunje, in a book titled 'Regional mobility and resource development in West Africa', takes a determinedly historical view and argues that movements have benefited not only the participants but also their host and home communities. He stresses the need

> . . . to dispel popular misconceptions, not only do they distort reality, but they also *prevent an appreciation of historical continuity in the development of West Africa. The time perspective allows us to see many of the consequences of regional mobility as a continuous adaptation and modification of traditional norms and institutions to meet new social and economic needs* (Mabogunje 1972:7; Italics mine).

With this perspective, Mabogunje considers the colonial period as one which 'saw the consolidation of situations that stimulated trade, travels and migrations among different ethnic groups to an unprecedented level' (p. 8). He is also explicit that many of those moving away return eventually to their home places: 'mobility need not imply any intention of abandoning former homes. Indeed if any intention can be deciphered it is that of making the former home a better place to return to periodically or at the end of a foreign sojourn' (pp. 15-16).

Nowadays, seasonal or short-term flows of labour tend to distract attention from the long-established, more diverse, and directionally more complex movement of traders, farmers and fishermen, which still continue. Mabogunje notes these latter as being associated with seasonality and distance, both given factors of the west African environment. Peoples of the interior, located within environments of marginal potential and subject to high transport costs that limited their agricultural production, were economically disadvantaged. To correct this, 'a large number find recourse in migrating to the more prosperous farms and plantations of the south' (Mabogunje 1972:53). The distinct seasonal contrasts of the west African interior, a short wet and long dry season

> . . . made such movement easy to contemplate for most of the people . . . it involved minimal disruption of the basic economic activity in their home area . . . The long slack period . . . from December to March provides an invaluable opportunity to earn extra income each year by migration . . . (ibid, p. 53).

In general, Mabogunje is emphatic about the positive advantages of regional mobility during the colonial period as a decisive factor in resource development.

413

A quite different ideological position is taken by Amin (1974), in a long essay that introduces 'Modern migrations in West Africa'. His concern is with recent migratory movements, but the historical context within which these are considered is very limited. 'Migrations have already passed very largely over a first phase, characterized by the preponderance of migrations of short periods (under one year) to a mature phase characterized by permanent migration' (Amin 1974:69). For the greater part of West Africa, he maintains there has been a massive draining of population from the interior and estimates that 4.8 million persons represent the likely 'demographic contribution of the interior to the coast during fifty years of migration' (p. 72). To these 'permanent migrants' must be added seasonal movers, who previously were much more numerous.

In his analysis of both permanent and temporary mobility, Amin strongly criticizes

> The environmental approach to migratory phenomena . . . carried out within a theoretical framework based on the hypothesis that the 'factors' of production (labour, capital, natural resources and land) are given *a priori* and geographically distributed unequally, the latter itself being taken *a priori* (p. 85).

The distribution of these factors is 'the result of the strategy of development', so that 'economic choice (so-called 'rational') and notably the decision of the migrant to leave his region of origin is then completely pre-determined . . .' (pp. 88-9). In other words migrants had no alternative but to respond to the constraints of the colonial capitalist system, which in turn deliberately developed some parts of West Africa but not others, consequently these latter 'have been conditioned more than others to meet the need for proletariat'.

These two very different views of population movement, one essentially negative and the other positive, provide further background for the following discussion. Amin, who makes only passing reference to the work of Mabogunje, sees most migrants today as dispossessed, and therefore oppressed by an economic system established throughout most of West Africa in the early decades of this century. Mabogunje, while not claiming that all mobility is good, inevitably of benefit to those who move and to all sources of origin and destination, sees it as a phenomenon which existed in the past, is rooted in traditional custom, and has changed in sensitive response to opportunities and potentialities — but without major social and/or economic disruption and in general with substantial benefit.

Circulation in West Africa

Social and behavioural scientists have given limited explicit attention to circulatory movements in West Africa. Taking as a point of reference the different kinds identified by a time-space typology (Table 19.1), little is known about the dimensions of daily circulation in rural areas associated with economic practice and domestic need. Rather more information is available about pastoralists than other groups (e.g. Stenning 1957) and of the circulation between residence and farms (daily, periodic, seasonal) in southwestern Nigeria (Ojo 1970, 1973). Of the daily movement to and from wage work in urban areas, there is awareness of its growing volume and of the problems created for over-burdened transport services, but few quantitative details of the numbers involved and their space-time parameters. Investigation of the much wider range of daily, intraurban movements is only beginning (Main 1981).

The situation is very similar for most forms of periodic, seasonal, and long-term circulation. Trading movements have received some attention, although largely indirect through research on market period-icity (e.g. Meillasoux 1971; Smith et al. 1972). The tantalizing but limited insights revealed by these studies have only emphasized how much more needs to be discovered, not only for its intrinsic interest but also for its relevance to important issues of socioeconomic change. Undoubtedly, best documented is the circulation of labourers, who travel varying distances in search of wage work and stay away from their home places for varying periods. The rest of this paper concen-trates on these movements, because of their historical depth, contem-porary relevance, and comparative interest for both tropical Africa and Melanesia.

Circulation studies made in West Africa during the 1950s and 1960s did not appreciate sufficiently the continuity of mobility from past to present, nor the significance of links maintained between home places and destinations set within the particular regional context. The ideas applied in West Africa were derived from classic work on 'labour migration' begun in the 1940s, mainly by British-trained anthropolo-gists, and continuing through to the 1950s and 1960s. Their regional focus was southern Africa, notably the plural societies and dual econ-omies of what were then Northern and Southern Rhodesia, together with South Africa. A recurrent theme, that the circulation of labour was a consequence of socioeconomic disequilibrium, had been articu-lated by Godfrey Wilson (1941–42) in his pioneer study of migrant labour in Broken Hill (now Kabwe, Zambia; Mitchell 1985). These ideas were important for West African research in pointing to common features of labour movements for all of tropical Africa. They did result, however, in too little notice being taken of the particular socioecon-omic, cultural, and political settings of wage-labour circulation in West Africa and how these differed from elsewhere in the continent (for southeast Asia, cf. Goldstein this volume).

From south to north, West Africa exhibits a succession of east-west belts of environments that range from tropical rainforest through savanna to desert (Fig. 19.1). The environmental features of each belt are given, cannot be ignored, were relevant in the past, and will continue to exert a major influence into the foreseeable future. Viewed negatively, the zone of the Sahel immediately south of the desert, and the adjacent northern parts of the Sudan savanna zone, offer limited potentials and great risks for agriculture. Viewed positively, this east-west zonation contributed to the exchange of agricultural products through trade. Consequently, both negative and positive circumstances encouraged human interchange between these various zones long before Europeans first penetrated West Africa, not only before colonial rule and economy were introduced but also before wage-labour movements evolved (e.g. Meillasoux 1971).

Indigenous urban places were long established in West Africa, in distinct contrast with towns created more recently by and for Europeans in south-central and southern Africa. Traditionally, in the former, urban and rural areas were closely interactive in their society, economy, and polity: rural and urban places were extensions of one another and mutually dependent, so that rural-urban relationships were unified and not dichotomous (e.g. Goddard 1965; Miner 1965; Mortimore 1972). While the new towns founded by colonial powers in West Africa exhibited different features from the indigenous ones, nonetheless these introduced centres could be quite easily comprehended within traditional concepts and people's experience.

Except for plantations in southern parts of the Ivory Coast and in the Cameroon (Fig. 19.1), throughout West Africa little land was alienated to Europeans. Producing crops for cash and export thus remained mostly in indigenous rather than expatriate hands, although rising demands for their cultivation were stimulated directly by the external market. The output of traditionally-cultivated crops, like cotton and groundnuts, was increased through extending and intensifying production, and the cultivation of introduced ones, such as cocoa and coffee, expanded with the response of indigenous entrepreneurs to market demands. The latter process was especially pronounced in Ghana and Nigeria (e.g. Hill 1963).

With these very important facts in mind, population mobility in West Africa may be characterized as involving continuity between times past and more recent. Of major and continuing importance for circulation is the maintenance of links between areas of origin and destination of those who move. Based on research for other parts of the continent, scholars have been inclined to assume that over time a break occurred in mobility behaviour in West Africa. Recent studies and some reassessment of earlier work underline the need to acknowledge evidence that indicates continuity rather than simply to assume discontinuity.

'Strange farmers' in Senegambia

In research undertaken in the early 1970s, Swindell (1977, 1978) considered the movements of the *navétanes*, or 'strange farmers' of Senegal, who replace and supplement family labour in the groundnut-growing areas of the Gambia (Fig. 19.1). They are involved in a host-client relationship, working for Gambian farmers for two to four days each week and growing groundnuts for the rest of the time on plots made available to them. Numbering about 30,000, excluding dependants, they provide labour at critical points in the farming schedule. They are needed notwithstanding mechanization which, while enlarging the area under cultivation, has also increased labour demands for weeding and harvesting. Holdings with *navétanes* cultivate on average about one-third or more land under groundnuts than those without.

Swindell (1979) notes the persistence of this system, providing evidence for its existence over 150 years, thereby preceding the main impact of colonial rule together with the establishment of the colonial-capitalist economy. In a special survey of 'strange farmers' undertaken during 1974–75 as part of the Census of World Agriculture, about a quarter of those interviewed (250 of 1,024) were Gambians, the remainder principally from Senegal, Mali and Guinea, and smaller numbers from Guinea Bissau, Mauritania and Upper Volta (Fig. 19.1). Their mobility reflects primarily economic factors, being variously influenced by pressures in home areas (high population density, arduous agricultural work, an ailing economy) and attractions of the coastal regions, of which Gambia is part: the efficient marketing system, favourable exchange rates, and availability of goods for purchase. During the late 1960s, the Mali government mounted an *Operation Arachide* to extend and improve production of groundnuts, in a positive desire to correct some of the economic imbalance between its land-locked country and the coast; but even this failed to stem movement to Gambia. A decade earlier, comparable attempts in Guinea had been similarly unsuccessful.

While farming work in Gambia extends from March to the year's end and is concentrated in the wet season, many *navétanes* have been away from their home areas for more than one season. Having made an initial move, they then alternate between wet-season agriculture in Gambia and various forms of dry-season employment in Gambia and adjacent parts of Senegal. This alternation involves what might be termed 'circulation within circulation', since most migrant workers maintain close links with their natal communities. They remit money and goods to support families (just over 50 per cent), save for bride price (29 per cent), and accumulate capital for trade or other activities of the informal sector (14 per cent).

Swindell's research is important in demonstrating the complexity of a mobility system that has operated over a long period of time, has continued despite administrative attempts to make the areas from which migrant farmworkers come economically more attractive, and has withstood such changes as the widespread use of oxen for the ploughing of groundnut holdings. While evidence exists of a shift from seasonal to long-term circulation, only one-fifth of those questioned said they wished to remain permanently in Gambia should the opportunity occur. *Navétanes* do not seem trapped by the constraints of a broader ecological and economic system that offers no flexibility. In the wet season they can work at groundnut cultivation, and with alternative wage employment in the dry season they have 'an ideal method for maximizing opportunities for income formation'. This is their desire, for 'the Gambia is a relatively prosperous economic enclave which attracts migrants and which presents an escape from bureaucratic situations *for a potentially highly mobile population, who are voting with their feet*' (Swindell 1977: italics mine).

Labour circulation in Upper Volta

This poor land-locked country has been the single most important source of labour migrants in West Africa. These are mostly young men, who leave to take up employment available in cocoa production, share cropping, construction work and forestry in Ivory Coast and Ghana (Fig. 19.1). Field research on these movements in the 1960s and 1970s suggested that, through political and economic measures (forced labour, military service, taxation), the French colonial system deliberately developed the Upper Volta as a labour reserve (e.g. Skinner 1960, 1965; Deniel 1968, 1974; Songré et al. 1974; Gregory 1974; Gregory and Piché 1978; Piché, Gregory, and Coulibaly 1980). Some men were recruited for wage-work in the Ivory Coast, but most moved there spontaneously, as well as to Ghana.

Skinner (1965:67), an anthropologist who made one of the earliest studies with reference to the Mossi of Upper Volta, wrote: 'Migrants try to ascertain whether more jobs are available in Ghana or in the Ivory Coast before they leave home, but few of them know exactly when they go off what kind of job they will obtain'. Overall, he considered that

> . . . labour migration in Africa is most properly seen as a post-
> European phenomenon. In most cases the first labour migrations
> were involuntary either because Africans were forcibly displaced to
> work on European projects, or because they went to work centres
> in response to the introduction of taxes which had to be paid in
> European currencies (Skinner 1960:376–7).

However, he did acknowledge some element of continuity between past and present by noting that 'these Africans associated with traditions of compulsory or tributary labour for the chief or specific types of social institutions often accepted forced labour and labour migration quite easily' (ibid:377).

By the second decade of this century, according to Skinner, the Mossi had evolved a seasonal pattern of absence – for most, an established practice of circulation with '. . . the failure of about 20 per cent of the migrants to return home in any one year' (Skinner 1960:381). Such wage-labour mobility was clearly ordered by economic factors and in 'itself brings little lasting prestige. It is not seen as a *rite de passage* or even an unusual act. The men who go away are not considered brave but poor . . .' (ibid:385).

In a more recent study in Upper Volta Finnegan, also an anthropologist, takes a much broader view of wage-labour circulation set within the context of a particular Mossi community. He identifies many forms of movement in precolonial and contemporary times, and relates the recurrent flow of labour to these. Finnegan (1976:9) makes plain that such an approach

> can provide data to support a different view of Mossi labor migration and its effect than has previously been expressed. My hypothesis is the following: international labor migration is only one manifestation (however striking) of how a Mossi village has used spatial deployment of its people to adapt to varying economic, social, and political conditions.

He argues, as does Mabogunje, for the 'too-little-recognized ability of people's subject to 'modern' forces to respond actively and creatively rather than passively' (p. 11), because their responses '. . . have clear roots in their adaptations to the demands and pressures of living in precolonial state society' (p. 13).

In precolonial times, Mossi states were comprised of local, semi-autonomous communities that often varied in their ethnic and linguistic origins. Given such autonomy and heterogeneity among the Mossi population, the membership of states themselves, as well as of local communities and their constituent lineages, was always open to change. Customary, even intrinsic forms of population movement included all or part of a lineage departing to found a new settlement when the old became overcrowded, as well as individuals as clients of chiefs and kings, as involuntary captives, as refugees, and in trade. At marriage, wives usually transferred to their husband's village, adults sought sociopolitical advantage by relocating and transferring their allegiance to particular kin, and children were exchanged between related households to ensure that each could function as an economic unit. Local society evolved means of dealing with all 'such movements, of preventing them from disrupting the village, and even of turning them to the advantage of the villagers as individuals and/or the total community' (Finnegan 1976:218).

Thus when the French colonial system subjected the Mossi to administrative and economic pressures, people were able to respond on the basis of established experience. Since at no level was their society rigid or fixed, its social field was extended with increasing transport (road, rail) and more efficient means of communication (radio) to include Upper Volta, Ivory Coast, and Ghana. Territorial mobility within this enlarged field was still with reference to the village whose members thereby obtained access to resources not locally available: alternative political systems and jurisdictions, access to jobs and opportunities for trade, education, and religious training. This situation continues today. Population movement is 'a positive institution enabling individuals and their villages to exploit resources otherwise unavailable to them' and, far from having a destructive impact on the local community, its main effect on 'social structure is to reinforce and extend it' (ibid:225).

Labour movements from northwest Nigeria (Sokoto province)

In light of further insight and experience, review of what was written twenty-five years ago (Prothero 1957, 1959) about 'migrant labour' from Sokoto Province, northwest Nigeria (Fig. 19.1), suggests that some points require different emphasis. Seasonal movements from areas in the northwest, where opportunities to earn money are limited, to those of more intensive indigenous agriculture and urban expansion within both Nigeria and beyond, can be seen as more closely related to traditional movements of *masu cin rani* (people who eat away during the dry season) that predate the establishment of the colonial capitalist economy. In environmentally marginal areas, in particular, the long dry season was a period of little agricultural activity. *Masu cin rani*, often involving whole families, certainly helped conserve food supplies in home areas through lessened demand by absence for several months each year, but the evidence is unclear to what extent this was a forced response to the meagre physical environment and low levels of agricultural production. By the third and fourth decades of this century, the combination of natural population growth and immigration from Niger (Fig. 19.1) to the north increased the pressure on some of these marginal areas in Sokoto province. Consequently there was greater impetus for dry-season circulation. Trade contacts were long-established with the south, where cash cropping was expanding which in turn produced demands for labour that coincided largely with the slack dry season in the agricultural calendar. In response, circulatory *cin rani* over relatively short distances came to extend much farther afield and could only be undertaken by active, young, adult males: *yan tuma da gora* (young men jumping with a gourd), so named from some of the limited equipment which they carried.

Unlike the French territories of West Africa, there was in Nigeria much less administrative pressure on people to move. Forced labour was limited and military service not compulsory; in addition, no major plantation areas existed in the then Gold Coast (Ghana) or in Nigeria for which labour had to be provided. Taxes were levied in Nigeria during the earliest phase of colonial occupation, the first years of this century in the northwest, but available documentation reveals no evidence that this created pressures to move. Colonial administrative reports for Sokoto province, for example, make no reference to long-distance seasonal transfers until the 1930s. These contrasts with francophone West Africa need to be pursued, provided that records can be discovered to sustain detailed enquiry.

Economic motives − seeking money and food and engaging in trade − undoubtedly were important to the emergence of long-distance movements from northwest Nigeria. But other factors also were involved and operated to both encourage people to return to their home places in circulatory fashion and prevent many from relocating permanently through migration. The precise nature of these influences during the 1950s remains unknown and unfortunately was not much investigated during the Sokoto research. In the southeast of the province, opportunities for resettlement existed in areas suitable for the cash-cropping of cotton and groundnuts, but villagers were little interested in these. The provision by the colonial capitalist system of both the incentive and the infrastructure to reduce the need for circulatory movement over long distance met with very limited positive response and thus paralleled the experience of Mali and Guinea administrations in respect of movement to the Gambia.

Since then, some changes have been noted about mobility within and from northwest Nigeria, albeit in rather fragmentary fashion. Research undertaken in the 1960s (Goddard 1971), but only partly published indicates that movement continued in large volume but over reduced distances; a high proportion of migrant labourers found work within Sokoto province but relatively few went beyond northern Nigeria. Such reorientation doubtless reflects changed circumstances in Nigeria − political and ethnic unrest leading in 1967 to civil war − as well as farther afield: economic depression, currency devaluation, and a rising antipathy to foreigners in Ghana.

Although severe drought in northwest Nigeria during the 1970s has had a major impact on the source areas of labour circulation, the environmental pressures and their socioeconomic consequences have not been monitored sufficiently to suggest how this has affected the nature and pattern of people's mobility. While longer-distance journeys to find employment have become reestablished, only limited information exists on the number and characteristics of those involved and it remains unclear whether this change represents a response to drought, the more stable political condition of Nigeria, or both. The nature of Sokoto movement will not remain static in the future, especially given the introduction of universal primary education and agricultural expansion through major projects of water-control, both underway or proposed.

These fragments of information underline the simple but important fact that conclusions about mobility cannot be drawn from time-specific data. Movement behaviour is so dynamic, so responsive and adaptive to changing circumstances, that meaningful statements can be drawn only from evidence collected over time. Unfortunately, there is very little such evidence presently available for any part of tropical Africa (Gould 1977b).

West African and Melanesian comparisons

The circumstances in West Africa are sufficiently different – politically, economically, culturally – from those in the southern parts of the continent to warrant different expectations of how and why various kinds of circulation evolved. Continuity in people's mobility from past to present is more likely, more evidence for which may be found if it is actively sought. Whilst there were changes in the nature of movement with the introduction of colonial rule and of elements of a modern capitalist economy, no major and concomitant hiatus occurred as in the plural societies of southern Africa. Throughout West Africa, indigenous political, social, and economic institutions of great longevity were affected by externally-generated developments but also ameliorated their impacts. Long-distance contacts linked north and south, for example, in complex marketing systems of trade and exchange. These and other traditional forms of mobility were sensitive and responsive to broader socioeconomic and political change in some instances adapting to new opportunities and contributing importantly to the expansion of a money economy. This ability of people's movement, in the aggregate, to meld both old and new still continues throughout West Africa – to the benefit of both their home areas and their destinations. Mabogunje's (1972) views thus accord more closely with the realities of existing evidence and environmental givens than do those of Amin (1974), although both interpretations need to be tested more rigorously and extensively in controlled and comparative research.

Regional variations in mobility, both traditional and contemporary, make any generalizations subject to qualification. More careful examination is needed, for instance, of how the differing colonial policies in French and British territories influenced the dynamics of movement. Since international boundaries are recent, one approach would be to look more closely at the linkages existing between places before and after they came under different political jurisdiction – for example, Upper Volta and Ghana, Niger and Nigeria – in the way Swindell

(1977, 1979) has done for the Gambia and Senegal, Mali and Guinea (Fig. 19.1). Mobility in West Africa cannot be considered meaningfully in terms of a particular country, even though recently in each the government has tended to emphasize its own distinctiveness, thereby causing movement between countries to become more restricted than in previous decades. Studies of migrant labour in the 1950s crossing between Niger, Nigeria, Upper Volta, and Ghana (e.g. Rouch 1957; Prothero 1957) dealt with a regional process of circulation that nowadays is termed 'international labour migration'.

Given the importance of recognizing differences in mobility behaviour between various parts of tropical Africa, and in West Africa in particular, great care is needed in attempting cross-cultural comparisons with Melanesia. There are basic differences between these two areas that inevitably affect the circumstances of population movement, of which scale is undoubtedly major. The land areas of West Africa are immense compared with those of island Melanesia, giving rise in the former to broad and conspicuous environmental contrasts. Even so, ecological differences between coast and highlands or wet and dry areas can be discerned on the larger parts of the Melanesian landmass, on the Papua New Guinea mainland and the biggest islands of Fiji, New Caledonia, and the Solomons. Size influences distance and therefore accessibility: the distances over land in West Africa parallel those by sea between the islands of Melanesia.

The effect of such distances on accessibility has varied over time, and although apparently more restricting in Melanesia has not markedly inhibited interisland mobility within the past century. Until very recently, movement in West Africa was almost entirely on foot, subject to the similar constraints as travelling beyond the coastal littoral of the larger islands, but not as absolute as with the sea barriers of Melanesia. More efficient forms of transport, particularly by road, are nowadays more widespread throughout West Africa and have increased both the ease and volume of movement. With reference to the Frafra of northeast Ghana, Hart (1974:329) writes vividly of the 'erosion of distance between Ghana's regions . . . visits in either direction are commonplace and Frafra hop around the country more often than many citizens of developed nations'. At the same time, ties are maintained with home places, 'mainly because they conceive of themselves as temporary absentees, with a long-term stake in joint property and the village community' (ibid). This contrasts with the Frafra situation three decades previously, when there was 'the traumatic severance of communications and relationships which seem to have characterized prewar ·mobility . . . The whole process of physically uprooting oneself seems so casual these days' (p. 331). In short, with improvements in transport and communications, attachment to places and communities of origin can be maintained more strongly because movement nowadays can be accomplished far more easily.

In Melanesia, facilities for movement also have improved between larger and smaller islands with the establishment of regular shipping to augment local travel by canoe, complemented more recently by the construction of airstrips and the introduction of air services. Where, as in the peripheral islands of Fiji, this improved accessibility is not sustained, movement away to the central and more developed islands continues but frequent visits to maintain contact with places of origin become more difficult. Thus Nair (this volume) reports that the largest group of Fijians resident in the capital of Suva came from Lau province, comprising the eastern outer islands, but those in salaried employment will not risk losing their jobs through being unable to connect with transport while visiting natal villages (Bedford, this volume).

Given the constant flows of individuals between rural and urban environments, Hart (1974) notes that the traditional lineages of the Frafra have been modified, expanded to include the whole of Ghana, and now constitute 'a single integrated field'. This is similar to the situation Finnegan (1976) describes for the Mossi and echoes the discussion of bilocality and multilocality among the Toaripi and the Kuni of Papua (Ryan, Vele, this volume). Such field evidence points to how advantage can be taken of widened opportunities for mobility, while at the same time maintaining links with natal places, so that the varied destinations of village movers become a sociospatial extension of the community of origin. This is akin to Mitchell's (1961a:279) notion of 'social field', the diverse social forces that connect places of origin and destination, which arose from studies of labour circulation in south-central Africa.

In some of the smaller islands of Melanesia the amount of outmovement has begun to produce conditions of depopulation, with inevitable demographic, social, and economic consequences (Bedford, this volume). Within the vastness of West Africa, areas of declining population are much more difficult to identify, given the problems of obtaining census data for small units over time. Undoubtedly such areas exist but, overall, out flows of people from interior to coastal regions and from rural to urban places still comprise a minority of the total population. Rapid growth continues to compensate for losses which occur through emigration from rural towns and villages, with annual rates of natural increase in them exceeding 2.5 per cent.

Contemporary migration and circulation in Melanesia are more effectively documented than for West Africa, with data from successive censuses, formal surveys, and field studies available to facilitate the study of mobility either for its own sake or as a major component in wider investigations (Conroy and Curtain, this volume). In part, the smallness of scale has made it easier to organize such data collections among less sizable communities in the more limited areas of Melanesia. A recent general study of migration in West Africa, based on census data and conducted by the World Bank, illustrates the paucity of material available relative to area and population involved (Zachariah and Condé 1981). Nigeria, the most populous country in the region, with a total population many times greater than those of all Melanesia, is not included, because the first Nigerian census conducted since the war (1952–53) yielded no data directly related to movement and all those undertaken since have been politically unacceptable. Ghana has the most frequent succession of postwar censuses in West Africa (in 1948, 1960, 1970) but they provide relatively limited information on mobility, although it was the location for a major and much-quoted sample survey of rural-urban movement (Caldwell 1969). Among francophone countries, Upper Volta has been most effectively studied, being the major source area for those going to the Ivory Coast and Ghana (Fig. 19.1).

Influences on population movement and redistribution in Melanesia, through administrative, religious, and economic contact with Europeans, appear far more direct and visible than in West Africa and make it possible to identify the decade when specific forms of movement began. Village relocation occurred through the requirements of not only the colonial administration, as happened in some parts of West Africa, but also of proselytizing missionaries (Bonnemaison, Connell, Hamnett, this volume). In the earliest phases of labour recruitment in Melanesia, control could be tight and effective since arrangements were necessary to transport adult males both among islands and between them and Australia. Such procedures quickly became institutionalized, although the initial period of outright coercion lasted perhaps as little as five years. There is similarity here with the control of wage-labour under contract to the gold mines of South Africa (Wilson 1972; Prothero 1974), but there was no comparable system in West Africa. Some contemporary circulation by the educated elite in both Melanesia and West Africa contains an involuntary element, as when teachers and health workers are posted to a sequence of locations throughout their professional careers (Racule, this volume; for southeast Asia, Goldstein this volume).

The great range of spatial and temporal attributes in West African circulation are far less well known than for Melanesia. In the past labour flows undoubtedly were strongly seasonal, being conditioned by the physical environment and by developments in cash cropping, which in turn were influenced although not determined by their environmental contexts. Seasonality of movement has been noted in Melanesia but is far more localized, reflecting the fact that the broad ecological zones of tropical Africa are compressed on the largest islands, from mountain crest to coastal seaboard – as manifest in customary trade and exchange between adjacent niches (Bonnemaison, Watson, this

425

volume). While seasonal labour circulation has probably diminished in West Africa, it has neither ceased nor been largely replaced by permanent migration as Amin (1974) would have us believe. As a characteristic of movement, seasonality will likely continue and be relatively stronger than in other parts of Africa.

The views on Melanesian mobility expressed in this volume are both varied and reveal the difficulties of meaningful generalization from particular instances:

> Mobility in Vanuatu is thus dominantly circular and shall remain so long as the actual relationship between people and territory endures (Bonnemaison).

> ... evidence from eastern Fiji suggests that circulation from a rural base has been replaced by a massive exodus from the villages (Bedford).

> Circulation over the lifecycle of the individual remains the dominant form of population movement throughout Papua New Guinea and will remain so in the foreseeable future for most people from most places (Conroy and Curtain).

> ... the longer-term plans of most urban dwellers still are dominated by a basically rural orientation (Conroy and Curtain).

> Circulation and migration ... co-exist, are to some extent contingent, and often substitute for each other (Nair).

> Even with the advantages To'ambaita migrants now enjoy in town, they are not totally committed to working there and do not regard Honiara as their permanent place of residence (Frazer).

Nonetheless, it is clear that circulation continues to exist or has been present in all localities, despite an apparent trend towards lengthier periods of absence from natal places to the point where ties become redefined and migration may be said to have occurred. Often relocation was not the intent of those who moved and is appreciated only in retrospect.

Doubtless as varied a selection of views could be compiled for West Africa. There, however, most people remain rural based, circulating within and from such areas. The large numbers who have gone to the towns continue these circuits of movement within and between urban settlements, and with the towns as bases they return frequently to the countryside. The concept of rural and urban places constituting a 'single integrated field' has existed in much of West Africa over many generations, certainly predates European contact (Goddard 1965;

(Mortimore 1972), and is valid irrespective of whether scholarly enquiry takes a rural or an urban focus. African social scientists, moreover, stress such continuing links, examining them in their research and maintaining them in their own lives.

Adepoju (1974), for instance, in a study of migrants in the town of Oshogbo, west Nigeria, found that seventy-eight per cent of those interviewed visited their natal communities at least twice a year and forty per cent between four and twelve times. He observes: 'Most town-dwellers do not become completely urban in their outlook. Frequent home visits serve as channels of cultural diffusion' (Adepoju 1974:135). As Mabogunje (1972) notes, these links also serve as important conduits for the diffusion of socioeconomic change. Among those in Oshogbo town, the strength of 'village patriotism' manifests itself in various ways. Organizations with such titles as 'Home Improvement Unions' and 'Sons Abroad' are established in the towns, not only within Nigeria but also in other parts of West Africa and overseas: as among Nigerians in the United Kingdom. The contributions made by these unions to finance schools, clinics, and local electricity in their places of origin are in fact institutionalized remittances that are reported also for Melanesia.

Town dwellers return to their home villages for family reunions and periodic festivals, as well as to maintain rights to land. Among railway workers, market traders, and senior civil servants surveyed in Enugu, southeast Nigeria, Gugler (1971:415) found that 'Everyone claims land rights at home; great majorities visit at least once a year; and nearly everyone, including 84 per cent of the senior civil servants, wants to be buried at home'. In Nigeria, at the time of the 1962/63 census, what have been termed 'census migrations' (more correctly, circulations) occurred back to the natal villages, to help swell their populations and thus ensure increased support from the government revenues in a federal system (Udo 1970). Town and city dwellers similarly return to their local communities for electoral registration and to vote. There could be no stronger affirmations of continuing links.

Conclusion

In the future, more so than has occurred with previous enquiries, there is a need to look more closely at links between past and recent mobility behaviours, at the nature of contemporary circulation and the means by which it links diverse locations and widespread communities. Among third world societies, changes undoubtedly are occurring in the patterns and processes of population movement in general and of circulation in particular, but these will be properly identified only when methods are devised for their ongoing study and precise monitoring. Only with more adequate understanding and explanation of the dynamics of people's mobility may the bases be provided for more effective policy decisions.

Writing of the Frafra, Hart (1974:336) concludes his study by noting: 'When we turn to specific theories of economic development and underdevelopment, we find a never-never land of ideological polarisation, conceptual sterility and empirical ignorance'. In both west Africa and other parts of the third world, the same may be said of mobility research in general and that of circulation in particular, although with less justification for Melanesia. Many authors in this volume, and Bedford in particular, note that the results obtained from mobility studies are highly contingent on definitions and parameters used. As Swindell (1979) has written, it is necessary in research on population movement to recognize the importance not only of whom questions are asked but also what kinds of questions and how. To this may be added the importance of who is posing these questions and what motives, purposes, and preconceptions underlie their asking. Not only is the mobility of people a diverse and complex subject but so also are those involved in its study.

Acknowledgment

An earlier version of this paper, and without the Melanesia comparisons, appeared in *Population Geography*, a journal of the Association of Population Geographers of India (Prothero 1979).

20 Me go 'walkabout'; you too?

Murray Chapman

Lela tells of a man who got time to go 'walkabout'. He [does] not
do any work, he just 'walkabout' . . . to see and hullo to everyone.
A man go[es] *lela* for 'kaikai' [food, particularly at feasts]; to go
and talk with friends; to have peace from his troubles or from his
rowing wife; to talk business; to see some of his good friends. You
always go *lela* in your bus and landrover. This is just the same! (The
late Marcus Pipisi of Duidui, south Guadalcanal, Solomon Islands,
24 January 1967).

Among the greatest faults of which we are guilty in migration research
is being locked into the same kinds of questions related to the same
concepts of migration that were developed years ago for a particular
setting at a particular time (Goldstein 1976:428).

By examining people's mobility on the ground and interpreting that
behaviour through constructs of limited domain, authors in this volume
are able to convey the depth and volatility of movement for Melanesian
life. Essays detail the intricate transformations within society, economy,
and polity that have occurred over generations and how population
movement both mirrors and has shaped those broader contexts. Conversely, they provide neither simple answers for mobility specialists nor
standard recipes for Melanesian planners, apart from a shared philosophy
that intensive studies of people's behaviour represent the most effective
means to achieve various levels of analysis along the continuum between
the individual mover and the aggregate population. Thus an argument
eloquently put in one chapter is likely to be just as firmly contradicted by
its neighbour. Three of the papers concerned with Fiji demonstrate that
neither parallel experience nor a common literature necessarily lead to
comparable conclusions, for their authors — a Fijian, an Indo-Fijian, and a
european New Zealander — put quite different interpretations on basically
the same words written by the noted scholar Rusiate Nayacakalou.

Such differences apart, these essays articulate points of view. The act of movement and the actions of movers are considered in terms of an intrinsic and interactive system; as a means to link complementary yet diverse places and circumstances; as an unfolding and transformation of age-old practice and habit through to the present; and as a strategy by which families minimize risk within communities that lie at the margins of the world capitalist system. Throughout, there is an abiding concern within the dialectic between family and community, region and nation, society and polity. Consequently the ladder of enquiry must be scaled both up and down; understanding derives from a range of data, approaches, and constructs rather than from the elegance and exclusivity of a powerful determinism. In this and other senses, the essays in this volume document the intellectual history of mobility research undertaken amongst the peoples of Melanesia and for which commentaries on West Africa and southeast Asia provide broader context.

From this collective whole, agreement often occurs at only the broadest level. Far more challenging than unanimity, there are delineated basic issues in concept and method and differences in analytical perspective and scale of inquiry. As with the 1978 seminar, from which this book derives, there emerges instead five major themes: continuities and discontinuities in people's mobility over time; variety in the conception of movement both through and across time; the definition of different kinds of mobility given ranges of scale and socioeconomic condition; and problems of measurement and technique in the field study of recurrent mobility.

Unlike the early ethnographers, authors accept that movement was a basic feature of Melanesian life in pre- and immediately postcontact times. Societies could be envisaged as a 'moving sea of more or less rooted, more or less rootless folk', to use Watson's phrase, and people circulated for many reasons: wives, garden land, ceremonials, politics, warfare, barter, and asylum. Whether the sociopolitical context within which these circuits traditionally occurred has much, some, or little significance for the contemporary scene underlies the range of opinion expressed about long-term continuities and discontinuities in Melanesian mobility. A sequence of historical experiences — sandalwood traders, labour recruiting, proselytizing missionaries, alien administrative structures, and world trade conditions — has resulted in a complex reality that will support practically any analytic perspective. Over 150 years, circuits between a village and other communities, between points of origin and destination, have intensified, until today the overall impression about people's mobility is of great flexibility and thus heightened ambiguity. Observed Hamnett at the Honolulu seminar: 'The Eivo of central Bougainville would like to maintain the ambiguity of where they live'. Some fieldworkers also note their own ambivalence to movement: when resident in villages circulation seemed more basic, but when in town migration appeared paramount. In broad terms, whether more continuities than discontinuities are seen lurking in the Melanesian case depends on whether long-term changes in movement behaviour are conceived primarily as ones of degree or of kind.

Such ambiguity and volatility in people's movement on the ground emphasizes the need for flexible interpretation and seriously questions some cherished conventions. Usually, as stated many times by authors in this volume, 'migration' and 'circulation' are distinguished as moves that are permanent or quasipermanent as against those that are intentionally impermanent. Implicit in this formulation is the view that individuals or groups move from their natal communities or ancestral places, sometimes to return and sometimes not. But in this context, what is 'permanent' and what 'impermanent'? What is the meaning of 'home' and of its abandonment? The social sciences are littered with such dualities: rural versus urban, core set against periphery, peasantry counterposed with proletariat. Avoidance of such polar opposites is a more fruitful strategy when seeking to generalize about complex reality in the observable world. Population movement, when defined by time and space, can be viewed in terms of 'movement systems' and 'movement fields'; it is, in Hägerstrand's (1963:67) words, 'the geographical distribution of all movements . . . which can be attributed to a delimited area'. Alternatively, 'social fields' and 'social networks' refer to the complex set of relationships (like genealogy and affinity) within which individuals are enmeshed and as a result of which some decide to move. Although the same terms have different meaning across disciplines, such systems, fields, and networks generally articulate a context for the actions of individuals or groups and make more comprehensible their ebb and flow from one location to another.

Some social structures, as among the Toaripi and Deva Deva Kuni (Gulf and Central provinces, Papua New Guinea), have been described as bi- or multi-local and the varied destinations of village movers are today a sociospatial extension of the origin community. Wage-labour circuits between village and work place may represent a calculated strategy of risk minimization, just as elites may reside for years in town but refuse to surrender title to local land. Ambiguity also exists in people's ideologies of movement. In the past, the flexibility and pragmatism inherent within Melanesian concepts permitted a reconciliation of two apparently contradictory ideologies: the desire for territorial fixity and the temptation of the 'journey' (Bonnemaison, this volume). Often, nowadays, young adult males do not disclose to fellow villagers their likelihood of return, perhaps because they are unsure, perhaps because they do not wish to specify their plans, but most importantly because they do not desire to foreclose their options for future action.

The incorporation of such ambiguities into definition and concept, which in themselves ought to be as unambiguous as possible, is one of

431

the current challenges in mobility research. Existing typologies anchored in coordinates of time and space — as in commuting, circulation, and migration — imply that any one form of movement is discrete from all others (cf. Chapman and Prothero 1983). The advantages of such definitional simplicity are quickly weakened if criteria are added to encompass the functional significance of moves made and the perspectives of people involved in them. Demonstrably different kinds of mobility are neither discrete nor mutually exclusive, even when the usual criteria of time and space are applied. For example, with the establishment of rural feeder roads and with the increasing availability of wheeled transport, villagers in Viti Levu (Fiji), Guadalcanal (Solomon Islands), and New Britain (Papua New Guinea) are now able to commute to work each day or every week whereas a decade ago they could have only circulated or migrated to earn cash. Distance, time, and effort are not in themselves critical, but rather the nature of opportunities available and the risks involved in grasping them. 'Circulation', observed Breman during the Honolulu discussion, 'permits people to be both peasant and proletariat'. Thus attempts to capture the full meaning of recurrent mobility might delineate for individuals or groups the degrees of involvement with (action) and commitment to (attitude) specified places and communities. The variability of allegiance might be indicated, for example, in the primary location of household members and significant kin as distinct from mobile individuals; in the different sites of land and other property owned or held for use; the proportion of income or savings and the frequency of gifts sent by the mover to variously located family members, relatives, or kin; the incidence of return visits made by absentees; and the primary orientation of informal and formal associations with which those who have moved affiliate (cf. Hugo 1983:87-9).

To facilitate cross-cultural inquiry, some authors express a strong desire for precise definitions of movement based upon strictly specified criteria. Although unresolved, much of this genuine concern could be met if definitions were acknowledged to reflect the nature of the research problem and consequently to vary with level of investigation. Minimal and unambiguous criteria of time and space may be the most appropriate when quantitative data are to be collected through census questions, but they should be much augmented if prospective registers and formal surveys are the main field instruments. Such data, when available, can be grouped at various levels of aggregation to permit cross-regional or cross-national comparison. At these intermediate levels of resolution, the potential also exists to incorporate qualitative details about the social significance of who moves and who does not. Finally, concepts and theories of limited domain must apprehend the various contexts within which territorial mobility occurs and incorporate the functional meaning of that behaviour for the participants involved. Thus more attention must be focused upon not only the social but also the economic and political structures of which potential movers are part, upon what has been called 'the cultural meaning of spatial mobility' (Schneider 1970: personal communication) and upon what, in another dimension, has been termed 'the "language" of those who move' (Chapman and Prothero 1977b). As Germani (1965) once observed, to address these needs requires that field investigations and analyses proceed at three levels: the objective, the normative, and the psychosocial.

Such variety of definition and perspective serves only to emphasize the key importance of measurement and technique in making detailed studies of population movement and the critical need for longitudinal as well as cross-sectional data. Mitchell's (1954) index of stabilization, which summarizes the time spent in town in person-months and years, has been employed in Melanesia but the results, as noted in the original work, are greatly influenced by the age and position in the lifecycle of those interviewed. Other indices have been developed to delineate the formal pattern of mobility between village nodes and various points of destination (Bedford 1971) and to describe the ratio of return circuits to all moves made from a given place within a specified period (Young 1977b). A desirable goal, the use of life-table procedures to compute the probabilities for different kinds of circulation among various cohorts in a population, rarely has been attempted.

The difficulties of recall, always present in the retrospective collection of mobility as with other data, can be alleviated by use of the life-history matrix. For this method (Perlman 1976:263–78), unstructured discussion about a person's lifecourse proceeds according to changes in critical events (birth, marriage, education, occupation, land ownership), which may be related to moves between various places at different times. Urban and rural research in Mexico, Brazil, Colombia, and Thailand has demonstrated that this technique requires a sensitive understanding of the people under study if it is to capture the details of mobility intertwined with other life events (e.g. Lauro 1979a; Singhanetra-Renard 1981). Otherwise, the field account quickly reduces to a standard retrospective record of more permanent and long-distance movements elicited without reference to their varied contexts. Again, intent or expectation to return to a community of origin may be judged a poor indicator of future acts of circulation, but such a view overlooks the need to distinguish clearly between structural and behavioural data. Research in some African cities (Mitchell 1969b) has revealed the degree of fit between what is said and what subsequently done to be surprisingly close. This relationship, in short, still constitutes an empirical question.

Flux, ambiguity, and system

Flux and ambiguity, so much a feature of Melanesian mobility, at first glance imperils the search for broader patterns in human behaviour and for systematic links between the actions of individuals, families, or small groups and the structural contexts within which they occur. One answer to this apparent contradiction is to begin holistically and, following Hägerstrand's (1963) example, try to delineate the field of movement for a given people or society. This was done years ago for two village communities in south Guadalcanal, Solomon Islands (Chapman 1970: chapter 5; 1976) and is summarized here, partly to demonstrate the method and partly to indicate some further implications that went unrecognized at the time.

The field investigation, conducted from October 1965 to February 1967, focused on the coast settlement of Duidui and the inland village of Pichahila. Population movement was considered from these local points of reference and incorporated journeys involving an absence of twenty-four hours on the one hand to the relocation of entire communities on the other. Both cross-sectional and longitudinal data were collected, for which a *de jure* census was the baseline and to which was related information from a mobility register maintained over ten months, a two-year record of absences for wage-labour, conjugal histories for all married persons or formerly married persons, a handful of migrant life histories, and oral accounts from senior men about village relocation since the turn of the century (Chapman 1971:10–20).

From this information there emerged three clusters of population movement.

1   Shifts in village locations, generally though not inevitably from the interior to the coast: a customary practice in response to natural hazards, epidemics, land exhaustion, warfare, and sorcery that, since about 1900, has been reinforced by administration and mission efforts to consolidate settlement on the coasts or in the river valleys.

2   Moves by younger persons, and predominantly of adult males, that are of at least ten days' duration but seldom for more than one year: to earn money in the main town, at district centres, on European-owned plantations, or from Solomon Island entrepreneurs; to visit Honiara, district and mission stations for retail, educational, medical, and administrative services; to leave for other villages to go to school, or because of a serious dispute.

3   Short-term, mainly familial moves of a highly spontaneous nature that usually involve absences of less than eight days: to 'go walkabout' to other villages and visit kin, discuss clan and church business, attend a feast; to live temporarily in the garden shelter; to hunt wild pig or trap fish; to quit the community briefly out of shame or by way of protest.

The clusters of mobility identified for the people of both Duidui and Pichahila have two crucial features: periodicity (time) and oscillation (area). Movements for particular objectives occur at varying rates; they similarly involve different kinds of people acting as individuals, as a member of a small group like the family, or as inhabitant of a community. Thus village shifts since 1900 have averaged one every sixteen (Pichahila) and twenty (Duidui) years, going away to school or wage labour follow 10 or 11 and 3 or 6 month cycles respectively, and the ephemeral absences to attend a feast, visit kin, or seek out both customary and introduced medical help rarely span more than eight days.

These markedly different patterns may be viewed as strata within a system of total movement, the lowest stratum of which is complete stability and denotes the periods an individual is physically present in the natal or ancestral community (Fig. 20.1). Everyone, within their life span, oscillates between the village and one or more of these strata. From the standpoint of Duidui and Pichahila, it would be exceptional for a person never to have left the local community for at least twenty-four hours but far less rare for someone, particularly if senile or very young, to have been continuously resident the entire year. Over time, an individual's ebb and flow is consequently the sum total of oscillations between village household and one or more of the mobility strata (Fig. 20.1). To outside and often bemused observers, it is this constant coming and going that summarizes the flexibility and ambiguity of Melanesian movement.

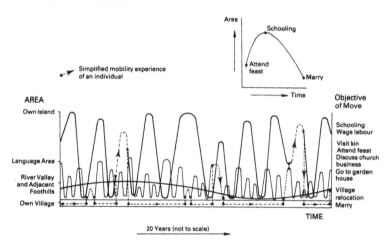

FIGURE 20.1 *Time-area relationships in a village mobility system*

Such focus upon the time dimension ignores the relatedness of movement to area, except for the obvious fact that a village's site is place specific. In Hägerstrand's (1963:67) term, the 'field' — or range — of the various acts of mobility subsumed within and illustrated by each stratum vary appreciably. Nowadays, only wage labour and formal schooling draw people beyond the confines of their own island, Guadalcanal, and the extent to which this occurs has diminished since the second world war. The administrative areas within which Duidui and Pichahila are located have long been renowned as a source of unskilled workers and in the early 1950s patrol reports noted considerable numbers of adult males away in the central and western Solomons (Fig. 5.1). During 1966, by contrast, a Duidui man on Vella Lavella was the solitary instance of wage employment beyond Guadalcanal. At the same time, three pupils from Duidui whose parents are South Seas Evangelicals received their senior primary schooling on the nearby island of Malaita, but this practice seemed unlikely to continue with the upgrading of several primary schools on north Guadalcanal.

Short-term moves involving absences of less than eight days infrequently penetrate the language boundary — Duidui:Poleo, Pichahila: Birao — and even more rarely extend beyond south Guadalcanal. This pattern reflects partly the fact that such oscillation involves mainly family groups rather than persons as individual actors, partly that genealogical ties become more attenuated and less meticulously upheld beyond one's own language area. As distance away from the local community increases, so more and more does it become a matter of personal choice whether a feast ought to be attended, relatives visited, or clan business shared. Thus the incidence of the decison to go varies inversely with distance but directly with status.

Territorially, the relocation of villages is even more highly localized, since inhabitants must be genealogically related and at least one of them own land that can be allocated to each and every householder for food gardens. The land upon which the village and gardens of Pichahila are situated belongs to one man and efforts during 1966 to establish communal coconut groves downriver were thwarted when his ownership of the area chosen was strenuously disputed by several coastal villages. Since the turn of the century, eighteen identified shifts of Duidui's antecedents have been confined to the fan and adjacent foothills of the Haligecha river, the present site, while seventeen relocations of settlements that preceded Pichahila have been similarly concentrated within the parallel river valleys of the Alualu (present site) and the Sabahalava immediately to the west (Chapman 1970:53–8, 69–76). For each community, the area traversed was a mere twenty-three square kilometres.

Broadly conceived, the movement fields of the three strata range from the river valley and adjacent foothills in village shifts to the home island and beyond for wage labourers and school pupils. The territorial manifestations of each cluster or set of mobility, like the ripples from a pebble in a pool, are finely graded and may be taken as the basis for ordering each stratum with respect to the other two (Fig. 20.1). Thus longer-term, but highly-localized village transfers are placed closest to, and annual but more distant moves of wage labourers farthest from, residential stability. On average, eligible women leave to marry no more than twice during their lifetime but these are seldom shifts beyond a neighbouring village. Conversely, the paucity of local schools annually draws children far beyond south to north Guadalcanal as well as to the island of Malaita (Fig. 5.1). From these two examples it can be seen that, over time, the area traversed during a specified move is poorly correlated with the frequency at which such shifts occur. Put another way, the time-area relationships for the various components of a village mobility system are neither simple nor linear but rather complex and curvilinear (Fig. 20.1:inset). In this respect, these field data from south Guadalcanal confirm what has been summarized about other cultural contexts (Morrill 1963).

It is inevitable, when delineating these three mobility strata, that time should have been stressed more than place, since movement analyses are heavily if decreasingly cross-sectional and thus capture the experience of given communities at a particular instant. The misrepresentation that can occur is well demonstrated by arbitrarily selecting any cross-section from Figure 20.1:the upswings of one stratum rarely coincide with the other two and any results, no matter how elegantly presented, are simplistic distortions of reality.

Thus far, delineation of a village mobility system has not addressed two key questions raised by several authors in this volume. To what extent, for the people involved, is this model congruent with their conception of movement, not to mention the values and orientations that underpin that conception; and to what extent does this construct document and reflect continuities and transformations in people's behaviour? To take an example from one stratum, if local visiting is viewed as recurrent and time-honoured behaviour, then villagers nowadays are constantly traversing the territory inhabited by closely related kin. A three-hour walk to the local health clinic is unlikely to exceed the range of long-established travel patterns and may thus be viewed as a contemporary adaptation of customary practice. On the other hand, a facility located a hard day's walk from the community will tend to be ignored unless some special emergency arises or, as in the case of Pichahila, it forms part of an institution like a mission that performs more than a single function. In these situations, the reaction of villagers is as much traditional as modern. Why should a person leave one's land, one's relatives, and the protection of one's ancestral spirits to travel to a foreign place, located beyond the territorial core of the lineage, unless for some particular and demonstrable reason?

In this context, the willingness for countless generations of individuals to shift their house sites and of hamlet communities to relocate themselves within clearly bounded limits becomes less surprising. It might even be speculated that, during the decade 1870–80 and given this lifestyle, men from south Guadalcanal were more apt to be enticed from the beaches for consignment as indentured labourers to the canefields of Fiji and Queensland. From the first decade of this century, when this nefarious practice was concluded, it is similarly understandable that south Guadalcanal became known throughout the protectorate as one assured source of able-bodied labourers who, in order to earn money, would leave their village homes for periods of up to two years. Again, with the expansion of primary education during the 1950s, Duidui and Pichahila families might be said to have been more willing to tolerate, if reluctantly, the annual absence of those teenage children who continued to higher levels.

Such interpretations are congruent with the field data, but demand the more detailed and tougher examination of local concepts, as well as of continuities and discontinuities in mobility behaviour, that in this volume Bathgate has provided for the Ndi-Nggai of northwest Guadalcanal and Bonnemaison for the Tannese of Vanuatu. Certainly the potential exists for such a systemic model simultaneously to articulate mobility behaviour from the perspective of both the village participant and the outside observer. But does such a construct also have the ability to address yet another crucial dimension: the wider institutional context within which movement occurs and the particular socioeconomic and political constraints that help shape the mobility behaviour of people acting independently or in groups?

In the original study (Chapman 1976:133–9), this question was addressed in ways intrinsic to the communities of Duidui and Pichahila and, in particular, to the roles performed by villagers. The combination of being a parent, spouse, blood relative, clan member, food producer, church leader, wage worker, entrepreneur, and political figure was closely related to the incidence of movement among different categories of people ('big man', adult married male, young single female). This relationship, moreover, was confirmed for the two strata of the mobility system that described villagers acting independently or in small groups rather than as members of a whole community (Fig. 20.1). For circuits of movement examined in sequence, the combination of roles thereby involved was a simple but powerful index of the likely rate of mobility for specified individuals.

Although not emphasized at the time, the close links between the rate of mobility and the combined roles thus performed also pointed to a complex of cultural, socioeconomic, and political constraints: genealogy, affinity, social organization, subsistence production, local-level politics, introduced religion, colonial economy, and extralocal administration. Some of these factors were intrinsic to the communities of Duidui and Pichahila and yet others extrinsic but, to the people, they constituted the collective context of mobility behaviour and influenced, both in general and in particular, the decision whether to go or to stay. To villagers, genealogy, social organization, food production, and local-level politics are important yet differential constraints on movement, notwithstanding the fact that structuralists prefer to emphasize others of equal but not greater relevance: systems of colonial administration, cash-crop estates, big farms and multinational corporations. In structural analysis, the result is a persuasive mode of explanation but one unable to incorporate the flux, flexibility, ambiguity, and contradiction that is so much part of Melanesian mobility; and the scholarly response has been to ignore such volatility on the pretext that it is sheer noise, unimportant, irrelevant, or perhaps does not even exist.

But is, in fact, this volatility in movement behaviour so irrelevant? The geographer Bonnemaison (1984:117) notes that, on the island of Tanna in Vanuatu, man 'is a tree that must take root and stay fixed in its place' whereas the local group 'is a canoe that follows "roads" and explores the wider world'. Thus is reconciled the dual metaphor of the tree, symbol of rootedness and stability, and the canoe, symbol of journeying and unrestricted wandering – which in turn is so critical to an understanding of Tannese identity and Tannese mobility. In another island realm, the anthropologist Carnegie (1982) conceives movement throughout the Caribbean in terms of 'strategic flexibility', or as an expression of cultural ideals that emphasize the building of multiple options. *Mantche shien, pwen shat*, 'if you lose the dog, grab the cat', is a St. Lucian expression that captures this willingness to move in response to unexpected opportunities rather than holding doggedly to a single, and perhaps fruitless option. The fact of Caribbean islanders in constant movement thus constitutes an eloquent statement on the permeability of the nation state. Finally, the economic historian Kussmaul (1981) describes the mobility of English farm servants in another era (1692-1843) as being inherently ambiguous. 'Ambiguity abounds . . . Half family members, half wage earners, the chimerical farm servants are inherently ambiguous. They are not born into a farmer's family, nor married into it, but hired into it; they were wage earners, but not propertyless adult proletarians. It is only fitting that their mobility also be ambiguous' (Kussmaul 1981:234).

When scholars of such different persuasion discover congruent concepts and report convergent results about societies spread so widely in time and over space, then the paradox, flux, and ambiguity in movement should be confronted rather than so assiduously avoided. To do so would lead, ironically, to the other side of the structuralist coin and to an incorporation of various levels of context for and different kinds of constraints upon the potential act of moving or staying. To abstract, by way of example, leaving for wage labour from its particular stratum of the village-based model of mobility, it is possible to isolate the influence of missions, government, and expatriate entrepreneurs and to incorporate that modification into the original. How adult men and women from Duidui and Pichahila initially secure a job will be reflected not only in kinds of available employment but also in their subsequent experience of different occupations and of skills acquired onsite. In research still in progress. Friesen (1983) demonstrates the potential of such an approach by examining the life histories of Mbambatana men and women from south Choiseul in the western Solomons (Fig. 5.1). In work histories drawn from a single village, thirty-five out of forty-two men had engaged in wage labour at some stage of their lives.

The Methodist mission, present in the area since 1902, was the first institution to offer paid employment for Choiseulese with some education, dominated until the 1940s, and was not eclipsed by government as employer of educated labour until the 1950s (Fig. 20.2). Those recruited as 'preacher-teachers' had to have completed several years of formal schooling, then received further training from the mission at a rural institution, and subsequently were posted to a local community, from which they were transferred every four or five years until returning to the village on retirement (cf. Racule this volume). The importance of independent entrepreneurs, notably the chinese owners of local trading vessels, as sources of jobs derives from the fact that they have been operating around Choiseul for many decades and consequently recruit seamen directly from Mbambatana communities. For males to be eligible requires 'little more than the ability to lift heavy sacks of copra and a willingness to accept low wages' (Friesen 1983:17) – attributes that subsequently translate into short interludes of wage labour (three to thirty-six months) broken by longer periods back in the village, and a total working life of no more than ten years. The government as a vehicle for career employment is a far more recent institutional prospect. Candidates usually are identified while attending secondary school, undergo further training at either a rural institution or in the capital of Honiara, spend longer periods in gainful employment, and reside in a greater range of places beyond Choiseul.

Viewed retrospectively, the initial act of taking up employment with the Methodist mission, a chinese entrepreneur, or a colonial or post-colonial government constitutes what Young (1982) terms a 'critical move' – one that predisposes towards a particular kind of wage work and increasingly channels the subsequent choices made from a range of mobility options. Within any run of years, variation in periods of employment, in the ratio of time spent physically in the Mbambatana village as against beyond it, and in different places of residence, together summarize which of the three institutional contexts exert the more powerful influence on a given number of villagers. In turn, these contrasting sequences of circulation for paid employment disaggregate the wage-labour stratum of the original model. Such chains of mobility experience, when added to and depicted on the village-based diagram, provide graphic comparison to yet other chains, whose broader context will have been defined by variable sets of differential circumstances: some intrinsic and some extrinsic to the community of reference; some ecological or cultural, others sociological or economic, yet others political. Melanesian mobility on the ground, the domain of the real world, is not quite as simple and uncomplicated as our univariate constructs might have us to believe.

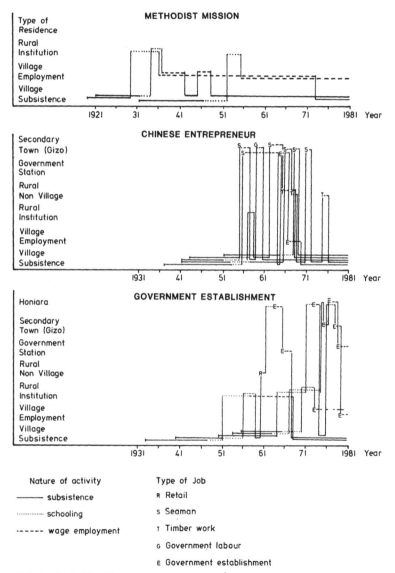

FIGURE 20.2 *Mobility contexts for some Choiseul wage-workers,*
1921–81

441

Toward what kind of future?

To propose in North America, in the early 1960s, a study of population movement from the standpoint of the village was viewed by university departments and funding agencies as quaint, if not downright idiosyncratic. By the 1980s the Americans, along with other superpowers, had rediscovered the island Pacific as a vast arena within which to peddle their ideological wares. Scholars prefer to distance themselves from such political insensitivity, but their intellectual constructs are similarly governed as much by whim and fashion as by their inherent worth. At the turn of this century, visiting ethnographers described the small communities of precontact Melanesia as indelibly fixed to a particular residential site; conversely, when viewed through a different set of philosophical lenses during the 1970s, these same local groups were considered so mobile as to seem unhitched from both space and time.

Despite an intellectual history at times unimpressive, there is gradually emerging a convergence of interest about mobility between the societies of Melanesia and the scholars — both local and foreign — who cross their beaches and walk their mountain trails. For Melanesians, what matters is how mobility can be deployed for a better world, a less arduous life, and a more secure future more definitively their own; for theoreticians, how an understanding of Melanesian mobility can inform a scholarly world dominated by western precepts and manners of thought. The anthropologist Lystad, when introducing a survey of social research in Black Africa, has argued for the latter kind of contribution:

> The feeling is strong that the distortion of concrete African social realities is too great, the omission of relevant, even crucial, data too likely when the data are injected into categories predefined on the basis of non-African materials and from 'Western' models. Writers in nearly all disciplines have encountered these conceptual difficulties on the level at which raw data are minimally organised and at the next highest levels on which relationships among minimum categories are being determined. At the same time, the need to define new (or to modify 'Western') categories has the salutary effect of forcing a discipline's non-Africanists to review and revise those categories that fail adequately to comprehend African data. This, more than perhaps any other, is the principle service African . . . studies can provide to theoreticians in any discipline working towards general models and theories (Lystad 1965:2).

Melanesians define their arena of interest far more broadly. The movement of people is part of life itself, with antecedents stretching back into mythology and with dimensions an inherent part of cosmology. What is discerned and articulated by indigenous scholars consequently will proceed from vastly different frames of reference. Witness

442

this final word from John Waiko, a Melanesian historian rather than a historian of Melanesia:

> I do not think that we can create writers just out of the colonial experience because it is too short a time and indeed too shallow a basis: our great writers are the masters of our own culture who are living in the villages, and who unfortunately are old. The sad fact is that every day the old people die and they are buried with 'wealth of libraries' of every branch of knowledge in their minds beyond recovery. If this continues it will be only a matter of a generation before this country will have lost its identity based on the people's culture. In its stead we will have a 'national literature' with no roots (Waiko 1978:24).

Acknowledgment

Appreciative thanks go to Ward Friesen, University of Auckland, for permission to use and to quote from field research in the Solomon Islands that is still in progress and subject to final analysis.

Bibliography

ABU-LUGHOD, J. (1975), 'Comments: the end of the age of innocence in migration theory', in *Migration and Urbanization: Models and Adaptive Strategies* (eds.) B.M. Du Toit and H.I. Safa:201-6, The Hague: Mouton.

ADEPOJU, A. (1974), 'Rural-urban socio-economic links: the example of migrants in south-west Nigeria', in *Modern Migrations in Western Africa* (ed.) S. Amin:127-37, London: Oxford University Press for International African Institute.

ADEPOJU, A. (1976), 'Internal migration in Nigeria', report of Seminar on Internal Migration in Nigeria held at the University of Ife, Nigeria, May 1975.

ALLEN, M.R. (1968), 'The establishment of Christianity and cash-cropping in a New Hebridean community', *Journal of Pacific History*, 3:25-46.

ALLEN, M.R. (1972), 'Rank and leadership in Nduindui, Northern New Hebrides', *Mankind*, 8:270-82.

ALLEN, J. and C. HURD (1963), *Languages of the Bougainville District*, Ukarumpa, Papua New Guinea: Summer Institute of Linguistics.

AMIN, S. (ed.) (1974), *Modern Migrations in Western Africa*, London: Oxford University Press for International African Institute.

ARNOLD, F. and S. BOONPRATUANG (1976), *1970 Population and Housing Census: Migration*, Subject Report 2, Bangkok: National Statistical Office, Office of the Prime Minister.

ARONSON, E. (1968), 'Dissonance theory: progress and problems', in *Theories of Cognitive Consistency: A Source Book* (eds.) R.P. Abelson, E. Aronson, W.J. McGuire, T.M. Newcomb, M.J. Rosenberg and P.H. Tannenbaum: 5-27, Chicago: Rand McNally.

Aruliho School (1970), 'Vocational and career preferences', Summary report of thought-papers written by students on the theme of Christian dedication, Aruliho: St. Paul's High School.

AVOSA, M. (1977), 'Lese Avihara', in *The Rural Survey 1975, Supplement to Vol. 4 of Yagl-Ambu* (eds.) J. Conroy and G. Skeldon:

140–60, Port Moresby: Institute of Applied Social and Economic Research.

BALAN, J., H.L. BROWNING, E. JELIN, and L. LITZLER (1969), 'A computerized approach to the processing and analysis of life histories obtained in sample surveys', *Behavioral Science*, 14:105–19.

BARTH, F. (1966), *Models of Social Organization*, Occasional Publication 23, London: Royal Anthropological Institute of Great Britain and Ireland.

BARTON, F.R. (1910), 'The annual trading expedition to the Papuan Gulf', in *The Melanesians of British New Guinea* (ed.) C.G. Seligmann, M.D.:96–120, Cambridge: University Press.

BASTIN, R.R. (1980), 'Cash, calico, and christianity: Individual strategies of development on Tanna, New Hebrides', PhD thesis, University of Sussex, Falmer, Brighton.

BASTIN, R.R. (1981), 'Economic enterprise in a Tannese village', in *Vanuatu: Politics, Economics, and Ritual in Island Melanesia* (ed.) M. Allen:337–55, New York: Academic Press.

BATHGATE, M.A. (1973), *West Guadalcanal Report: A Study of Economic Change and Development in the Indigenous Sector, West Guadalcanal, B.S.I.P.*, Wellington: Department of Geography, Victoria University.

BATHGATE, M.A. (1975), '*Bihu matena golo*: a study of the Ndi-Nggai of West Guadalcanal and their involvement in the Solomon Islands cash economy', PhD thesis, 2 vols., Victoria University of Wellington.

BATHGATE, M.A. (1978a), 'Marketing distribution systems in the Solomon Islands: the supply of food to Honiara', in *Food, Shelter, and Transport in Southeast Asia and the Pacific* (eds.) P.J. Rimmer, D.W. Drakakis-Smith and T.G. McGee:63–100, Department of Human Geography Publication HG/12, Canberra: Research School of Pacific Studies, Australian National University.

BATHGATE, M.A. (1978b), *The Structure of Rural Supply to the Honiara Market in the Solomon Islands*, Occasional Paper 11, Canberra: Development Studies Centre, Australian National University.

BAXTER, M.W.P. (1973), *Migration and the Orokaiva*, Department of Geography, Occasional Paper 3, Port Moresby: University of Papua New Guinea.

BAYLISS-SMITH, T.P. (1977), *Koro in the 1970's: Prosperity through Diversity?*, UNESCO/UNFPA, Fiji Island Reports 2, Canberra: Australian National University for UNESCO.

BAYLISS-SMITH, T.P. (1978), 'Batiki in the 1970s: satellite of Suva', in *The Small Islands and the Reefs* (ed.) H.C. Brookfield:67–128, UNESCO/UNFPA Fiji Island Reports 4, Canberra: Development Studies Centre, Australian National University for UNESCO.

BEDFORD, R.D. (1971), 'Mobility in transition: an analysis of population movement in the New Hebrides', PhD thesis, Australian National University, Canberra.

BEDFORD, R.D. (1973a), *New Hebridean Mobility: A Study of Circular Migration*, Department of Human Geography Publication HG/9,

Canberra: Research School of Pacific Studies, Australian National University.

BEDFORD, R.D. (1973b), 'A transition in circular mobility: population movement in the New Hebrides, 1800–1970', in *The Pacific in Transition: Geographical Perspectives on Adaptation and Change* (ed.) H. Brookfield:187–227, Canberra: Australian National University Press.

BEDFORD, R.D. (1978), 'Rural Taveuni: perspectives on population change', in *Taveuni: Land, Population, and Production* (eds.) H.C. Brookfield, B. Denis, R.D. Bedford, P.S. Nankivell and J.B. Hardaker: 85–244, UNESCO/UNFPA Fiji Island Reports 3, Canberra: Australian National University for UNESCO.

BEDFORD, R.D. (1981), 'The variety and forms of population mobility in Southeast Asia and the Pacific: the case of circulation', in *Population Mobility and Development: Southeast Asia and the Pacific* (eds.) G.W. Jones and H.V. Richter:17–49, Development Studies Centre Monograph 27, Canberra: Research School of Pacific Studies, Australian National University.

BEDFORD, R.D. and M. BROOKFIELD (1979), 'Population change 1946–1976: perspectives on fertility and migration', in *Lakeba: Land Resources, Population and Economy* (ed.) H.C. Brookfield: UNESCO/UNFPA Fiji Island Reports, 5, Canberra: Development Studies Centre, Australian National University for UNESCO.

BEDFORD, R.D. and A.F. MAMAK (1976), 'Bougainvilleans in urban wage employment: some aspects of migrant flows and adaptive strategies', *Oceania*, 46:169–87.

BEDFORD, R.D., McLEAN, R.F., and J. MACPHERSON (1978), 'Kabara in the 1970's: home in spite of hazards', in *The Small Islands and the Reefs* (ed.) H.C. Brookfield:9–65, UNESCO/UNFPA Fiji Island Reports 4, Canberra: Development Studies Centre, Australian National University for UNESCO.

BELLAM, M.E.P. (1963), 'The Melanesian in town: a preliminary study of adult male Melanesians in Honiara, British Solomon Islands', MA thesis, Victoria University of Wellington.

BELLAM, M.E.P. (1970), 'The colonial city: Honiara, a Pacific Islands' case study', *Pacific Viewpoint*, 11:66–96.

BELSHAW, C.S. (1954), *Changing Melanesia: Social Economics of Culture Contact*, London: Oxford University Press.

BERG, E.J. (1961), 'Labour supply functions in dual-economies – the African case', *Quarterly Journal of Economics*, 75:468–92.

BERG, E.J. (1965), 'The economics of the migrant labor system', in *Urbanization and Migration in West Africa* (ed.) H. Kuper:160–81, Los Angeles: University of California Press.

BERG, E.J. (1966), 'Backward-sloping labor supply functions in dual economies – the Africa case (1961)', in *Social Change: The Colonial Situation* (ed.) I. Wallerstein:114–36, New York: John Wiley & Sons.

VAN BINSBERGEN and H.A. MEILINK (eds.) (1978), *Migration and the Transformation of Modern African Society: African Perspectives*, Leiden: Afrika-Studiecentrum.

BONNEMAISON, J. (1972), 'Système de grades et différences régionales en Aoba (Nouvelles-Hébrides)', *Cahiers ORSTOM Série Sciences Humaines*, 9:87–108.

BONNEMAISON, J. (1974), 'Changements dans la vie rurale et mutations migratoires aux Nouvelles-Hébrides', *Cahiers ORSTOM Série Sciences Humaines*, 11:259–86.

BONNEMAISON, J. (1976), 'Circular migration and uncontrolled migration in the New Hebrides', *South Pacific Bulletin*, 26:7–12.

BONNEMAISON, J. (1977a), 'The impact of population patterns and cash-cropping in urban migrations in New Hebrides', *Pacific Viewpoint*, 18:119–32.

BONNEMAISON, J. (1977b), *Système de Migration et Croissance Urbaine à Port Vila et Luganville (Nouvelles-Hébrides)*, Travaux et Documents 60, Paris: Office de la Recherche Scientifique et Technique Outre-Mer.

BONNEMAISON, J. (1979), 'Les voyages et l'enracinnement: formes de fixation et de mobilité dans les sociétés traditionelles des Nouvelles-Hébrides', *L'Espace Géographique*, 4:303–18.

BONNEMAISON, J. (1984), 'The tree and the canoe: roots and mobility in Vanuatu societies', *Pacific Viewpoint*, 25:117–51.

BREESE, G. (1966), *Urbanisation in Newly Developing Areas*, Englewood Cliffs, New Jersey: Prentice-Hall.

BREMAN, J. (1978–79), 'Seasonal migration and cooperative capitalism: the crushing of cane and of labour by the sugar cane factories of Bardoli, South Gujurat', *Journal of Peasant Studies*, 6(1):41–70, (2):168–209.

BREWER, K.R.W. and R.W. MELLOR (1973), 'The effect of sample structure on analytical surveys', *Australian Journal of Statistics*, 15:145–52.

British Solomon Islands (1970), *Annual Abstract of Statistics*, Honiara: Statistical Office.

BROOKFIELD, H.C. (1960), 'Population distribution and labour migration in New Guinea', *Australian Geographer*, 7:233–42.

BROOKFIELD, H.C. (1961), 'The highland peoples of New Guinea: a study of distribution and localization', *Geographical Journal*, 127:436–48.

BROOKFIELD, H.C. (1962), 'Local study and comparative method: an example from central New Guinea', *Annals Association of American Geographers*, 52:242–54.

BROOKFIELD, H.C. (1970), 'Dualism and the geography of developing countries', Presidential address to Section 21 (Geography), Conference of Australian and New Zealand Association for the Advancement of Science, Port Moresby.

BROOKFIELD, H.C. (1977), 'The effect of transport arrangements on the human geography of small islands', in *Selected Papers Prepared for an Expert Group on Inter-Island Services by Sea and Air for Developing Island Countries* (ed.) E. Dommen: 57–88, Geneva: United Nations Conference on Trade and Development.

BROOKFIELD, H.C. (ed.) (1980), *Population-Environment Relations in Tropical Islands: The Case of Eastern Fiji,* Man and the Biosphere Technical Notes 13, Paris: UNESCO.

BROOKFIELD, H.C. and D. HART (1971), *Melanesia: A Geographical Interpretation of an Island World,* London: Methuen.

BURAWOY, M. (1976), 'The functions and reproduction of migrant labor: comparative material from Southern Africa and the United States', *American Journal of Sociology,* 81:1050–87.

BURNS, T., M. COOPER and B. WILD (1972), 'Melanesian big men and the accumulation of power', *Oceania,* 43:104–12.

BUTTIMER, A. (1976), 'Grasping the dynamics of lifeworld', *Annals Association of American Geographers,* 66:277–92.

CALDWELL, J.C. (1969), *African Rural-Urban Migration: The Movement to Ghana's Towns,* New York: Columbia University Press.

CAMPBELL, J.R. (1977), 'Hurricanes in Kabara', in *The Hurricane Hazard: Natural Disaster and Small Populations,* (eds.) R.F. McLean, T.P. Bayliss-Smith, M. Brookfield, and J.R. Campbell:149–75, UNESCO/UNFPA Fiji Island Reports 1, Canberra: Australian National University for UNESCO.

CARNEGIE, C.V. (1982), 'Strategic flexibility in the West Indies: a social psychology of Caribbean migration', *Caribbean Review,* 11:1–13, 54.

CAVALLI-SFORZA, L. (1963), 'The distribution of migration distances: models and applications to genetics', in *Human Displacements: Measurements, Methodological Aspects* (ed.) J. Sutter:139–58, Monaco: Centre International d'Etude des Problèmes Humains.

CHAGNON, N.A. (1974), *Studying the Yanomamo,* New York: Holt, Rinehart and Winston.

CHALMERS, J. (1886), *Adventures in New Guinea,* London: Religious Tract Society.

CHALMERS, J. (1887), *Pioneering in New Guinea,* London: Religious Tract Society.

CHALMERS, J. (1898), 'Toaripi', *Journal of the Anthropological Institute of Great Britain and Ireland,* 27:326–34.

CHAMRATRITHIRONG, A. (1979), *Recent Migrants in Bangkok Metropolis: A Follow-up Study of Migrants' Adjustment, Assimilation, and Integration,* Bangkok: Institute of Population and Social Research, Mahidol University.

CHAPMAN, M. (1969), 'A population study in south Guadalcanal: some results and implications', *Oceania,* 40:119–47.

CHAPMAN, M. (1970), 'Population movement in tribal society: the case of Duidui and Pichahila, British Solomon Islands', PhD thesis, University of Washington, Seattle.

CHAPMAN, M. (1971), *Population Research in the Pacific Islands: A Case Study and Some Implications,* Working Paper 17, Honolulu: East-West Population Institute.

CHAPMAN, M. (1975), 'Mobility in a non-literate society: method and analysis for two Guadalcanal communities', in *People on the Move: Studies on Internal Migration* (eds.) L.A. Kosiński and R.M. Prothero: 129–45, London: Methuen.

CHAPMAN, M. (1976), 'Tribal mobility as circulation: A Solomon Islands example of micro/macro linkages', in *Population at Microscale* (eds.) L.A. Kosiński and J.W. Webb: 127–42, Christchurch: New Zealand Geographical Society and the Commission on Population Geography, International Geographical Union.

CHAPMAN, M. (1977), 'Circulation studies at the East-West Population Institute, Honolulu', *Indonesian Journal of Geography*, 7:1–4.

CHAPMAN, M. (1978), 'On the cross-cultural study of circulation', *International Migration Review*, 12:559–69.

CHAPMAN, M. (1979), 'The cross-cultural study of circulation', *Current Anthropology*, 20:111–14.

CHAPMAN, M. and R.M. PROTHERO (1977a), 'Circulation between home places and towns: a village approach to urbanization', paper presented at Working Session on Urbanization in the Pacific, Annual Meeting, Association for Social Anthropology in Oceania, Monterey, California, March.

CHAPMAN, M. and R.M. PROTHERO (1977b), 'Conceptualising mobility: the "language" of those who move', paper presented at Developing Areas Study Group Meeting on Internal Migration in Less Developed Countries, Institute of British Geographers, Newcastle, England, January.

CHAPMAN, M. and R.M. PROTHERO (1983), 'Themes on circulation in the third world', *International Migration Review*, 17:597–632.

CHO, L.J. (ed. with assistance of D.B. Johnson and M. Gannaway) (1976), *Introduction to Censuses of Asia and the Pacific, 1970–74*, Honolulu: East-West Population Institute.

CHURCHMAN, W.B. (1888), *Blackbirding in the South Pacific or The First White Man on the Beach*, London: Swan Sonnenschein.

CLUNIES ROSS, A. (1977a), 'Motives for migration', in *The Rural Survey 1975, Supplement to Vol. 4 of Yagl-Ambu* (eds.) J. Conroy and G. Skeldon:16–45, Port Moresby: Institute of Applied Social and Economic Research.

CLUNIES ROSS, A. (1977b), 'Village migration patterns: notes towards a classification', in *The Rural Survey 1975, Supplement to Vol. 4 of Yagl-Ambu* (eds.) J. Conroy and G. Skeldon:3–15, Port Moresby: Institute of Applied Social and Economic Research.

CLUNIES ROSS, A. (1984), *Migrants from Fifty Villages*, Research Monograph 21, Port Moresby: Institute of Applied Social and Economic Research.

CLUNIES ROSS, A., R.L. CURTAIN, and J.D. CONROY (1975), *The 1974/75 Rural Survey: Aims, Scope, and Methods*, Discussion Paper 7, Port Moresby: New Guinea Research Unit.

COALE, A.J. and P. DEMENY (1966), *Regional Model Life Tables and Stable Populations*, Princeton: University Press.

CONNELL, J. (1977), *The People of Siwai: Population Change in a Solomon Island Society*, Working Paper 8, Canberra: Department of Demography, Australian National University.

CONNELL, J. (1978), *Taim Bilong Mani: The Evolution of Agriculture in a Solomon Island Society*, Development Studies Centre Monograph

449

12, Canberra: Research School of Pacific Studies, Australian National University.

CONNELL, J. (1979), 'Temporary townsfolk? Siwais in urban Bougainville', manuscript, Department of Geography, University of Sydney.

CONNELL, J. (1981), 'Migration, remittances and rural development in the South Pacific', in *Population Mobility and Development: Southeast Asia and the Pacific* (eds.) G.W. Jones and H.V. Richter:229–55, Development Studies Centre Monograph 27, Canberra: Research School of Pacific Studies, Australian National University.

CONNELL, J., B. DASGUPTA, R. LAISHLEY and M. LIPTON (1976), *Migration from Rural Areas: The Evidence from Village Studies*, New Delhi: Oxford University Press.

CONROY, J.D. (1972), 'School leavers in Papua New Guinea: expectations and realities', in *Change and Development in Rural Melanesia* (ed.) M.W. Ward: 355–73, Canberra: Research School of Pacific Studies, Australian National University.

CONROY, J.D. (1975), 'A model of educational expansion and employment in Papua New Guinea', *Manpower and Unemployment Research in Africa*, 8:49–66.

CONROY, J.D. (1977), 'A longitudinal study of school leaver migration', in *Change and Movement: Readings on Internal Migration in Papua New Guinea* (ed.) R.J. May:96–115, Canberra: Australian National University Press.

CONROY, J.D. and R. CURTAIN (1973), 'Migrants in the urban economy: case studies of rural leavers in Port Moresby', *Oceania*, 44: 81–95.

CONROY, J. D. and R. CURTAIN (1982), 'Circular migration and the emergence of permanent urban populations', in *Essays on the Development Experience of Papua New Guinea*: 28–48, Research Monograph 17, Port Moresby: Institute of Applied Social and Economic Research.

CONROY, J. and G. SKELDON (eds.) (1977), *The Rural Survey 1975, Supplement to Vol. 4, of Yagl-Ambu*, Port Moresby: Institute of Applied Social and Economic Research.

CORRIS, P. (1970), 'Pacific Island labour migrants in Queensland', *Journal of Pacific History*, 5:43–64.

CORRIS, P. (1973), *Passage, Port and Plantation: A History of Solomon Islands Labour Migration 1870–1914*, Melbourne: University Press.

CURTAIN, R. (1975), 'Labor migration in Papua New Guinea: primary school leavers in the towns – present and future significance', in *Migration and Development: Implications for Ethnic Identity and Political Conflict* (eds.) H.I. Safa and B.M. du Toit:269–93, The Hague: Mouton.

CURTAIN, R. (1980a), 'Dual dependence and Sepik labour migration', unpublished PhD thesis, Australian National University, Canberra.

CURTAIN, R. (1980b), 'The structure of internal migration in Papua New Guinea', *Pacific Viewpoint*, 21:42–61.

DAKEYNE, R.B. (1967), 'Labour migration in New Guinea: a case study from Northern Papua', *Pacific Viewpoint*, 8:152–58.

DEACON, A.B. (1934), *Malekula: A Vanishing People in the New Hebrides*, London: George Routledge.

DENIEL, R. (1968), *De la savanne à la ville*, Paris: Aubier.

DENIEL, R. (1974), 'Mesures gouvernementales et/ou intérêts divergents des pays exportateurs de main d'oeuvre et des pays hôtes Haute Volta et Côte d'Ivoire, in *Modern Migrations in Western Africa* (ed.) S. Amin: 215-25, London: Oxford University Press for International African Institute.

DOUGLAS, M. (1970), *Natural Symbols: Explorations in Cosmology*, London: Barry and Rockliff.

East-West Population Institute (1978), *Summary Report: International Seminar on the Cross-Cultural Study of Circulation, 3-7 April*, Honolulu: East-West Center.

ELDRIDGE, H.T. (1965), 'Primary, secondary, and return migration in the United States, 1955-1960', *Demography*, 2:445-55.

ELKAN, W. (1960), *Migrants and Proletarians: Urban Labour in the Economic Development of Uganda*, London: Oxford University Press for the East African Institute of Social Research.

ELKAN, W. (1967), 'Circular migration and the growth of towns in East Africa', *International Labour Review*, 96:581-9.

ELKAN, W. (1976), 'Is a proletariat emerging in Nairobi?', *Economic Development and Cultural Change*, 24:695-706.

FEINDT, W. and H.L. BROWNING (1972), 'Return migration: its significance in an industrial metropolis and an agricultural town in Mexico', *International Migration Review*, 6:158-65.

FESTINGER, L. (1964), *Conflict, Decision, and Dissonance*, Stanford: University Press.

FINNEGAN, G.A. (1976), 'Population movement, labor migration, and social structure in a Mossi village', PhD thesis, Brandeis University, Waltham, Massachussetts.

FINNEY, B.R. (1973), *Big-Men and Businesss: Entrepreneurship and Economic Growth in the New Guinea Highlands*, Honolulu: University of Hawaii Press.

FIRTH, R. (1954), 'Social organization and social change', *Journal of the Royal Anthropological Institute of Great Britain and Ireland*, 84:1-20.

FIRTH, R. (1955), 'Some principles of social organization', *Journal of the Royal Anthropological Institute of Great Britain and Ireland*, 85:1-18.

FIRTH, R. (1959), *Social Change in Tikopia: Re-study of a Polynesian Community After a Generation*, London: George Allen & Unwin.

FIRTH, S.G. (1973), 'German recruitment and employment of labourers in the Western Pacific before the First World War', PhD thesis, Oxford University.

FORBES, D. (1978), 'Urban-rural interdependence: the trishaw riders of Ujung Pandang', in *Food, Shelter and Transport in Southeast Asia and the Pacific* (eds.) P.J. Rimmer, D.W. Drakakis-Smith and T.G. McGee: 219-36, Department of Human Geography Publication 12, Canberra: Research School of Pacific Studies, Australian National University.

BIBLIOGRAPHY

FOSTER, L.R. (1956), *Survey of Native Affairs, Port Moresby Area*, Mimeograph, Port Moresby: Department of Native Affairs.

FOX, R. (1967), *Kinship and Marriage: An Anthropological Perspective*, Middlesex: Penguin.

FRAZER, I.L. (1973), *To'ambaita Report: A Study of Socioeconomic Change in Northwest Malaita*, Wellington: Department of Geography, Victoria University.

FRAZER, I.L. (1981a), 'Man long taon: migration and differentiation amongst the To'ambaita, Solomon Islands', PhD thesis, Australian National University, Canberra.

FRAZER, I.L. (1981b), 'Trade unions and industrial relations in the Solomon Islands: the early 1960s', manuscript, Department of Anthropology, University of Otago, Dunedin.

FREMONT, A. (1976), *La région, Espace vécu*, Paris: Presses Universitaires de France.

FRIESEN, W. (1983), 'Accessibility and circulation in the western Solomon Islands', paper presented to Session on Accessibility, Circulation and Spatial Integration in Pacific Island Nations, Section C (Geography), Fifteenth Pacific Science Congress, Dunedin, New Zealand, February.

GALLAIS, J. (1976), 'De quelques aspects de l'éspace vécu dans les civilisations du monde tropical', *L'Espace Géographique*, 5:5-10.

GARBETT, G.K. (1960), *Growth and Change in a Shona Ward*, Occasional Paper 1, Salisbury: Department of African Studies, University College of Rhodesia and Nyasaland.

GARBETT, G.K. (1975), 'Circulatory migration in Rhodesia: towards a decision model', in *Town and Country in Central and Eastern Africa* (ed.) D. Parkin: 113-25, London: Oxford University Press for International African Institute.

GARBETT, G.K. and B. KAPFERER (1970), 'Theoretical orientations in the study of labour migration', *New Atlantis*, 2:179-97.

GARNAUT, R., M. WRIGHT, and R. CURTAIN (1977), *Employment, Incomes and Migration in Papua New Guinea Towns*, Research Monograph 6, Port Moresby: Institute of Applied Social and Economic Research.

GEORGE, P. (1959), *Questions de géographie de la population*, Institut national d'études démographiques, Travaux et Documents 34, Paris: Presses Universitaires de France.

GERMANI, G. (1965), 'Migration and acculturation' in *Handbook for Social Research in Urban Areas* (ed.) P.M. Hauser:159-78, Paris: UNESCO.

GIPEY, G.J. (1978), *A Personal History of Education and Mobility: Morobe, Papua New Guinea*, Occasional Paper 3, Port Moresby: Institute of Applied Social and Economic Research.

GODDARD, A.D. (1974), 'Population movements and land shortages in the Sokoto close-settled zone, Nigeria', in *Modern Migrations in Western Africa* (ed.) S. Amin:258-80, London: Oxford University Press for International African Institute.

GODDARD, S. (1965), 'Town-farm relationships in Yorubaland: a case study from Oyo', *Africa*, 35:21-29.

GODFREY, M. (1975), 'A note on rural-urban migration: an alternative framework of analysis', *Manpower and Unemployment Research in Africa*, 8:9–12.

GOLDSTEIN, S. (1964), 'The extent of repeated migration: an analysis based on the Danish population register', *Journal of the American Statistical Association*, 59:1121–32.

GOLDSTEIN, S. (1976), 'Facets of redistribution: research challenges and opportunities', *Demography*, 13:423–34.

GOLDSTEIN, S. (1977), 'Urbanization, migration, and fertility in Thailand', report prepared for Center for Population Research, National Institutes of Child Health and Human Development, Washington, DC.

GOLDSTEIN, S. (1978), *Circulation in the Context of Total Mobility in Southeast Asia*, Papers of the East-West Population Institute 53, Honolulu: East-West Center.

GOLDSTEIN, S. and A. (1979), 'Types of migration in Thailand in relation to urban-rural residence', in *Economic and Demographic Change: Issues for the 1980s*: 351–75, Proceedings, Helsinki, 1978, vol. 2, Liege: International Union for the Scientific Study of Population.

GOLDSTEIN, S. and A. (1985), 'Differentials in repeat and return migration in Thailand, 1965–70' in *Circulation in Third World Countries* (eds.) R.M. Prothero and M. Chapman: 380–413, London: Routledge and Kegan Paul.

GOULD, P. (1970), 'Tanzania, 1920–63: the spatial impress of the modernization process', *World Politics*, 22:149–70.

GOULD, W.T.S. (1977a), *A Bibliography of Population Mobility in Tropical Africa*, African Population Mobility Project Working Paper 31, Liverpool: Department of Geography, University of Liverpool.

GOULD, W.T.S. (1977b), *Longitudinal Studies of Population Mobility in Tropical Africa*, African Population Mobility Project Working Paper 29, Liverpool: Department of Geography, University of Liverpool.

GOULD, W.T.S. and R.M. PROTHERO (1975a), 'Population mobility in tropical Africa', in *The Population Factor in Tropical Africa* (eds.) R.P. Moss and R.J.A.R. Rathbone:95–106, London: University Press.

GOULD, W.T.S. and R.M. PROTHERO (1975b), 'Space and time in African population mobility', in *People on the Move: Studies on Internal Migration* (eds.) L.A. Kosiński and R.M. Prothero: 39–49, London: Methuen.

GRAVES, T.D. and N.B. GRAVES (1976), 'Demographic changes in the Cook Islands: perception and reality; Or, where have all the *Mapu* gone?', *Journal of the Polynesian Society*, 85:447–61.

GREGORY, J.W. (1974), 'Development and in-migration in Upper Volta', in *Modern Migrations in Western Africa* (ed.) S. Amin: 305–20, London: Oxford University Press for International African Institute.

GREGORY, J.W. and V. PICHÉ (1978), 'African migration and peripheral capitalism', in *Migration and the Transformation of Modern African Society: African Perspectives* (eds.) W.M.J. van Binsbergen and H.A. Meilink:37–50, Leiden: Afrika-Studiecentrum.

GROENEWEGEN, K. (nd 1972), *Report on the Census of the Population, 1970*, Honiara: Government of the British Solomon Islands Protectorate.

GROVES, M. (1973), 'Hiri', in *Anthropology in Papua New Guinea* (ed.) I. Hogbin:100–5, Melbourne: University Press.

GUGLER, J. (1968), 'The impact of labour migration on society and economy in SubSaharan Africa: empirical findings and theoretical considerations', *African Social Research*, 6:463–86.

GUGLER, J. (1969), 'On the theory of rural-urban migration: the case of SubSaharan Africa', in *Migration* (ed.) J.A. Jackson:134–55, Cambridge: University Press.

GUGLER, J. (1971), 'Life in a dual system: eastern Nigerians in town, 1961', *Cahiers d'Etudes Africaines*, 11:400–21.

GUIART, J. (1956), 'Unité culturelle et variations locales dans le Centre Nord des Nouvelles-Hébrides', *Journal de la Société des Océanistes*, 12:217–25.

HÄGERSTRAND, T. (1963), 'Geographic measurements of migrations: Swedish data', in *Human Displacements: Measurement, Methodological Aspects*, (ed.) J. Sutter:61–83, Monaco: Centre Internationale d'Etude des Problèmes Humains.

HAMNETT, M.P. (1977), 'Households on the move: settlement among a group of Eivo and Simeku speakers in central Bougainville', PhD thesis, University of Hawaii, Honolulu.

HAMNETT, M.P. and J. CONNELL (1981), 'Diagnosis and cure: the resort to traditional and modern medical practitioners in the North Solomons, Papua New Guinea', *Social Science and Medicine*, 15B: 489–98.

HANCE, W.A. (1970), *Population, Migration and Urbanization in Africa*, New York: Columbia University Press.

HARRÉ, J. (ed.) (1973), *Living in Town: Problems and Priorities in Urban Planning in the South Pacific*, Suva: South Pacific Social Sciences Association and the School of Social and Economic Development, University of the South Pacific.

HARRIS, G. (1974), 'Internal migration in Papua New Guinea: a survey of recent literature', *Yagl-Ambu* (Papua New Guinea Journal of the Social Sciences and Humanities), 1:154–81.

HARRIS, G.T. (1977), *Forces of Change in Subsistence Agriculture in Papua New Guinea*, Discussion Paper 31, Port Moresby: Economics Department, University of Papua New Guinea.

HART, J.K. (1974), 'Migration and the opportunity structure: a Ghanaian case-study', in *Modern Migrations in Western Africa* (ed.) S. Amin:321–42, London: Oxford University Press for International African Institute.

HAYS, T.E. (1976), *Anthropology in the New Guinea Highlands*, New York: Garland Publishing.

HEANEY, W.H. (1977), *Preliminary Findings and Implications of a Study of Rural Migration and Social Change in the Wahgi Valley, Western Highlands Province*, report to the Western Highlands Provincial Government, Honolulu: East-West Population Institute.

HENDERSHOT, G.E. (1971), 'Cityward migration and urban fertility in the Philippines', *Philippine Sociological Review*, 19:183–91.

HILL, P. (1963), *The Migrant Cocoa Farmers of Southern Ghana: A Study in Rural Capitalism*, Cambridge: University Press.

HOCART, A.M. (1929), *Lau Island, Fiji*, Bulletin 62, Honolulu: Bernice P. Bishop Museum.

HOGBIN, H.I. (1938), 'Social advancement in Guadalcanal, Solomon Islands', *Oceania*, 8:289–305.

HOGBIN, H.I. (1939), *Experiments in Civilization: the Effects of European Culture on a Native Community of the Solomon Islands*, London: George Routledge and Sons.

HOWARD, A. (1963), 'Conservatism and non-traditional leadership in Rotuma', *Journal of the Polynesian Society*, 72:65–77.

HOWLETT, D. (1973), 'Terminal development: from tribalism to peasantry', in *The Pacific in Transition: Geographical Perspectives on Adaptation and Change* (ed.) H. Brookfield:249–73, Canberra: Australian National University Press.

HOWLETT, D. (1980), 'When is a peasant not a peasant? Rural proletarianisation in Papua New Guinea', in *Of Time and Place: Essays in Honour of OHK Spate* (eds.) J.N. Jennings and G.J.R. Linge: 193–210, Canberra: Australian National University Press.

HUGO, G.J. (1975), 'Population mobility in West Java, Indonesia', PhD thesis, Australian National University, Canberra.

HUGO, G. (1978), *Population Mobility in West Java*, Yogyakarta: Gadjah Mada University Press.

HUGO, G. (1983), 'New conceptual approaches to migration in the context of urbanization: a discussion based on Indonesian experience', in *Population Movements: Their Forms and Functions in Urbanization and Development* (ed.) P. Morrison:69–113, Liege: Ordina Editions for International Union for the Scientific Study of Population.

HUGO, G. (1985), 'Circulation in West Java, Indonesia', in *Circulation in Third World Countries* (eds.) R.M. Prothero and M. Chapman: 75–99, London, Routledge and Kegan Paul.

HUGHES, A.V. (1969), 'Low-cost housing and home ownership in Honiara, British Solomon Islands Protectorate', *South Pacific Bulletin* 19:19–26.

INGLIS, J. (1851), *Report of a Missionary Tour in the New Hebrides . . . in the Year 1850 on Board HMS 'Havannah'*, Auckland: Williamson and Wilson.

IVENS, W.G. (1930), *The Island Builders of the Pacific*, London: Seeley, Service and Company.

JACKSON, R.T. (1975), 'A R.I.P. off?', *Yagl-Ambu* (Papua New Guinea Journal of the Social Sciences and Humanities), 2:285–94.

JACOBSON, D. (1973), *Itinerant Townsmen: Friendship and Social Order in Uganda*, Menlo Park, California: Cummings.

455

JELLINEK, L. (1978), 'Circular migration and the *pondok* dwelling system: a case study of ice-cream traders in Jakarta', in *Food, Shelter and Transport in Southeast Asia and the Pacific* (eds.) P.J. Rimmer, D.W. Drakakis-Smith and T.G. McGee:135–54, Department of Human Geography Publication HG/12, Canberra: Research School of Pacific Studies, Australian National University.

KASSMAUL, A.S. (1981), 'The ambiguous mobility of farm servants', *Economic History Review*, 34:222–35.

KEENY, S.M. (1973), 'East Asia review, 1972', *Studies in Family Planning*, 4:101–31.

KEESING, R.M. (1967), 'Christians and pagans in Kwaio, Malaita', *Journal of the Polynesian Society*, 76:82–100.

KEESING, R.M. (1976), *Cultural Anthropology: a Contemporary Perspective*, New York: Holt, Rinehart and Winston.

KEESING, R.M. (1976), 'Politico-religious movements and anticolonialism on Malaita: Maasina rule in historical perspective', *Oceania*, 48:241–61, 49:46–73.

KEOWN, P.A. (1971), 'The career cycle and the stepwise migration process', *New Zealand Geographer*, 27:174–84.

KERR, C. (1960), 'Changing social structures', in *Labor Commitment and Social Change in Developing Areas* (eds.) W.E. Moore and A.S. Feldman: 348–59, New York: Social Science Research Council.

KUANGE, P. (1977), 'Omkalai', in *The Rural Survey 1975, Supplement to Vol. 4 of Yagl-Ambu* (eds.) J. Conroy and G. Skeldon:121–39, Port Moresby: Institute of Applied Social and Economic Research.

LAURO, D.J. (1979a), 'The demography of a Thai village: methodological considerations and substantive conclusions from field study in a Central Plains community', PhD thesis, Australian National University, Canberra.

LAURO, D.J. (1979b), 'Life history matrix analysis: a progress report', in *Residence History Analysis* (ed.) R.J. Pryor: 134–54, Studies in Migration and Urbanization 3, Canberra: Department of Demography, Australian National University.

LEA, D.A.M. and H.C. WEINAND (1971), 'Some consequences of population growth in the Wosera area, East Sepik District', in *Population Growth and Socio-Economic Change* (ed.) M.W. Ward:122–36, Boroko and Canberra: New Guinea Research Unit and Australian National University.

LEAHY, M. and M. CRAIN (1937), *The Land that Time Forgot: Adventures and Discoveries in New Guinea*, London: Hurst and Blackett.

LEE, E.S. (1966), 'A theory of migration', *Demography*, 3:47–57.

LEVINE, H.B. and M.W. LEVINE (1979), *Urbanisation in Papua New Guinea: A Study of Ambivalent Townsmen*, Cambridge: University Press.

LIPTON, M. (1977), *Why Poor People Stay Poor: A Study of Urban Bias in World Development*, Cambridge, USA: Harvard University Press.

LIVINGSTONE, F.B. (1973), 'Gene frequency differences in human populations: some problems of analysis and interpretation', in

Methods and Theories of Anthropological Genetics (eds.) M.H. Crawford and P.L. Workman:39–66, Albuquerque: University of New Mexico Press.

LODHIA, R.N. (1977), *Report on the Census of the Population 1976, Volume 1*, Parliamentary Paper 13 of 1977, Suva: Government Printer.

LUANA, C. (1969), 'Buka! A retrospect', *New Guinea and Australia, the Pacific and South-east Asia*, 4:15–20.

LYSTAD, R.A. (1965), 'Introduction', in *The African World: A Survey of Social Research* (ed.) R.A. Lystad:1–7, New York: Praeger.

MABOGUNJE, A.L. (1972), *Regional Mobility and Resource Development in West Africa*, Toronto: McGill-Queen's University Press.

MAIN, H.A.C. (1981), 'Time-space study of daily activity in urban Sokoto, Nigeria', PhD thesis, University of Liverpool.

MAMAK, A. and R. BEDFORD (1977), 'Inequality in the Bougainville copper mining industry: some implications', in *Racism: The Australian Experience: Volume 3, Colonialism and After* (ed.) F.S. Stevens and E.P. Wolfers:427–55, Sydney: Australia and New Zealand Book Company.

MANTRA, I.B. (1978), 'Population movement in wet rice communities: a case study of two *dukuh* in Yogyakarta Special Region', PhD thesis, University of Hawaii.

MANTRA, I.B. (1981), *Population Movement in Wet Rice Communities: A Case Study of Two Dukuh in Yogyakarta Special Region*, Yogyakarta: Gadjah Mada University Press.

MASSER, I. and W.T.S. GOULD (with the assistance of A.D. Goddard) (1975), *Inter-Regional Migration in Tropical Africa*, Special Publication 8, London: Institute of British Geographers.

MAUDE, A. (1979), 'How circular is Minangkabau migration?', *Indonesian Journal of Geography*, 9:1–12.

MAYER, P. (1961), *Townsmen or Tribesmen: Conservatism and the Process of Urbanization in a South African City*, Capetown: Oxford University Press for the Institute of Social and Economic Research, Rhodes University.

McARTHUR, N. (1966), 'The demographic situation', in *New Guinea on the Threshold* (ed.) E.K. Fisk:103–14, Canberra: Australian National University Press.

McARTHUR, N. and J.F. YAXLEY (1968), *Condominium of the New Hebrides: A Report on the First Census of the Population 1967*, Sydney, New South Wales: Government Printer.

McKAY, J. and J.S. WHITELAW (1977), 'The role of large private and government organizations in generating flows of inter-regional migrants: the case of Australia', *Economic Geography*, 53:28–44.

MEILLASOUX, C. (ed.) (1971), *L'évolution du commerce Africain depuis le XIX siècle en Afrique de l'Ouest*, London: Oxford University Press for the International African Institute.

MEINKOTH, M.R. (1962), 'Migration in Thailand with particular reference to the northeast,' *Economics and Business Bulletin of the School of Business and Public Administration* (Temple University), 14:2–45.

457

MILLER, J.C. (ed.) (1980), *The African Past Speaks*, Folkestone, Kent: William Dawson Sons.

MINER, H.M. (1965), 'Urban influences on the rural Hausa', in *Urbanization and Migration in West Africa* (ed.) H. Kuper:110–30, Los Angeles: University of California Press.

MINTZ, S.W. (1974), 'The rural proletariat and the problem of rural proletarian consciousness', *Journal of Peasant Studies*, 1:291–325.

MITCHELL, J.C. (1954), 'Urbanization, detribalization, stabilization and urban commitment in Southern Africa: a problem of definition and measurement', in *Social Implications of Industrialization and Urbanization in Africa South of the Sahara* (ed.) C.D. Forde:693–711, Paris: UNESCO.

MITCHELL, J.C. (1959), 'The causes of labour migration', *Bulletin of the Inter-African Labour Institute*, 6:12–46.

MITCHELL, J.C. (1961a), 'The causes of labour migration', in *Migrant Labour in Africa South of the Sahara*:259–80, Abidjan: Commission for Technical Cooperation in Africa South of the Sahara.

MITCHELL, J.C. (1961b), 'Wage labour and African population movements in Central Africa', in *Essays on African Population* (eds.) K.M. Barbour and R.M. Prothero:193–248, London: Routledge and Kegan Paul.

MITCHELL, J.C. (1969a), 'Structural plurality, urbanization and labour circulation in Southern Rhodesia', in *Migration* (ed.) J.A. Jackson: 156–180, Cambridge: University Press.

MITCHELL, J.C. (1969b), 'Urbanization, detribalization, stabilization and urban commitment in Southern Africa: a problem of definition and measurement:1968', in *Urbanism, Urbanization, and Change: Comparative Perspectives* (eds.) P. Meadows and E.H. Mizruchi: 470–93, Reading: Addison-Wesley Publishing Company.

MITCHELL, J.C. (1985), 'Towards a situational sociology of wage-labour circulation', in *Circulation in Third World Countries* (eds.) R.M. Prothero and M. Chapman:30–54, London: Routledge and Kegan Paul.

MOORE, W.E. (1951), *Industrialization and Labour: Some Aspects of Economic Development*, New York: Cornell University Press.

MOORE, W.E. (1967), *Order and Change: Essays in Comparative Sociology*, New York: John Wiley and Sons.

MOORE, W.E. and A.S. FELDMAN (eds.) (1960), *Labor Commitment and Social Change in Developing Areas*, New York: Social Science Research Council.

MORAUTA, L.M. (1979), *Facing the Facts: The Need for Policies for Permanent Urban Residents*, Discussion Paper 26, Port Moresby: Institute of Applied Social and Economic Research.

MORAUTA, L.M. (1981), 'Mobility patterns in Papua New Guinea: social factors as explanatory variables', in *Population Mobility and Development: Southeast Asia and the Pacific* (eds.) G.W. Jones and H.V. Richter:205–28, Development Studies Centre Monograph 27, Canberra: Research School of Pacific Studies, Australian National University.

MORAUTA, L.M. and M. HASU (1979), *Rural-Urban Relationships in*

Papua New Guinea: Case Material from the Gulf Province on Net Flows, Discussion Paper 25, Port Moresby: Institute of Applied Social and Economic Research.

MORAUTA, L. and D. RYAN (1982), 'From temporary to permanent townsmen: migrants from the Malalaua district, Papua New Guinea', *Oceania*, 53:39–55.

MORRILL, R.L. (1963), 'The development of models of migration and the role of electronic processing machines', in *Human Displacements: Measurement, Methodological Aspects* (ed.) J. Sutter:213–30, Monaco: Centre International d'Etude des Problèmes Humains.

MORTIMORE, M.J. (1972), 'Some aspects of rural-urban relations in Kano, Nigeria', in *La Croissance Urbaine en Afrique noire et à Madagascar* (ed.) P. Vennetier: 871–88, Paris: Centre National de la Recherche Scientifique.

VAN DEN MUIJZENBERG, O.D. (1973), *Horizontal Mobility in Central Luzon, Characteristics and Background*, Publication 19, Amsterdam: Center for Anthropological and Sociological Studies, University of Amsterdam.

VAN DEN MUIJZENBERG, O.D. (1975), 'Involution or evolution in Central Luzon?', in *Current Anthropology in the Netherlands* (eds.) H.J.M. Classen and P. Kloos: 141–55, The Hague: N.S.A.V./Staatsdrukkerj.

MURDOCK, G.P., (1949), *Social Structure*, New York: Macmillan.

MURDOCK, G.P. and S. WILSON (1972), 'Settlement pattern and community organization: cross-cultural code 3', *Ethnology*, 11:254–95.

NAGATA, J.A. (1974), 'Urban interlude: some aspects of internal Malay migration in West Malaysia', *International Migration Review*, 8:301–24.

NAIR, S. (1980), *Rural-born Fijians and Indo-Fijians in Suva: A Study of Movements and Linkages*, Development Studies Centre Monograph 24, Canberra: Research School of Pacific Studies, Australian National University.

NANKIVELL, P.S. (1978), 'Income inequality in Tavenui district: a statistical analysis of data from the resource base survey', in *Taveuni: Land, Population and Production* (eds.) H.C. Brookfield, B. Denis, R.D. Bedford, P.S. Nankivell, and J.B. Hardaker:245–98, UNESCO/UNFPA Fiji Island Report 3, Canberra: Australian National University for UNESCO.

NASH, J. (1974), *Matriliny and Modernisation: the Nagovisi of South Bougainville*, New Guinea Research Bulletin 55, Port Moresby: New Guinea Research Unit and Canberra: Australian National University.

NAYACAKALOU, R.R. (1963), 'The urban Fijians of Suva', in *Pacific Ports and Towns: A Symposium* (ed.) A. Spoehr:33–41, Honolulu: Bishop Museum Press.

NAYACAKALOU, R.R. (1975), *Leadership in Fiji*, Melbourne: Oxford University Press in association with The University of South Pacific.

NELSON, J.M. (1976), 'Sojourners versus new urbanities: causes and consequences of temporary verus permanent cityward migration in developing countries', *Economic Development and Cultural Change*, 24:721–57.

ODONGO, J. and J.P. LEA (1977), 'Home ownership and rural-urban links in Uganda', *The Journal of Modern African Studies*, 15:59–73.

OESER, L. (1969), *Hohola: The Significance of Social Networks in Urban Adaptation of Women in Papua New Guinea's First Low-cost Housing Estate*, New Guinea Research Bulletin 29, Port Moresby: New Guinea Research Unit and Canberra: Australian National University.

OGAN, E. (1972), *Business and Cargo: Socio-Economic Change Among the Nasioi of Bougainville*, New Guinea Research Bulletin 44, Port Moresby: New Guinea Research Unit and Canberra: Australian National University.

OJO, G.J.A. (1970), 'Some observations on journey to agricultural work in Yorubaland, southwestern Nigeria', *Economic Geography*, 46:459–71.

OJO, G.J.A. (1973), 'Journey to agricultural work in Yorubaland', *Annals Associations of American Geographers*, 63:85–96.

OLIVER, D.L. (1955), *A Solomon Island Society: Kinship and Leadership Among the Siuai of Bougainville*, Cambridge, Mass.: Harvard University Press.

ORAM, N.D. (1968a), 'Culture change, economic development and migration among the Hula', *Oceania*, 38:243–75.

ORAM, N.D. (1968b), 'The Hula in Port Moresby', *Oceania*, 39:1–35.

ORAM, N.D. (1976), *Colonial Town to Melanesian City: Port Moresby 1884–1974*, Canberra: Australian National University Press.

Papua New Guinea, Central Planning Office (1976), *The Post-Independence National Development Strategy*, Port Moresby.

Papua New Guinea, Committee on Review on Housing (1978), *Report*, Port Moresby: Government Printer.

Papua New Guinea, National Planning Office (1977), *Managing Urbanisation in Papua New Guinea*, Waigani.

PARKIN, D. (ed.) (1975), *Town and Country in Central and Eastern Africa*, London: Oxford University Press for International African Institute.

Parliament of the Commonwealth of Australia (1924), *Report to the League of Nations on the Administration of the Territory of New Guinea from 1 July 1922 to 30 June 1923*, Melbourne: Government Printer.

PERLMAN, J.E. (1976), *The Myth of Marginality: Urban Poverty and Politics in Rio de Janeiro*, Berkeley: University of California Press.

PERNIA, E.M. (1975), *Philippine Migration Streams: Demographic and Socioeconomic Characteristics*, Research Note 51, Manila: Population Institute, University of the Philippines.

PICHÉ, V., J. GREGORY, and S. COULIBALY (1980), 'Vers une explication des courants migratoires Voltaiques', *Land, Capital and Society*, 13:77–103.

PRACHUABMOH, V. and P. TIRASAWAT (1974), *Internal Migration in Thailand, 1947–1972*, Paper 7, Bangkok: Institute of Population Studies, Chulalongkorn University.

PROTHERO, R.M. (1957), 'Migratory labour from northwestern Nigeria',

Africa, 27:251–61.
PROTHERO, R.M. (1959), *Migrant Labour from Sokoto Province,*
Northern Nigeria, Kaduna: Government Printer.
PROTHERO, R.M. (1961), 'Population movements and problems of
malaria eradication in Africa', *Bulletin of the World Health Organ-*
ization, 24:405–25.
PROTHERO, R.M. (1963), 'Population mobility and trypanosomiasis in
Africa', *Bulletin of the World Health Organization*, 28:615–26.
PROTHERO, R.M. (1964), 'Continuity and change in African population
mobility', in *Geographers and the Tropics: Liverpool Essays* (eds.)
R.W. Steel and R.M. Prothero:189–213, London: Longman.
PROTHERO, R.M. (1965), *Migrants and Malaria*, London: Longmans.
PROTHERO, R.M. (1968), 'Migration in tropical Africa', in *The Popula-*
tion of Tropical Africa (eds.) J.C. Caldwell and C. Okonjo:250–63,
London: Longmans.
PROTHERO, R.M. (1973), 'Migrant adjustment: spatial perspectives', in
Proceedings, International Population Conference, Vol. 1:305–13,
Liege: International Union for the Scientific Study of Population.
PROTHERO, R.M. (1974), 'Foreign migrant labour in South Africa',
International Migration Review, 8:383–94.
PROTHERO, R.M. (1975), *Mobility in North Western Nigeria: Perspec-*
tive and Prospects, African Population Mobility Project Working
Paper 23, Liverpool: Department of Geography, University of
Liverpool.
PROTHERO, R.M. (1977), 'Disease and mobility: a neglected factor in
epidemiology', *International Journal of Epidemiology*, 6:259–267.
PROTHERO, R.M. (1979), 'The context of circulation in West Africa',
Population Geography, 1:22–40.
RACULE, R.K. (1974), 'Mere Usakinakula in Suva: a migrant case
study', *Basic Research in Pacific Islands Geography 1*, Suva: School
of Social and Economic Development, University of the South Pacific.
RAISON, J-P. (1976), 'Espaces significatifs et perspectives régionales à
Madagascar', *L'Espace Géographique*, 5:189–203.
READ, K.E. (1954), 'Cultures of the Central Highlands, New Guinea',
Southwestern Journal of Anthropology, 10:1–43.
REW, A.W. (1974), *Social Images and Process in Urban New Guinea: A*
Study of Port Moresby, Monograph 57, Seattle: American Ethno-
logical Society.
RICHARDS, A.I. (1954), *Economic Development and Tribal Change:*
A Study of Immigrant Labour in Buganda, Cambridge: University
Press for the East African Institute for Social Research.
VAN RIJSWIJCK, O. (1967), 'Bakoiudu: resettlement and social change
among the Kuni of Papua', PhD thesis, Australian National Univer-
sity, Canberra.
RODMAN, M. (1979), 'Following peace: indigenous pacification of a
northern New Hebridean society', in *The Pacification of Melanesia*
(ed.) M. Rodman and M. Cooper: 141–60, ASAO Monograph 7,
Ann Arbor, Michigan: University Press.
RODMAN, W.L. (1973), 'Man of influence, man of rank: leadership

and the graded society on Aoba', PhD thesis, University of Chicago.

DEL ROSARIO, J., M. LOURDES and Y. KIM (1977), 'Migration differentials by migration types in the Philippines: a study of migration typology and differentials, 1970', paper presented at the General Conference, International Union for the Scientific Study of Population, Mexico City, August.

ROSEMAN, C.C. (1971), 'Migration as a spatial and temporal process', *Annals, Association of American Geographers*, 61:589–98.

ROSS, H.M. (1973), *Baegu: Social and Ecological Organization in Malaita, Solomon Islands*, Urbana: University of Illinois Press.

ROUCH, J. (1957), 'Migration au Ghana', *Journal de la Société des Africanistes*, 26:33–196.

ROUCH, J. (1960), 'Problèmes relatifs à l'étude des migrations traditionelles et des migrations actuelles en Afrique occidentale, *Bulletin de l'Institut Français d'Afrique Noire*, 22:369–78.

ROWLEY, C.D. (1965), *The New Guinea Villager: A Retrospect from 1964*, Melbourne: F.W. Cheshire.

ROYALL, R.M. (1970), 'On finite population sampling theory under certain linear regression models', *Biometrika*, 57:377–87.

RYAN, D. (1965), 'Social change among the Toaripi of Papua', MA Thesis, University of Sydney.

RYAN, D. (1968), 'The migrants', *New Guinea and Australia, the Pacific and Southeast Asia*, 2:60–6.

RYAN, D. (1970), 'Rural and urban villagers: a bi-local social system in Papua', PhD thesis, University of Hawaii.

SAHLINS, M.D. (1963), 'Poor man, rich man, big man, chief: political types in Melanesia and Polynesia', *Comparative Studies in Society and History*, 5:285–303.

SAHLINS, M.D. (1972), *Stone Age Economics*, Chicago: Aldine.

SALISBURY, R.F. (1964), 'Despotism and Australian administration in the New Guinea Highlands', in *New Guinea: The Central Highlands* (ed.) J.B. Watson:225–39, Special Publication, Menasha, Wisconsin: American Anthropological Association.

SALISBURY, R.F. and M.E. SALISBURY (1970), 'Siane migrant workers in Port Moresby', *Industrial Review*, 8:5–11.

SCARR, D. (1967), *Fragments of Empire: A History of the Western Pacific High Commission, 1877–1914*, Canberra: Australian National University Press.

SEILER, D. (1974), *Gulf Report – Development in the Eastern Papuan Gulf, Papua New Guinea: A Case Study of the Cultural Conditions and Needs of the Eastern Elema People in the Malalaua Sub-district and in Port Moresby*, Joint program of studies in the transport process, Port Moresby: Department of Transport, Government of Papua New Guinea.

SELIGMANN, C.G. (1910), *The Melanesians of British New Guinea*, Cambridge: University Press.

SEVELE, F. (1973), *The Importance of Agriculture in Pacific Island*

Development, New Zealand: Episcopal Conference of the Pacific and CORSO.

SHINEBERG, D. (1967), *They Came for Sandalwood: A Study of the Sandalwood Trade in the South-west Pacific, 1830-1865*, Melbourne: University Press.

SIAOA, C. (1977), 'Lese Oalai', in *The Rural Survey 1975, Supplement to Vol. 4 of Yagl-Ambu* (eds.) J. Conroy and G. Skeldon:161–79, Port Moresby: Institute of Applied Social and Economic Research.

SILVERMAN, M.G. (1971), *Disconcerting Issue: Meaning and Struggle in a Resettled Pacific Community*, Chicago: University Press.

SINGHANETRA-RENARD, A. (1981), 'Mobility in north Thailand: a view from within', in *Population Mobility and Development: Southeast Asia and the Pacific* (eds.) G.W. Jones and H.V. Richter:137–66, Development Studies Centre Monograph 27, Canberra: Research School of Pacific Studies, Australian National University.

SINGHANETRA-RENARD, A. (1982), 'Northern Thai mobility 1870–1977: a view from within', PhD thesis, University of Hawaii.

SKELDON, R. (1976), *The Growth of Goroka: Towards an Interpretation of the Past and a Warning for the Future*, Discussion Paper 6, Port Moresby: Institute of Applied Social and Economic Research.

SKELDON, R. (1977a), 'The evolution of migration patterns during urbanization in Peru', *Geographical Review*, 67:394–411.

SKELDON, R. (1977b), *Internal Migration in Papua New Guinea: A Statistical Description*, Discussion Paper 11, Port Moresby: Institute of Applied Social and Economic Research.

SKELDON, R. (1978a), *Evolving Patterns of Population Movement in Papua New Guinea with Reference to Policy Implications*, Discussion Paper 17, Port Moresby: Institute of Applied Social and Economic Research.

SKELDON, R. (1978b), *Recent Urban Growth in Papua New Guinea*, Discussion Paper 21, Port Moresby: Institute of Applied Social and Economic Research.

SKINNER, E.P. (1960), 'Labour migration and its relationship to sociocultural change in Mossi society', *Africa*, 30:375–401.

SKINNER, E.P. (1965), 'Labor migration among the Mossi of the Upper Volta', in *Urbanization and Migration in West Africa* (ed.) H. Kuper: 60–84, Los Angeles: University of California Press.

SMITH, D.W. (1975), *Labour and the Law in Papua New Guinea*, University Monograph 1, Canberra: Development Studies Centre, Australian National University.

SMITH, R.H.T. (ed.) (1972), 'Markets in West Africa', *Economic Geography*, 48 (Special number):299–355.

Solomon Islands, Statistics Division (1980), *Report on the Census of Population 1976*, vol. 1, Honiara: Ministry of Finance.

SONGRE, A., J-M. SAWADOGO, and G. SANOGOH (1974), 'Réalités et effets de l'emigration massive des Voltaiques dans le contexte de l'Afrique Occidentale', in *Modern Migrations in Western Africa* (ed.) S. Amin: 384–406, London: Oxford University Press for International African Institute.

SOPE, B. (1976), *Land and Politics in the New Hebrides*, Suva: South Pacific Social Sciences Association.

SOUTHALL, A.W. and P.C. GUTKIND (1957), *Townsmen in the Making: Kampala and its Suburbs*, East African Studies 9, Kampala: East African Institute of Social Research.

SPATE, O.H.K. (1959), *The Fijian People: Economic Problems and Prospects*, Legislative Council Paper 13 of 1959, Suva: Government Printer.

STAMPER, B.M. (1973), 'Population policy in development planning', *Reports on Population/Family Planning* 13:1–30.

STENNING, D.J. (1957), 'Transhumance, migratory drift, migration: patterns of pastoral Fulani nomadism', *Journal of the Royal Anthropological Institute*, 87:57–73.

STRANGE, H. (1976), 'Village paths and city routes', paper prepared for the Conference on Woman and Development, Wellesley, Massachusetts, June.

STRATHERN, M. (1972), 'Absentee businessmen: the reaction at home to Hageners migrating to Port Moresby', *Oceania*, 43:19–39.

STRATHERN, M. (1975), *No Money on our Skins: Hagen Migrants in Port Moresby*, New Guinea Research Bulletin 61, Port Moresby: New Guinea Research Unit and Canberra: Australian National University.

STRATHERN, M. (1977), 'The disconcerting tie: attitudes of Hagen migrants towards home', in *Change and Movement: Readings on Internal Migration in Papua New Guinea* (ed.) R.J. May:247–66, Canberra: Australian National University Press and Port Moresby: Institute of Applied Social and Economic Research.

STRAUCH, J. (1977), 'Cyclical labor migration among Chinese in Hong Kong and Malaysia: social and political implications for rurally-based workers, their families, and their home communities', proposal, Department of Anthropology, Harvard University.

STRAUCH, J. (1980), 'Circulation in Malaysia and Hong Kong: linkages between town and villages', paper presented at Intermediate Cities in Asia Meeting, Population Institute, East-West Center, Honolulu, July.

STRAUCH, J. (1984), 'Women in rural-urban circulation networks: implications for social structural change', in *Women in the Cities of Asia: Migration and Urban Adaptation* (eds.) J.T. Fawcett, S-E. Khoo, and P.C. Smith:60–77, Boulder, Colorado: Westview Press.

STRETTON, A. (1978), 'Independent foremen and the construction of formal sector housing in the Greater Manila area', in *Food, Shelter and Transport in Southeast Asia and the Pacific* (eds.) P.J. Rimmer, D.W. Drakakis-Smith and T.G. McGee:155–69, Department of Human Geography Publication HG/12, Canberra: Research School of Pacific Studies, Australian National University.

STRETTON, A.W. (1979), *Urban Housing Policy in Papua New Guinea*, Research Monograph 8, Port Moresby: Institute of Applied Social and Economic Research.

STRETTON, A.W. (1981), 'The building industry and urbanization in Third World countries: a Philippine case study', *Economic Development and Cultural Change*, 29:325–39.

SUTLIVE, V.H. Jr. (1977), 'A comparison of urban migration in Sarawak and the Philippines', *Urban Anthropology*, 6:355–69.

SWINDELL, K. (1977), ' "Strange Farmers" of the Gambia', paper presented at International Seminar on Migration and Rural Development in Tropical Africa, Afrika-Studiecentrum, Leiden, May.

SWINDELL, K. (1978), 'Family farms and migrant labour: the strange farmers of the Gambia', *Canadian Journal of African Studies*, 12:3–17.

SWINDELL, K. (1979), 'Sera Woollies, Tillibunkas and Strange Farmers: the development of migrant groundnut farming along the Gambia river', *Journal of African History*, 21:93–104.

TEDDER, J.L.O. (1966), 'Honiara (Capital of the British Solomon Islands Protectorate)', *South Pacific Commission Quarterly Bulletin*, 16:36–43.

TEXTOR, R.B. (1956), 'The northeastern samlor driver in Bangkok', in *The Social Implications of Industrialization and Urbanization* (ed.) P.M. Hauser, Calcutta: UNESCO.

Thailand National Statistical Office (1978), *The Survey of Migration in Bangkok Metropolis, 1977*, Publication Series N-Rep., 1–78, Bangkok: National Statistical Office, Office of the Prime Minister.

THOMPSON, L. (1940), *Fijian Frontier*, San Francisco: American Council, Institute of Pacific Relations.

TODARO, M.P. (1969), 'A model of labor migration and urban unemployment in less developed countries', *American Economic Review*, 59:138–48.

TODARO, M.P. (1976), *Internal Migration in Developing Countries: A Review of Theory, Evidence, Methodology and Research Priorities*, Geneva: International Labour Office.

UDO, R.K. (1970), 'Census migrations in Nigeria', *Nigerian Geographical Journal*, 13:3–8.

ULLMAN, E.L. (1956), 'The role of transportation and the bases for interaction', in *Man's Role in Changing the Face of the Earth* (ed.) W.L. Thomas: 862–80, Chicago: University Press.

United Nations (1970), *Methods of Measuring Internal Migration, Manual IV*, ST/SOA/Series A/47, New York: United Nations.

UNDAT (1974), *Report to the Independent Chairman on a Study of the Fiji Sugar Industry*, Suva.

UNESCO/UNFPA (1977), *Population, Resources and Development in the Eastern Islands of Fiji: Information for Decision-Making*, Canberra: Development Studies Centre, Australian National University for UNESCO.

UN Economic and Social Commission for Asia and the Pacific (1980), *National Migration Surveys*, Manuals I–VII, New York: United Nations.

VELE, M. (1977), 'A survey of four peri-urban settlements near Port Moresby', *Yagl-Ambu* (Papua New Guinea Journal of the Social Sciences and Humanities), 4:191–210.

VELE, M. (1978), *Rural Village and Peri-urban Settlement: A Case-study of Circulation from the Central Province*, Occasional Paper 2, Port Moresby: Institute of Applied Social and Economic Research.

VUSONIWAILALA, L. (1978), 'Stop the rock! Young islanders want to get off', *Pacific Islands Monthly*, 49:11-12.

WAIKO, J.D. (1978), 'The political responsibility of a writer', *Read* 13 (1):22-5.

WALSH, A.C. (ed.) (1976), *The characteristics of some urban households in Suva, Fiji*, Urban Report 5, Suva: School of Social and Economic Development, University of the South Pacific.

WALSH, A.C. (1977), 'Urbanization in Fiji', *Perspective 14*, Christchurch: Manawatu Branch, New Zealand Geographical Society.

WALSTER, E. and E. BERCHEID (1968), 'The effects of time on cognitive consistency' in *Theories of Cognitive Consistency: A Source Book* (eds.) R.P. Abelson, W.J. McGuire, T.M. Newcomb, M.J. Rosenberg, and P.H. Tannenbaum: 599-608, Chicago: Rand McNally.

WARD, R.G. (1961), 'Internal migration in Fiji', *Journal of the Polynesian Society*, 70:257-71.

WARD, R.G. (1965), *Land Use and Population in Fiji: A Geographical Study*, London: Her Majesty's Stationery Office.

WARD, R.G. (1967), 'The consequences of smallness in Polynesia', in *Problems of Smaller Territories* (ed.) B. Benedict:80-96, Institute of Commonwealth Studies Papers 10, London: Athlone Press.

WARD, R.G. (1971), 'Internal migration and urbanisation in Papua New Guinea', in *Population Growth and Socio-Economic Change* (ed.) M.W. Ward: 81-107, New Guinea Research Bulletin 42, Boroko: New Guinea Research Unit and Canberra: Australian National University.

WARD, R.G. (1980), 'Migration, myth and magic in Papua New Guinea', *Australian Geographical Studies*, 18:119-34.

WATSON, J.B. (1967), 'Tairora: the politics of despotism in a small society', *Anthropological Forum*, 2:53-104.

WATSON, J.B. (1970), 'Society as organized flow: the Tairora case', *Southwestern Journal of Anthropology*, 26:107-24.

WATSON, J.B. (1980), 'Crowded fields and the intensification of reciprocity', Typescript, Copyright registration number TXU 37-110, Washington: US Copyright Office.

WATSON, J.L. (1975), *Emigration and the Chinese Lineage: The Mans in Hong Kong and London*, Berkeley: University of California Press.

WATSON, W. (1958), *Tribal Cohesion in a Money Economy: A Study of the Mambwe People of Northern Rhodesia*, Manchester: University Press.

WATTERS, R.F. (1969), *Koro: Economic Development and Social Change in Fiji*, Oxford: Clarendon Press.

WEST, F.J. (1958), 'Indigenous labour in Papua New Guinea', *International Labour Review*, 77:89-112.

WHEELER, G.C. (1943), *The Mono-Alu People of Bougainville Strait*, Microfilm FM 4/2701, Sydney: Mitchell Library.

WHITEMAN, J. (1973), *Chimbu Family Relationships in Port Moresby*,

New Guinea Research Bulletin 52, Port Moresby: New Guinea
Research Unit and Canberra: Australian National University.

WHITNEY, V.H. (1976), 'Population planning in Asia in the 1970s',
Population Studies, 30:337-51.

WIESENFELD, S.L. and D.C. GAJDUSEK (1976), 'Genetic structure and
heterozygosity in the Kuru region, Eastern Highlands of New
Guinea', *American Journal of Physical Anthropology*, 45:177-89.

WILMSEN, E.M. (1973), 'Interaction, spacing behaviour, and the organization of hunting bands', *Journal of Anthropological Research*,
29:1-31.

WILSON, F. (1972), *Labour in the South African Goldmines 1911-65*,
Cambridge: University Press.

WOLFERS, E.P. (1975), *Race Relations and Colonial Rule in Papua New
Guinea*, Sydney: Australian and New Zealand Book Company.

YOUNG, E.A. (1973), 'Population mobility in the Kainantu area:
patterns of movement of the Agarabi/Gadsup people from contact
until the present day', MA thesis, University of Papua New Guinea,
Port Moresby.

YOUNG, E.A. (1977a), 'Population mobility in Agarabi/Gadsup,
Eastern Highlands Province', in *Change and Movement: Readings
on Internal Migration in Papua New Guinea* (ed.) R.J. May: 173-202,
Canberra: Papua New Guinea Institute of Applied Social and Economic Research in association with Australian National University.

YOUNG, E.A. (1977b), 'Simbu and New Ireland migration', PhD thesis,
Australian National University, Canberra.

YOUNG, F.S.H. (nd 1925), *Pearls from the Pacific*, London: Marshall
Brothers.

YOUNG, M.L. (1982), 'Critical moves: a life history approach to the
structural analysis of migration', invited paper presented at Conference on Urbanization and National Development, East-West Center,
Population Institute, Honolulu, Hawaii, January.

ZACHARIAH, K.C. and CONDE, J., (1981), *Migration in West Africa:
Demographic Aspects*, New York: Oxford University Press for the
World Bank.

ZELINSKY, W. (1971), 'The hypothesis of the mobility transition',
The Geographical Review, 61:219-49.

ZELINSKY, W. (1979), 'The demographic transition: changing patterns
of migration', in *Population Science in the Service of Mankind*:
165-89, Liege: International Union for the Scientific Study of
Population.

ZWART, F.H.A.G. (1968), *Report on the Census of the Population
1966*, Legislative Council Paper 9 of 1968, Suva: Government
Printer.

Index

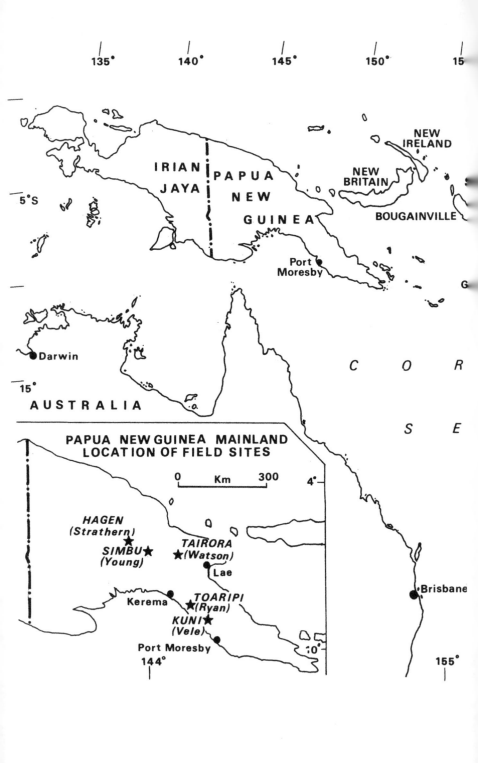

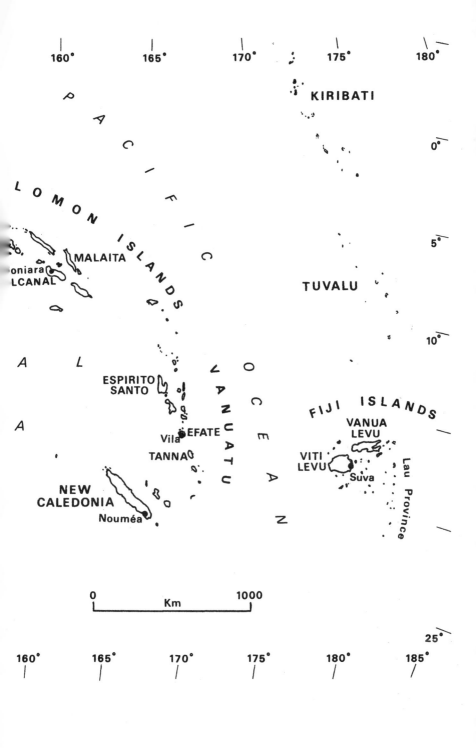